Digital Apollo

Digital Apollo:

Human and Machine in Spaceflight

David A. Mindell

The MIT Press
Cambridge, Massachusetts
London, England

First MIT Press paperback edition, 2011

For information about special quantity discounts, please email special_sales@mitpress.mit.edu

This book was set in Stone Serif and Stone Sans on 3B2 by Asco Typesetters, Hong Kong.
Printed and bound in the United States of America.

Library of Congress Cataloging-in-Publication Data

Mindell, David A.
Digital Apollo : human and machine in spaceflight / David A. Mindell.
p. cm.
Includes bibliographical references and index.
ISBN 978-0-262-13497-2 (hc. : alk. paper)—978-0-262-51610-5 (pb. : alk. paper) 1. Human-machine systems. 2. Project Apollo (U.S.)—History. 3. Astronautics—United States—History. 4. Manned spaceflight—History.
I. Title.
TA167.M59 2008
629.47′4—dc22 2007032255

20 19 18 17 16 15 14

For Pamela, who flies me to the moon.

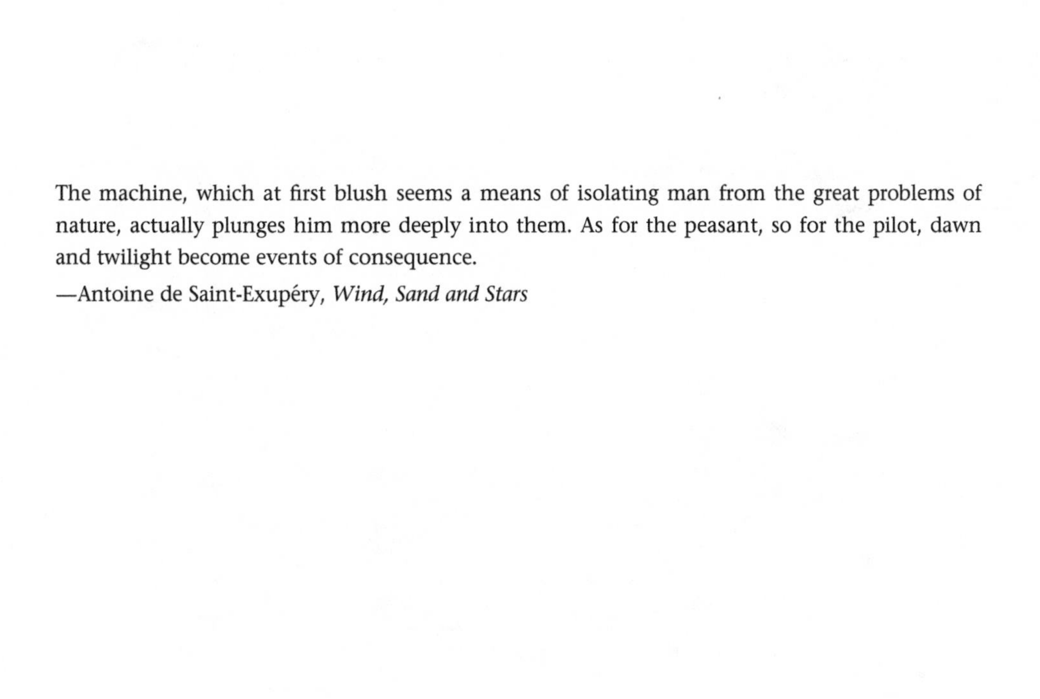

The machine, which at first blush seems a means of isolating man from the great problems of nature, actually plunges him more deeply into them. As for the peasant, so for the pilot, dawn and twilight become events of consequence.

—Antoine de Saint-Exupéry, *Wind, Sand and Stars*

Contents

Preface

A great deal has changed in the world of human spaceflight in the three years since the publication of the hardcover edition of *Digital*. The Obama administration canceled the Constellation program, a product of the Bush "vision" for space exploration, in early 2009 and replaced it with a sweeping, if as yet undefined, vision of a "flexible path" through Lagrange points, asteroids, and eventually to Mars. With a heavy emphasis on private contractors providing launch services for crews to and from low earth orbit and intense debate in Congress, the new landscape is uncertain and stressful to many, but vibrant and necessary as the country confronts the legacy of Apollo and the status of human spaceflight in a world of robotics and telepresence.

In the fall of 2008, before the presidential election, I and a group of MIT colleagues wrote a white paper, "The Future of Human Spaceflight," based in part on ideas from *Digital Apollo*, that rethought the purposes of human spaceflight in combination with remote presence and their policy implications. We briefed the paper to the Obama transition team for NASA in December 2008.

Digital Apollo, with its fundamental rethinking of the human role in spaceflight, has helped shape this conversation. In recognition of its contributions, *Digital Apollo* was awarded the Eugene Emme prize from the American Astronautical Association. It has been read by a number of the Apollo moonwalkers, leaders at NASA, and current astronauts (I've been told there's copy in the astronaut office in Houston that gets passed around). One Apollo commander recently told me he learned about some of the circumstances of his own landing from the book. Several current astronauts, including the commander of an upcoming mission on the International Space Station, approached me and asked, "How can we convince you to do a *Digital Apollo*-like analysis of our operations on the station?" Indeed if we can find the appropriate funding, we will study the relative roles of humans and machines as visiting vehicles dock to the station—some human-occupied, some remotely controlled, and some autonomous.

For me, as an historian/engineer, gratifying responses have come from pilots and other operators of heavily automated, complex real-time systems. Readers have recognized that *Digital Apollo* is a book not only about human spaceflight but about human operators' interactions with computers, automation, networks, and the people who design them in contemporary systems. I have heard from operators of remote exploration missions in space who compared the ground controllers' interactions with the Apollo computer to their own conversations with spacecraft computers orbiting other plants. The chief pilot from a major European airline wrote me that the book influenced its decision to include advanced heads-up displays rather than automatic landing systems in their new airliners. One engineer wrote that "although we are not designing spacecraft, the perspective your book offers is very relevant to the work we are doing designing analytic tools and information visualizations. Here, there is always a critical interplay between automation and human analysis/decision making." Similarly I have heard from embedded-systems engineers, designers of human-machine interfaces, even physicists designing remote experiments for earth orbit.

These responses have spawned a new research program at MIT, the Laboratory for Automation, Robotics, and Society, to investigate a similar set of issues across a variety of systems, including human spaceflight, commercial aviation, remote robotics in the U.S. Air Force, and undersea exploration. In each, human, remote, and autonomous systems coexist and work together, while changing the nature of the work, human roles, and professional identities. Eventually we will turn our attention to other domains such as surgery and automobiles. In some sense, Neil Armstrong's complex interaction with the Apollo Guidance Computer while landing Apollo 11 is repeated daily in each of these domains. New automated safety features, for instance, are rapidly appearing in our cars. Will they allow people to drive longer into old age? Will they encourage people to read or nap on the highway? How would you feel about driving a car with a computer and software that can slam on the brakes at any moment?

Preface to 2008 Edition and Acknowledgments

On June 14, 1966, a robotic spacecraft had just landed on the moon and begun transmitting images to NASA. Project Gemini was drawing to a close, Apollo hardware was beginning to emerge from factories, and Apollo software was experiencing a crisis. And on that day I was born.

I do not remember the first lunar landing of Apollo 11 or the drama of Apollo 13, but I do remember watching the later launches and landings on television. In that sense, I am among the first of a generation—those for whom lunar landings have always been a fait accompli—for whom the twentieth century's greatest technological spectacle was an accomplishment rather than a dream. Nevertheless, as a boy I was fascinated by images of Apollo. When my father brought home a book, *Apollo: Expeditions to the Moon*, filled with wonderful, complicated imagery, I pored through it hundreds of times. The book shaped my lifelong fascination with machinery and my later choice to become an engineer.

This book arises out of a later scholarly trajectory. It is the third in an unplanned trilogy, a series of books I have been writing since I chose to become a historian in 1991. For fifteen years, beginning with a study of the USS *Monitor* in the American Civil War, I have written about the relationship of humans and machines, the experiences of new technologies and their effects on human identity. My second book, *Between Human and Machine: Feedback, Control, and Computing before Cybernetics*, explored the history of human interfaces, control systems, and digital computing. That book included an episode during World War II when Charles Stark Draper (and his young associate Robert Seamans) collaborated on a gun sight project with James Webb, then a lawyer for the Sperry Gyroscope Company. These three men would play central roles in the Apollo program. As I completed that book I found more and more continuities between the earlier history and the lunar landings. A project funded by the Sloan Foundation and the Dibner Foundation on the History of Recent Technology on the World Wide Web provided early support for collecting documents and interviews (that website is available at http://digitalapollo.mit.edu). A senior research fellowship from the Dibner Institute supported early writing.

A number of colleagues, friends, and students have been patient interlocutors and have read manuscripts in various states: Alexander Brown, Stephen Cass, Paul Ceruzzi, Don Eyles, Slava Gerovitch, Jeff Hoffman, Thomas P. Hughes, Chihyung Jeon, Rich Katz, Alex Kosmala, Roger Launius, John Logsdon, Fred Martin, Larry McGlynn, Dava Newman, Jim Nevins, Chuck Oman, Wayne Ottinger, Philip Scranton, Sherry Turkle, John Tylko, and Rosalind Williams. Paul Fjeld made incisive readings and a variety of documents from the Grumman Archives available from his personal collection; John Knoll generously created the cover image with historical help from Fjeld. Eldon Hall also generously shared documents and photographs, as did Hugh Blair Smith and Jim Nevins. Victor McElheny, as always, proved an engaging friend and also provided access to his collection of documents from his reporting on Apollo for the *Boston Globe*. Sarah Fowler assisted the research for the final stages of the manuscript with energy and humor. Thanks to Jack Garman for granting permission to produce the image from his Apollo 11 program alarm "cheat sheet," and to the Society of Experimental Test Pilots for providing access to their back issues of journals and newsletters (some of them dug out of a closet). A number of Apollo participants generously gave their time for a series of group oral history interviews: Ramon Alonso, Dave Bates, Hugh Blair-Smith, Ed Blondin, Herb Briss, Ed Copps, Ed Duggan, Cline Frasier, Joe Gavin, John Green, Eldon Hall, Margaret Hamilton, David Hanley, David Hoag, Alex Kosmala, Dan Lickly, Fred Martin, Jim Miller, John Miller, Jack Poundstone, Herb Thaler, and Bard Turner.

While writing this book I created a course at MIT, "Engineering Apollo: The Moon Project as a Complex System," as an exploration of the project from numerous angles, from management techniques to software, from presidential policy to press coverage. I've been fortunate to teach in collaboration with Professor Larry Young from whom I have learned a great deal. We brought a wonderful mix of guests to class to create a unique educational experience for graduate students in engineering, management, and history: Buzz Aldrin, Dick Battin, Hugh Blair-Smith, Charlie Duke, Don Eyles, Joe Gavin, Eldon Hall, Sy Liebergot, John Logsdon, Victor McElheny, Ed Mitchell, Bob Parker, and Bob Seamans. A particularly interesting moment was attending a lunch with Chris Kraft, Bob Seamans, Aaron Cohen, and Jeff Hoffman. Each, through their generosity, memory, and insight, has contributed to my own thinking on Apollo.

I have also been fortunate to serve on the NASA Historical Advisory Committee and to work with the NASA historians, archivists, and librarians without whom books like this would not be possible. These include Nadine Andreassen, Steve Dick, Steve Garber, Christian Gelzer, Mike Gorn, Roger Launius, Peter Merlin, Jane Odom, Curtis Peebles, Jennifer Ross-Nazal, Rebecca Wright, and the numerous interviewers who have collected NASA history since the 1960s. The NASA History Office preserves and publishes a history that, in addition to being central to spaceflight, exemplifies the evolution of a large technological system. Open and well documented, NASA's systems are more accessible to scholars than military or corporate ones, and hence provide crucial material for understanding the human evolution of technology.

Only the second time I met my wife's family, they accompanied me to the Cradle of Aviation Museum in Long Island and enjoyed a detailed tour of a lunar module. Their genuine interest and excitement made a warm welcome into the Getnick family that I will always remember.

Pamela, whom I met and married while writing this book, sang me through it. I dedicate the book to her, for laughing with me through every day and into our future.

1 Human and Machine in the Race to the Moon

A July Day on the Moon

On a July day in 1969, after a silent trip around the far side of the moon, the two Apollo spacecraft reappeared out of the shadows and reestablished contact with earth. The command and service module (CSM) (sometimes simply "command module") was now the mother ship, the capsule and its supplies that would carry the astronauts home. The CSM continued to orbit the moon, with astronaut Michael Collins alone in the capsule. "Listen, babe," Collins reported to ground controllers at NASA in Houston, "everything's going just swimmingly. Beautiful." His two colleagues Neil Armstrong and Edwin "Buzz" Aldrin had just separated the other spacecraft, the fragile, spidery lunar module (LM, pronounced "lem"), nicknamed Eagle, from the command module. This odd, aluminum balloon, packed with instruments and a few engines, would carry the two men down to the lunar surface.

A rocket engine fired to slow the LM, causing it to fall out of orbit. Once on its way down, the spacecraft would either execute a dangerous abort or soon hit the moon. Whether the impact was a landing or a crash depended on the next ten minutes—the longest continuous series of critical events in the entire mission.

In the LM, weight was such a premium that seats had been eliminated altogether. The astronauts stood up, stabilized by tensioned cables connected to the floor. The spacecraft was an enclosing home, complementing the astronauts' bodies. It supplied their food, exchanged their gases, and collected their wastes. The human occupants, in turn, controlled flows of the spacecraft's numerous fluids, drinking some for hydration and carefully igniting others for propulsion. An inertial navigation system—accelerometers wedded to precisely spinning gyroscopes—measured the vehicle's motions. A radar reached an invisible beam down to sense the first approach of the moon's surface, like a blind man's stick tapping for a curb.

Tying the whole thing together was an embedded digital computer, made out of exotic devices called "integrated circuits"—silicon chips, running a set of esoteric programs. In the middle of the instrument panel, amid familiar dials and switches, stood the computer interface, a numeric keypad glowing with segmented digits. Throughout the mission the astronauts punched in numbers, ran programs, and read the displays. Much of the landing was under direct control of these programs. Neil Armstrong, when he did fly, did not command the spacecraft directly, but rather used two control sticks to command the computer, whose programs fired the thrusters to move the LM. Every move was checked and mediated by software, written by a group of young programmers half a world away.

Nor was the LM alone with its computer, for the command module had an identical computer of its own; both were linked to Houston's control center. Down in Houston, numerous experts monitored systems, offered advice, and controlled some parts of the flight (they even had a remote computer keypad for entering commands directly into the computer in either of the two spacecraft). Communication between the three nodes was calm, matter-of-fact conversation, the precise technical banter of professionals, with an audience of millions.

As the LM began to descend the ground controllers focused their attention. Mission Control locked its doors.

Suddenly, the LM lost contact with Houston. The main antenna that carried data and voice communications to NASA's control center in Houston was having problems. It had to point directly at the earth to work, but other parts of the spacecraft blocked the path, so a computer-controlled feedback loop was commanding the antenna to "hunt" around to seek a new orientation. Aldrin intervened, turned off the automatic control, and adjusted the antenna by hand. The imperfect communications now required Aldrin's attention to keep on track. Frustrated flight controllers in Houston strained through the noise to hear the astronauts, struggling to piece together a continuous story from intermittent bursts of data. "This is just like a simulation," one controller observed on the intercom. Indeed, the performance had been rehearsed, countless times in countless variations, in computer-controlled virtual simulations on the ground.

The astronauts stood in a high state of tension. Their attention was a scarce resource, and any increase in the "workload" could cause them to lose control of the situation. Indeed, for Armstrong the faulty communications detracted from the intense focus of his task.

Uncertainty and ambiguity: on one hand, the astronauts were in control, piloting an autonomous machine far from home; on the other hand, they were part of a network of communications channels, human experts, and control centers. Intermittent communications caused Aldrin to oscillate across this borderline: "You didn't know where you were—whether you were on your own, or whether you were still under the close supervision of ground control. And that sort of reality is rarely simulated in training." The engineers who designed the system (including the astronauts themselves) did not anticipate how electrical noise could interfere with this critical control loop, half a million miles long. Still, in the scheme of things, these problems were minor, easily handled by the conservative design of the LM and the calm professionalism of the astronauts.

At 50,000 feet above the moon's surface, the LM rocket engine fired again, the powered descent initiation (or PDI) that would bring the vehicle to the surface. "Throttle up. Looks good!" Aldrin radioed. They were going down. The computer was in control.

Then, as the spacecraft passed 35,000 feet above the moon, an unexpected light flashed on the computer display.

"Program alarm," Armstrong called out, with noticeable concern.

The computer was having a problem, calling the astronauts' attention. Like its users, the machine had a limited amount of workload and the processor was overloaded with data. This problem might not be serious, but at this moment even a benign distraction could cause trouble. The computer restarted itself, Aldrin punched some keys to inquire about the problem, and it indicated alarm code 1202.

"Give us a reading on the 1202 Program Alarm," Armstrong urgently asked Houston. With the push of a button, the astronaut could abort the landing, an action practiced countless times in simulation. Yet he held off. Armstrong later explained himself as a mechanism: "In simulations we have a large number of failures and we are usually spring-loaded to the abort position. And in this case in the real flight, we are spring-loaded to the land position."

Houston checked out the problem. Young engineers recognized it from a recent simulation, and conferred with their support teams in the back room. They quickly found the cause. The computer was overloading and restarting but not shutting down. It was ignoring low-priority tasks, but these were not critical for the mission. "We're go on that alarm," the ground controller replied, meaning the LM could proceed. For his role in clearing the landing, engineer Steven Bales later accepted a presidential award on behalf of the flight control team.

Armstrong surveyed the computer display; it had frozen. He checked the LM's systems. The vehicle seemed to be responding to his commands, meaning the computer was still running. So he continued. But these checks focused his attention inside the cockpit for critical moments when he should have been looking out the window for a landing site.

Now 2,000 feet above the surface, he again looked out the window. There he saw a potential disaster—a large crater stood where the landing area should have been, boulders surrounding its rim. In that moment Armstrong quit being a computer operator and became a pilot. He seized control of the spacecraft from the computer and flew the LM past the crater. The move took precious additional seconds, and ground controllers became concerned the LM would run out of fuel. Armstrong knew the limits, however, and guided the vehicle down. When the computer sensed the LM was a few feet off the moon Armstrong hit a button and shut off the engine. The LM fell the last few feet with a gentle thud. Aldrin called out the descent systems' shutdown sequence: "Mode control: both auto. Descent engine command override: off. 413 is in."

Relieved, Armstrong then chimed in with his definitive line of technical poetry: "Houston, Tranquility Base here. The Eagle has landed."

Human and Machine

The Eagle's landing is a familiar story, one of the great technological mythologies of the twentieth century.[1] I have retold it here by emphasizing elements that usually hide in the background: the interaction between human and machine, the role of the computer in mediating the astronauts' responses, the network of connections in space and on the ground (figure 1.1). Frequently mentioned but rarely analyzed, these relationships lie at the core of manned spaceflight since its inception, and they continue to frame questions surrounding our proposed future in space.

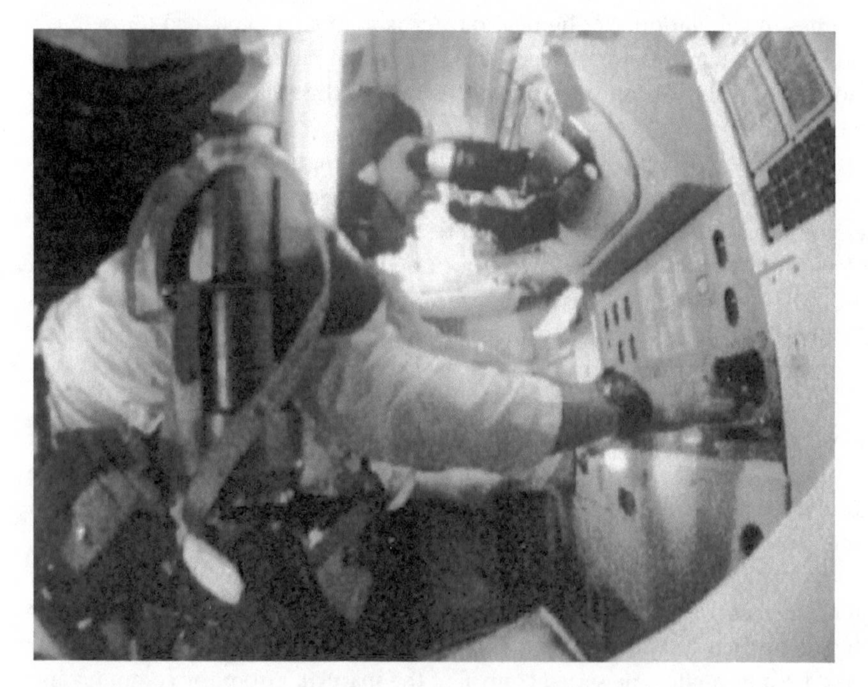

Figure 1.1
Jim Lovell on Apollo 8 aligning the optics for the Apollo guidance and control system. His left hand controls the spacecraft's attitude, while his right hand points the optics. (NASA JSC photo S69-35097.)

Human and machine: their relationship is not a new story. Indeed, it is one of the great narratives of the industrial world, from the mythical John Henry, who won a race with a steam drill at the cost of his life, to Charles Lindbergh, who used the word "we" to describe his partnership with his aircraft.[2] Even during the 1960s, scholars and philosophers debated the appropriate trade-offs between automatic systems and human skills. Yet the many accounts of space travel have failed to explore this profound part of the venture. This book tells the story of the relationship between human and machine in the Apollo project and how that relationship shaped the experience and the technology of flying to the moon. It is a story of human pilots, of automated systems, and of the two working together to achieve the ultimate in flight. It is also a story of public imagery, professional identities, and social relationships among engineers, pilots, flight controllers, and many others, each with their own visions of spaceflight.

To engage the nation, NASA's publicity machine drew on age-old American icons of control and mythologies of individuality and autonomy, from cowboys to sea captains.[3] Apollo's astronauts shared their title with the men who plied the great riverboats down the Mississippi: they were pilots. From the beginnings of aviation up through Apollo and the spaceflight of today, the identity of the aviator-pilot shaped, and was shaped by, technologies of flight.

For Apollo, NASA and its contractors built a "man-machine" system that combined the power of a computer and its software with the reliability and judgment of a human pilot. Keeping the astronauts "in the loop," overtly and visibly in command with their hands on a stick, was no simple matter of machismo and professional dignity (though it was that too). It was a well-articulated technical philosophy. It was also necessary to achieve the political goals of the space program and show that the classical American hero—skilled, courageous, self-reliant—had a role to play in a world increasingly dominated by impersonal technological systems (especially in contrast to the supposedly over-automated Soviet enemy).

That technical philosophy reflected policy making at the highest levels. When NASA administrator James Webb argued for the project, he cautioned that the decision "can and should not be made purely on the basis of technical matters," but rather on the "social objectives" of putting people into space. He and Secretary of Defense Robert McNamara argued that, "it is man, not merely machines, in space that captures the imagination of the world."[4] Presidential science advisor Jerome Wiesner famously opposed a manned lunar program because its scientific goals did not justify the cost. The debates leading up to Kennedy's decision distinguished between "exploration," which is manned, and "science," which has higher intellectual prestige value but is best conducted remotely.[5]

Yet the grammar of President Kennedy's 1961 call to action contained an ambiguity. "Achieving the goal, before this decade is out, of landing a man on the moon and returning him safely to the earth," made the astronaut a passive participant.[6] Indeed, NASA made an early, radical decision to use a digital computer, a "thinking machine," in the Apollo capsule that would control much of the flight. The computer design and its software then reflected a philosophy of automating the fights while not actually replacing the astronauts.

In the end, the astronauts "flew" a very small part of the total mission by themselves, but their control included critical moments of the lunar landing (as well as rendezvous and docking)—landing having long been the ultimate expression of a pilot's skill. Even then, the astronauts controlled the lander indirectly: unless in an emergency mode, their sticks actually commanded a software program, which then controlled the vehicle. Software, a concept barely understood at the start of Apollo, became critical during its development. The programs had the ability to bring the LM right down to the lunar surface under automatic control. They could also crash and kill the astronauts if they went wrong.

Despite the automation, on each of the six Apollo landings the astronaut in command took control and landed in a manual mode. This book explores why.

Chapter 2 begins by examining the anxieties surrounding the role of aircraft pilots in the 1950s. Professional test pilots debated human-machine interactions in the years just before human spaceflight as a host of new technologies—from electronic flight controls to computers in the cockpit—both mediated their control of the machine and gave them access to new realms. When the shock of Sputnik launched the space age in 1957, pilots pondered their potential role in this new era. The X-15 rocket-plane, the subject of chapter 3, sought to prove that human skill and judgment would be required for at least one phase of spaceflight: reentry from space back into the atmosphere. Pilots mastered reentry with the help of computers that augmented their skills and stabilized their flight.

In the wake of the Sputnik scare, NASA was created out of the National Advisory Committee on Aeronautics (NACA) and a variety of federal research groups, and chapter 4 follows pilots into the space age. X-15 test pilots like Neil Armstrong proclaimed that human operators could manually fly the huge new rockets directly off the launch pad and on toward the moon. The powerful Wernher von Braun had an alternative vision, of rockets as automata, carrying passive human cargo. His idea would overrule the astronauts' desire to fly off the pad.

Yet pilots had new powers of their own. The Project Mercury astronauts sparked intense public interest in human spaceflight and immediately pressed for their vision of professional identity, heroism, and control. Arguments over "spam in a can" and the amount of control appropriate for the human cargo ran throughout the Mercury project. Engineers working on the project, who would go on to form the core of the Apollo team, had long experience studying human-machine interactions in aircraft. They held pilots in high regard, relished a close collaboration, and spent years flying and testing dangerous machines. Project Mercury's successor, Project Gemini, epitomized the pilot-centered approach, enabling hands-on control of orbital maneuvers. The complexity of rendezvous operations, however, also called for computational aids, from paper charts to digital computers.

The Apollo program began with the new Kennedy administration and its recognition of the public, political impact of human spaceflight. Kennedy's speech launching the moon program came just weeks after the suborbital flight of Alan Shepard, who was hailed as a space-age Charles Lindbergh. Yet the first contract of the new moon program was let not for rocket engines or fuel tanks or launch pads, but for a computer, the subject of chapter 5. Engineers at MIT's Instrumentation Laboratory, who would build that computer, in the 1930s had helped change the nature of flight from "seat of the pants" intuition to numerical, instrument-based tasks. Their Apollo proposal derived from a Mars probe, designed but never built, and from the inertial guidance system for the Polaris nuclear missile. The MIT engineers valued accuracy and autonomy and studied how a set of gyroscopes could find its way to the moon and back.

But the Apollo system had a human user, someone who would require an "interface" to issue commands and requests to the computer and read out information. This requirement raised a series of difficult, interesting problems. The machine would have to be much more reliable for the two-week lunar missions than for a missile's short flight. It would interact with two planetary bodies instead of one, and two spacecraft instead of one. It would need to be calibrated against the stars by a human user. And if it failed, people might die. Chapter 6 follows the Apollo computer as it became operational hardware and recounts its designers' decisions about human interface, reliability, and manufacturing.

Exotic as it seemed, the hardware adapted for Apollo was relatively familiar in the world of military avionics. Radically new, however, was the software that would interact with the astronauts to control the mission, the subject of chapter 7. At first, programming was treated as a secondary, almost trivial task, but by 1966, it seemed the early Apollo flights might be delayed for lack of available computer programs. Only the 1967 Apollo 1 fire that killed three astronauts and NASA's management intervention into the programming team brought the software project under control and on schedule. Early, unmanned Apollo flights revealed the delicate mix of reliability, flexibility, and accountability that would surround these new, software-controlled systems.

The entire Apollo program culminated in the landing. The final ten or so minutes before touchdown formed the most critical period of the mission. The remainder of the book examines the design and execution of the lunar landings. Chapter 8 describes the plans for this phase of the flight, incorporating fundamental physics, lunar models, computers, human performance, and a host of uncertainties, including questions about the human role. From 50,000 feet down, the LM made the transition from purely inertial guidance to include radar and the human eye. To allow the astronauts time to visually assess their landing site, the maneuver's design had to incorporate detailed consideration of human capabilities. Automatic systems would fly the LM down to an altitude of a few hundred feet, where the commander could take over semi-automatic control and bring the LM down with his hands on a stick.

The final chapters go through each of the landings and the interactions between the astronauts and their machinery, and with their colleagues on the ground. The minute details afforded by transcripts and data telemetry allow a kind of real-time ethnography, a thick description of human-machine interactions and their cultural context during critical operations.

Each of the landings stood out in some dimension. Chapter 9 dissects Apollo 11 and the famous "Program Alarm" that began this chapter to examine risk, responsibility, and error in the distributed software-based system. Chapter 10 looks at the remaining landings. As they progressed, the technical task lost some of its challenge and uncertainty, while the scientific goals of the program took greater prominence—hence the later missions landed in geologically more interesting, but tactically more difficult, areas.

The final chapter of the book extends the analysis to the broader history of human spaceflight in America and follows some threads from the Apollo story into today's world. Human-machine relationships in Apollo had significant implications for the space shuttle, and hence for decades of American space policy. Reframing the "humans versus robots" debate into one that is richer and more forthright about both human and remote presence and their social implications is crucial for U.S. space policy as it faces the space shuttle's retirement and a possible return to the moon or human venture to Mars.

Rethinking Apollo

Flying to the moon was among the most notable technological achievements of the entire twentieth century, or at least the most noted. Even today, scan the culture for references to the Apollo program and you'll it everywhere: from clips in music videos (music television network MTV's first moments of broadcast in 1981 were a picture of Buzz Aldrin on Apollo 11 superimposed with the MTV logo; the icon remains the basis for the network's Video Music Award statuette) to repeated calls for a new Apollo project to solve one or another of society's ills. I recently looked up from my seat on a subway to see the Apollo motto, "For All Mankind," emblazoned on the derriere of a pair of designer jeans.

Even a casual bibliography has hundreds of entries. Ever conscious of its public image, NASA documented and wrote Apollo's own history in parallel with the project itself, producing numerous informative, if ponderous, volumes on everything from the spacecraft to the launch pads and the lunar science experiments.[7] NASA has also conducted hundreds of oral history interviews, from the beginnings of Apollo up to the present day. Those collected during the 1960s provide immediate, primary insights into the project, while those collected more recently document the participants' memories.

Of the twelve men who walked on the moon, at least eight have written, or have had ghostwritten, some kind of memoir, and numerous other Apollo crew members have chimed in as well.[8] Lately the ground controllers have gotten into the act, with similar, though delayed, levels of public interest and attention.[9] One account used interviews with engineers to tell the story of the technical people behind the scenes.[10] A few of Apollo's engineers have added their own stories as well.[11] Numerous other popular accounts cover the project from a variety of angles; one was even made into a TV miniseries, following on the successful feature film *Apollo 13*.[12]

With shelves straining from all this Apollo material, what could possibly be left to say? To begin with, histories of the Apollo program are nearly all project oriented—they begin at Apollo's beginning and end at its end. Other than personal background in memoirs, little is said about Apollo's connection to larger currents in the history of twentieth-century technology. The technical histories focus on hardware and description, with little broader analysis.

Those histories that provide "context" tend to be political or cultural, and don't delve into the machines themselves, the people who built them, or what they meant.[13] Hence they solidify the canonical narrative of the project around key themes and events: Kennedy's visionary decision, the frenetic engineering efforts, the triumphs of the astronauts, the tragic fire, the triumph of Apollo 11, the drama of Apollo 13, and so on.

Most Apollo histories adopt a heroic tone, retelling how the skill and cunning of the astronauts, ground controllers, and engineers brought the mission to a safe, successful conclusion, sometimes in the face of bureaucratic incompetence or technical failures. Heroic stories have been with us for a long time, at least since Homer, and play important cultural roles. Heroic narratives follow a prescribed "cycle," with a variety of stages in which the hero matures and proves himself—think of the labors of Hercules, or the island challenges of Odysseus.[14] The astronauts' stories, which vary widely in quality and interest, tend to follow a similar pattern—the boyish fascination with flight, early military service, transition to test piloting, miraculous selection by NASA, rigorous training, climactic moments of judgment and skill at critical points in the flight. Historian Asif Siddiqi calls them "nosecone histories" for their limited views of the projects they describe, yet they still carry cultural weight because of the "there I was..." character and the astronauts' lingering heroic public image.[15]

The human-machine relationships of Apollo both reflected and shaped the political and cultural goals of the program and the machines themselves. Sometimes technical decisions threatened the centrality of the human pilot. At other times they left key tasks to human skill and judgment. At no time, however, did some abstract set of technical requirements uniquely define the human role. Engineers and astronauts, as well as journalists and policymakers, constantly debated the appropriate tasks for the spacecrafts' human operators, continuing conversations rooted in the earliest days of aviation.

Aviation Heroism

Michael Collins, who orbited the moon on Apollo 11, remembered being inspired as a young man in the 1930s by the dashing figure of the barnstormer pilot Roscoe Turner. "Roscoe had flown with a waxed mustache and a pet lion named Gilmore," Collins remembered wistfully. "We flew with a rule book, a slide rule, and a computer." His comment captures in one sentence Apollo's relationship to aviation. Collins felt caught between "the colorful past I knew I had missed and the complex future I did not know was coming"[16] (figure 1.2).[17]

Roscoe Turner's career peaked just a few decades before Collins's, but the two seemed worlds apart. Dubbed "Aviation's Master Showman," in the 1920s and 1930s Turner barnstormed his way from rural America to Hollywood. He had little training and even less formal education. Yet he fashioned himself as a colorful character, sporting a waxed mustache and a made-up uniform from a nonexistent military in which he had never served. He was married in the cockpit of his Curtiss Jenny and flew his giant Sikorsky S-29 airplane, dressed up as a German bomber, in Howard Hughes's film *Hell's Angels*. As Collins noted, Turner flew with his pet lion Gilmore, named after the oil company that sponsored them. Turner embodied the showy, excited world of aviation in its "golden age" of transition from dangerous curiosity to commercial service.[18]

Collins was not alone in noting the passage from a hands-on past to a computer-controlled and rule-based future. In the mid-twentieth century, a host of professionals and craftspeople—from industrial managers to shop-floor machinists, from farmers to soldiers—reacted to the advent of computers and automated systems. Yet along with computers came new skills, work practices, and professional identities. Astronauts and their spacecraft were but the most visible manifestation of broad changes that raised fundamental questions: in a world of intelligent machines, who is in control? Can it be "manly" to control a machine by simply pushing buttons? How does software change the equation?

Figure 1.2
Michael Collins training in a command module simulator. Note the checklists in his left hand, the hand controller at his right, and the optical sighting equipment for the Apollo guidance and navigation system at his feet. June 1969. (NASA photo 69-H-978. Scan by Ed Hengeveld in *Apollo Lunar Surface Journal*, http://www.hq.nasa.gov/alsj/frame.html [accessed February 2007].)

Spacecraft and Symbolism

As different as they were, Michael Collins shared one characteristic with Roscoe Turner: both were on display. Turner flew in an age when aviation's commercial potential had yet to be realized, when the airplane remained a dazzling curiosity and most professional pilots earned a livelihood through entertainment. By Collins's day a pilot could make a living with more prosaic tasks like flying airliners; however, the astronauts, like Turner, worked with a technology of unclear civilian utility but whose imagery captivated the attention of the press, the public, and the state.

Put a human being inside a rocket, add the resonance of a journey into the blackness of space with all its allusions to the heavens and the long history of human fascination with the stars, and one has a technology that linked humankind's most earthly, practical endeavors (fuels, oxidizers, pipes, breathing, eating, shitting) to its most lofty ambitions.

None of the symbolic power of spaceflight was lost on the visionaries who promoted the space program, the politicians who supported it, the press who reported it, or the public who consumed the news about it. They very consciously built symbols as well as spacecraft.

In creating that symbolism, the Kennedy administration drew on American imagery of exploration, individualism, and geographical conquest to sell Apollo to the press and to Congress. Kennedy seized on the most powerful mythology in American history, the frontier narrative, and reopened it by aiming for the moon. Within this framing, the endeavor had all the elements of a classic frontier adventure: an unknown, but conquerable geography full of lurking dangers, even villainous antagonists—the competing Soviets.

Most important, the frontier narrative called upon heroic pioneers. The press may have been biased against large government projects (delighting in exposing waste and fraud), but they were heavily biased in favor of individual, human tales. Human presence made spaceflight into a story. For the American public, that story involved people who embodied American virtues, from humility and self-control to self-reliance and creativity, "part Davy Crockett and part Buck Rogers."[19] And for that story to be credible, the astronauts had to be in control. Frontiersmen could not be passengers.

Imagery of active pilots pervaded Apollo, but coexisted with another, subtler trend. The moon project resonated within a culture deeply concerned with the social implications of technology. It was conceived in the wake of Russia's Sputnik success and in the early Kennedy years when large-scale science and technical and managerial projects seemed to promise solutions to political problems. But Apollo unfolded in the era of Vietnam, 1960s counterculture, and increasing questioning of the social benefits of large technological systems. Commentators worried about the phenomenon of "deskilling" as computerized machine tools transformed work on the factory floor.[20] In his speeches and writings, for example, Martin Luther King frequently mentioned automation as a cause of the social displacements he was seeking to redress. Even NASA director James Webb suggested that the jobs generated by the Apollo program would help mollify unemployment created by automation.

The Apollo years spanned the release of Stanley Kubrick's *Dr. Strangelove* (1964), about an automated Soviet machine that triggers the end of the world, and his *2001: A Space Odyssey* (1968), in which an intelligent computer murders American astronauts. Also during Apollo, Jacques Ellul's book *The Technological Society* (published in 1965 in English) challenged the increasing dominance of "technique" in human culture. In 1967 Lewis Mumford named the "megamachine" as the aggregate of technology, social organization, and management that suppressed individual human values.

Philip K. Dick published *Do Androids Dream of Electric Sheep?* in 1968 (later made into the film *Blade Runner*), recasting traditional demarcations between humans and machines. Thomas Pynchon's *Gravity's Rainbow* (1973) took "the rocket" as its central literary figure, exploring the technical, psychological, and religious dimensions of a state that worshiped at the altar of technology, and the paranoia engendered by its invisible, clockwork plans.[21]

NASA and its astronauts faced such tensions in the daily engineering of their systems, questions with the potential to undermine the symbolic agenda of the program. Would the exigencies of rockets, supersonic flight, and split-second decisions, not to mention onboard computers, threaten the classical heroic qualities? What tasks were susceptible to human skill, and what was too fast, complex, or uncertain for a human to intervene? How were Apollo designers to engineer a system that had a place for a heroic operator? As Apollo's machines were designed, built, and operated they called the very nature of "heroism" into question. What did it mean to be in control?

Embedded Computers, Embedded Assumptions

A note on terminology raises the stakes. Essentially all of the sources from the 1950s and 1960s use the term "manned" for projects sending humans into space, so I use the term "manned," as the participants did. Similarly, I refer to "men" or "men versus machines" when referring to a particular historical group of men, namely, the Apollo astronauts, who were all men. I use the gender-neutral term "human," however, in reference to abstract notions of human-machine interaction. This approach avoids awkwardness and confusion in the text while highlighting the artificial nature of the "manned" terminology (NASA today uses the awkward and easily misheard term "crewed").

We now know how some scientists studied women's potential as astronauts, but NASA chose all of its early astronauts to be men. This decision may have countered engineering logic: when weight and space are premium, skilled women might make more sense than men, as they are smaller, lighter, and consume less. Yet U.S. experts cited the 1963 Soviet feat of putting a woman in space (two decades before an American woman flew) as evidence that the heavily automated *Vostok* spacecraft did not require a skilled operator. At Lyndon Johnson's suggestion, NASA insisted that astronauts be test pilots qualified in jets, which excluded women by definition. "The very qualifications required for NASA astronauts," argues historian Margaret Weitekamp, "proved the complexity of U.S. space achievements. Demonstrating that a woman could perform those tasks would diminish their prestige."[22]

The role and nature of "men" were very much at stake in the design of Apollo's control systems. Scholars have recently begun to outline the changing faces of masculinity during the last century and the public image of the astronaut certainly played a role in that evolution.[23] Astronauts' accounts continually reaffirm that what it means to be a man is related to control and interpret threats to pilots' control as threats to their manhood. Less recognized in the historical literature is how engineers responded to those changes. Did they design systems, knowingly or unknowingly, to leave the operators a sense of mastery? Was that mastery perceived differently when astronauts were pushing buttons and entering computer commands rather than having their hands on a control stick?

Sources and Implications

How did the Apollo engineers accommodate human beings in their machines? How did they build a computer that kept humans "in the loop" for the critical functions of the lunar landing? When were the human operators operating as skilled, intelligent beings, and when were they machine-like, following prescribed scripts? This borderline, *between human and machine,* reveals the human aspects of Apollo amid so many seemingly cold, technical calculations.

Much, if not all, of the engineering work was incredibly *mundane*: writing reports, holding meetings, testing machines, developing procedures, practicing pushing buttons, weaving hair-like wires through tiny magnetic cores thousands of times in mind-numbing succession. Human players interacted in ordinary ways: competition, collaboration, professional pride and anxiety, struggles to influence and define the project. Sources prove even more prosaic: project updates, status reports, interoffice memos, engineering drawings, test reports, logs of an astronaut's seemingly endless hours in a simulator, dry mission transcripts, technical debriefs, and mission reports. As participants often pointed out, the high abstractions of systems engineering frequently meant added layers of paperwork bureaucracy. Yet lurking within these ordinary documents are critical tensions and embedded assumptions whose explication makes the detail come alive.

One of my goals is to explain, really explain, how Apollo worked, and to make one of the most difficult engineering accomplishments of the twentieth century accessible and understandable. The story combines the intrigue and suspense of a group of engineers working at the cutting edge of technology with the drama and interest of spaceflight and the social importance of computers. Tracy Kidder's 1981 book about a group of engineers building a computer, *The Soul of a New Machine,* had the ironic result that the computer he focused on, a minor commercial machine, was forgotten, while his book is long remembered. This story has a similar cast of characters but in this case the computer and its task made history.

I hope that the interested, nontechnical reader will gain from this story insight and intuition into the thorny and fascinating engineering problem of how to fly to the moon, particularly how to land on its surface, and some understanding of the fundamental questions of machine control and human-machine interaction. These reappear in high-risk, high-reward technologies of today, from airline operations to nuclear power plants to proposals for a new era of space exploration.

A few words on what this book is not. It is not a reminiscence of NASA's glory days of Apollo, and it does not seek to explain what went wrong at NASA in the three decades since.[24] It does not repeat the numerous clichés of "we went to the moon with a computer that was less capable than a pocket calculator." That may be true if you measure a computer's capability in memory capacity or machine cycles alone. But if you consider interconnections, reliability, ruggedness, and documentation, the Apollo guidance computer is at least as impressive as the PC on your desktop, and the Apollo software an equally intricate ballet of many people's work and ideas.

Members of a video-game generation may find that Apollo makes sense when explained through stories of joysticks, cockpit displays, and hand-eye coordination. Indeed, the word *cyborg* was coined by NASA researchers studying bioastronautics in the 1950s.[25] The earliest video games appeared during the Apollo years, one of which was called "Lunar Lander" (with instructions that read: "You are landing on the moon and have taken over manual control 500 feet above a good landing spot..."). In the climactic moment of George Lucas's 1977 film *Star Wars*, the hero Luke Skywalker turns off his computerized sighting device and relies on the intuitive "Force" to help him destroy the enemy Death Star.

Still, I also do not contend that Apollo *caused* changes in human-machine relationships or that it created new technologies that altered those relationships. My argument is that Apollo exemplified broad changes in human-machine relationships, not that it caused them.

Human and Machine in the Future of Spaceflight

Yet Apollo's human-machine history does speak to the lasting debate over whether humans or robots should be flying into space and exploring the solar system.[26] Current polemics usually polarize around creative, flexible humans versus mindless automata, the former being capable of "exploration" and the latter collecting data for "science." Such rhetoric has arguably produced more heat than light in recent decades, although the stakes are high as NASA determines new policy directions. Yet the advocates for either side usually neglect or misunderstand the mixings and combinations of manual and automated, especially experiences made possible by communications links and remote controls. The Mars rovers named Spirit and Opportunity that captured public imagination in recent years, for example, are less "robots" acting as autonomous agents than "telerobots" responding to commands from the earth and providing data for ground controllers, scientists, and the public to experience a foreign world from afar. Similarly, the Apollo spacecraft and astronauts had tight connections to the ground and transmitted images, words, data, and experience through remote channels. No computers made decisions on their own; all were programmed by people, distanced in space and time from the landings, who embedded their own ideas, models, and assumptions into the machines.

In this vein, I do not take a position on the humans versus robots debate, but rather seek to clarify some of its terms. What, exactly, do people do in space? Which of their tasks require strict adherence to procedures? Which require subtle perceptions and skills? When do they use their judgment? When do they err? Less frequently debated than the humans versus robots question is the equally contentious: which people? What kinds of professionals? If a major goal of human spaceflight is inspiration, or expanding the realm of human experience, should we not consider selecting and training people to communicate those experiences? What follows comprises but a first look, and raises more questions than it answers. Still, the concluding chapter suggests that similar analyses applied to space shuttles, deep-space probes, or robotic missions could help redefine and advance a debate that has been stuck in circular argument for a generation. Informed public discourse on human spaceflight is essential for a successful, sustained human future in space, whether directly or remotely present.

2 Chauffeurs and Airmen in the Age of Systems

Test Pilots and Survival

The Society of Experimental Test Pilots (SETP) held its first annual awards banquet on October 4, 1957. These men sat at the top of the piloting profession, crossing the border between engineering and flying skills. They had been rocketed to fame by Chuck Yeager's epochal supersonic flight nine years before. The atmosphere in the banquet hall was electric as the group celebrated its new society epitomizing professional maturity. Six hundred and fifty people attended, many of them making the drive from the dry desolation and professional focus of Edwards Air Force Base, a few hours north in the Mojave desert, down to the cosmopolitan fashion of the new Beverly Hilton. This was Southern California at its 1950s best, as the serious, focused flyers enjoyed an evening in the heart of mid-century Hollywood glamour (figure 2.1).

The fledgling society was barely a year old. To bolster its reputation the SETP associated itself with great names in aviation. On this evening, the group awarded honorary fellowships to Air Force General James Doolittle, the Ph.D.-educated test pilot who led the famous raid over Japan, and to Howard Hughes, the eccentric aviator and manufacturer. Charles Lindbergh had been offered a similar honor, but declined it, perhaps because he had never heard of the society (he would eventually accept the award twelve years later).[1]

"Honored guests, ladies and gentleman," the evening's keynote speaker began, quieting the room. He was Richard Horner, assistant secretary of the air force for research and development. Horner was himself a pilot who had flown in North Africa in World War II, earned a graduate degree in aeronautical engineering at Princeton, and worked as a test pilot (two years later, Horner would become NASA's first associate administrator, the agency's highest civil service position).

Horner began with an anecdote. When he had been invited to speak, he asked a test pilot he knew what subject would be of most interest for the inaugural banquet. "His answer was one word—SURVIVAL—and it was so immediate as to leave no doubt in my mind that indeed was the uppermost thought in his mind," Horner said. He

Figure 2.1
SETP inaugural banquet, October 4, 1957. (SETP, *History*. Reprinted by permission.)

therefore thought it logical enough that the test pilots in the audience would be interested in ejection seats, escape capsules, pressure suits, goggles, helmets, and the like. Indeed the SETP charter listed safety and escape devices as the society's primary interests. As Horner relayed his thoughts on these devices to his test pilot acquaintance, however, he realized that they were talking at cross purposes. The young pilot had in mind a very different kind of survival—"the survival of the cockpit itself."

What could he mean? In this, the golden age of test flying, with test pilots energetically organizing for a professional society, what could threaten their existence? Horner immediately realized that "my opposite number in this conversation was thoughtfully considering such names as Bomarc, Matador, Snark, Thor, Atlas, Titan, and perhaps a little wistfully, Navaho."[2] These names referred to the variety of missiles then under development by the U.S. Air Force. Some were rockets; others had wings and flew like airplanes. Each had the dreaded modifier, "unmanned." The question of survival, then, was not simply for individual pilots, but for their very profession.

Usually in his speeches Horner extolled the virtues of automation. Here he was speaking to 650 people whose livelihoods would be determined by this question. For Horner, the U.S. national posture of deterrence, and the necessity for specialized weapons, virtually required unmanned systems for certain missions: "It is perfectly obvious that one of the pre-requisites for taking the man out of the systems operation must be the capability to describe very carefully, and in some detail, the characteristics of the operation before it starts. Of course, in some instances the man can be included by leaving him on the ground and providing him with necessary intelligence."[3]

Horner prophetically acknowledged the potential of remote control (which the pilots themselves rarely did), but he also tried to mollify the audience, declaring that "there is a place for manned aircraft in our military systems, now and as far as we can see into the future" and that "it is difficult to postulate a military engagement of any kind where the flexibility and discrimination of man's judgment and power of reasoning wouldn't be superior at some stage of the conflict.... The strongest advocates recognize missiles as complementary to, rather than a replacement for the manned aircraft."[4] To the SETP audience Horner presented an articulate, nuanced analysis of the engineering trade-offs between manned and unmanned systems, emphasizing that "the real justification for the inclusion of a man in a system is to capitalize on his reasoning, judgment, and flexibility of response."

Still, Horner also raised arguments against the presence of human pilots, and he did not dismiss them out of hand. "In manned vehicles, the same performance goals come easier in a system not handicapped by such idiosyncrasies of the human being as a desire to come home."[5] Technology would continue to progress, Horner concluded, but human operators would always be the same old folk: "We must recognize... that the one link in the manned system which we have that improves the least in successive generations, is the man himself." Horner notably avoided the position that the human presence in military aircraft was inevitable, claiming only that it would be a matter of engineering decision to incorporate human abilities into a particular mission.

The SETP banquet left its members with a new sense of professional camaraderie and growth, but it also left unresolved questions about their future. The next morning, as they read their morning newspapers, the world had changed. For on October 4, 1957, the day of that first SETP banquet, the Soviet Union launched the first artificial satellite, dubbed Sputnik, into orbit. The space race had begun, and its first hero was a machine.

Stability and Control

Horner's questions lay at the very heart of aviation, poised to take center stage in space. What would be the role of the human pilot as aircraft entered ever faster, higher, and more demanding flight regimes? How would the pilot share the cockpit, and control of the aircraft, with electronics and computers?

Beginning with the Wright brothers and continuing to this day, control is an essential feature of aviation. How should the pilot command an airplane? How should the pilot be trained? What devices should aid the pilot? In the early twentieth century, new personalities accompanied the new technology: the aristocratic aviator, the fighter pilot, the ace, the barnstormer, the stunt pilot, to name but a few. Since then, the technology of aviation and the profession of the pilot have evolved together.

Yet within this history lay a paradox, or at least an irony. As aviation matured, aeronautical science became increasingly adept at measuring and modeling the airflow around an aircraft and designing structures and devices to accommodate it. But the core of the aircraft was still the pilot, a human being, a subject that engineering has never fully mastered. Hence the pilot's importance: performing tasks that are difficult to measure or model.

At heart, debates about control and automation in aircraft are debates about the relative importance of human and machine. They go back to the very origins of powered flight, and with a few modifications they pervaded the Apollo program. From the Wright brothers' flying machines to the jets of the 1950s, technologies of control have evolved in parallel with the people who did the controlling. In aviation, as with all technologies, technical change and social change are intertwined.

First, a few technical terms. In aviation, people and machines come together through stability and control. *Stability* is the tendency of an aircraft to remain in straight and level flight, at a given airspeed, even without the pilot's inputs, and to return to that state after external disturbances like wind gusts. Stability is a feature of the general design of an aircraft—the placement of the wings and center of gravity, their subtle angles and orientations. Most modern airplanes are inherently stable—they will return to straight and level flight if the pilot takes his or her hands and feet off the controls. Similarly, an automobile is stable: on a flat road it should drive in a straight line if the driver's hands are removed from the wheel. An unstable aircraft with no pilot inputs, by contrast, will depart from straight and level flight, and eventually crash. A bicycle is unstable—it will not go straight, or even go at all, without some active participation from the rider. Note that instability does not make a bicycle unrideable or an airplane unflyable—it just takes more work, attention, and skill on the part of the operator.

Today it might seem obvious that stability should be built into an aircraft, and many engineers indeed share this view. But consensus on this point took decades to develop within aeronautics. At the heart of the debate was a tension between stability and *control*, which to some degree operate at cross purposes. The more stable an aircraft is, the more effort will be required to move it off of its point of equilibrium. Hence it will be less controllable. The opposite is also true—the more controllable, or maneuverable, an aircraft, the less stable it will be. A fighter plane is more responsive than an airliner, but also more difficult to fly.

Chauffeurs and Airmen

The tension between stable and unstable aircraft dates from the earliest days of aviation. As one observer put it in 1910: "Equilibrium [i.e., stability] has developed into a controversy dividing aviators into two schools. One school holds that equilibrium can be made automatic to a very large degree; the other... claims that equilibrium is a matter for the skill of the aviator."[6]

Already at this early date, just a few years after the first public demonstrations of powered flight, the debate took on a social dimension and created two distinct groups of people. The historian Charles Harvard Gibbs-Smith named these two schools: the *chauffeurs* who thought the aircraft should be inherently stable, and the *airmen,* who actually preferred the aircraft to be unstable. "The chauffeur attitude to aviation," wrote Gibbs-Smith, "regards the flying machine as a winged automobile, to be driven into the air by brute force of engine and propeller, so to say, and sedately steered about the sky as if it were a land—or even marine—vehicle."[7] Gibbs-Smith identified the chauffeur attitude with the Europeans, particularly the French, who were experimenting with flying machines around the turn of the twentieth century. These chauffeurs saw themselves as "outside" their machines, which had a certain stately autonomy; the human role was to guide, rather than direct.

By contrast, the airmen looked upon the early "aeroplane" as "as something to experience, to learn in, and to fly in."[8] Gibbs-Smith includes Otto Lilienthal and Octave Chanute among this school, but the ultimate airmen were, of course, the Wright brothers, who realized the critical importance of flight control to any workable airplane. "The true 'airmen's' attitude," wrote Gibbs-Smith, "was evident in the pilot's desire to identify himself with his machine ... or ride it like an expert horsemen." Indeed Wilbur Wright likened learning to fly to "learning to ride a fractious horse" and noted that "the problem of equilibrium constituted the problem of flight itself."[9] Bicycles and horses were the Wright brothers' key metaphors for flying aircraft. The Wrights trained extensively in gliders before attempting powered flight. It is well understood by historians, as it was by people at the time, that among the key contributions of the Wright brothers were their means of controlling the airplane, and indeed their very idea that an airplane should be controlled at all. Put another way, the Wright brothers invented not simply an airplane that could fly, but also the *very idea* of an airplane as a dynamic machine under the control of a human pilot.

Less recognized by historians are the social implications of the airmen's philosophy. First, if the airplane was inherently unstable, then it required a great deal of skill to fly. Second, a less stable aircraft was more dangerous, and put the pilot at risk. As Wilbur

Wright wrote to Octave Chanute in 1900: "What is chiefly needed is skill rather than machinery."[10] Inherent in their idea of a controllable aircraft was a new type of person: a skilled pilot.

Skill and Class

Let us pause for a moment to consider *skill*, a common enough notion in everyday life, but also a key to understanding the social dimensions of technology. On one hand, skill is highly personal. It is practical knowledge, it implies a certain amount of cleverness, perhaps expertise, and we often think about it as residing in our bodies, particularly our hands (e.g., "manual skills"). On the other hand, skill is also deeply social. It is not inborn, but acquired, as distinct from an innate quality like talent. Skill implies training, the time and effort to learn and master it, often with the help of another person. We also associate skill with pleasure—in acquiring it, exercising it, and observing it in others. Hence the joy in watching a musician, a figure skater, or a baseball pitcher (who are all probably talented as well) perform. Skill garners respect, and the more skill you are perceived to have, the more respect you seem to earn.

Skilled workers include surgeons, carpenters, and waiters. Obviously not all skills are equal. Some are more respected than others, and there tend to be social and economic differences among their practitioners. Skill also sets people apart. The word itself comes from an Old Norse word meaning *distinction* or *difference*, ideas that remain integral to today's meaning.[11] For any skill, some people have it and some people don't. The very notion of skill implies a social group, possibly even an elite.

When people with common skills come together, they often form societies, set standards, and create and uphold traditions. They also police the boundaries of who is in and who is out, and for high-status skills this makes them members of *professions* (those with more traditional skills belong to *crafts*).[12] Most would agree that surgeons are professionals, but are carpenters, or waiters?

The Wright brothers, by emphasizing the importance of skill, created not simply a controllable flying machine, but also its human counterpart—the pilot. From the moment Wilbur Wright first flew, this new professional was born.

In this light, the language of Charles Gibbs-Smith's chauffeur versus airmen distinction bears some examination. The term *chauffeur* connotes an ordinary skill, a person who is paid to perform a common task, driving an automobile, for another person of higher social status (the word *robot* has similar connotations). For the chauffeur school, flight is inherent in the machine, and hence the product of the engineer or designer.

The term *airmen*, however, suggests someone who flies for himself, who is part of a new profession, living in a new element, and it also identifies him as male. In the airmen school, the pilot's continuous, active involvement—his skill—is required to maintain stable flight. Flight itself is then a product of the pilot. If the pilot creates flight, then his status rises accordingly. The element of risk increases it further.

The terms *chauffeur* and *airmen* were Gibbs-Smith's from 1970, not those of the early aviators. Nor do we need to argue that the Wrights created these distinctions intentionally. Still, the dichotomy of chauffeurs and airmen makes it clear how high the stakes were in the debates over the stability of flight around the turn of the twentieth century. At issue was not only a technical question of aircraft design, but also the very professional status of the pilot: Were they to be mere engine drivers, like so many wage-earning machine operators before them? Or would they be independent professionals, masters of the new element?

Numerous models existed for pilots in the early decades of aviation: mechanic, tinkerer, engineer, sportsman, artist, aristocrat, and soldier, among others. Boosters saw aviators as mechanical angels, carrying the "winged gospel" of modernity, while futurists and Dadaists saw in them a modernist blurring of organic and mechanical.[13] Different notions of aviators would compete and evolve as aircraft changed, and the technology itself would adapt to suit the interests and dreams of its operators.

By phrasing the debate as *chauffeurs versus airmen*, the winning choice should have been clear, for who would choose to be subservient? The history of airmen seems a history of the victors. The Wrights triumphed, the airmen won. Or did they?

Stable or Unstable Aircraft

Indeed the Wrights' early opinion that aircraft should be inherently unstable became the consensus among aviators for several decades. The fighter aircraft of World War I were notoriously unstable and difficult to fly, but highly maneuverable. Learning to fly could be as dangerous (and possibly as heroic) as facing the enemy—during the war, more British pilots died in training than in combat.[14] And indeed with these aircraft matured a new cadre of skilled pilots, and a new breed of war hero, characterized as much by technical skill as by innate courage.[15] Jerome Hunsaker, one of the great early aeronautical engineers, referred to the "almost universal prejudice among accomplished fliers against so-called 'stable aeroplanes.'"[16]

But during the 1920s, the consensus among aviators changed from favoring unstable airplanes to favoring stable ones. By 1935 the soon-to-be-legendary aircraft designer Clarence "Kelly" Johnson was writing "the reasons why an airplane must be stable are more or less obvious"—indicating that engineers had changed their position. Another opinion captured the new consensus, that "the machine should be stable, but not *too* stable."[17] What had changed, argues the engineer and historian Walter Vincenti, was human fatigue—as aircraft acquired greater range, and pilots flew longer flights, the effort of constantly keeping the aircraft in level flight became tiring, and pilot opinion shifted in favor of stable aircraft (although it also split as aircraft types evolved, deeming that fighter aircraft should be less stable than bombers and transports).

Why would the pilots give up their attachment to total control, to creating flight itself with their skills? In new opportunities open to them—long-distance flight, air mail runs, commercial passenger flights, and increasing influence in military circles—professional development would not depend on so-called stick-and-rudder skills alone.

The Drawing Board and the Cockpit

In the 1930s writer-pilot Ernest Gann described himself as a "craftsman" and "a skilled artisan," with no interest in the machinery itself. "In the beginning many of us were scientific barbarians," Gann wrote, "we had neither the need nor the opportunity for technical culture."[18]

Yet during that decade an array of new devices emerged to aid pilots on long flights and in inclement weather. A variety of "instruments" began to populate the cockpit. Precision altimeters replaced crude air-pressure gauges, gyroscopic directional indicators complemented inaccurate magnetic compasses, and instruments dubbed "artificial horizons" replicated the reference points of the natural world inside the cockpit. These and similar devices allowed pilots to maintain their bearings during "blind flying," when they could not see out the window.[19] As historian Erik Conway notes, pilots had to learn to trust instruments more than themselves, making the transition from "natural" pilots who flew by sense, to "mechanical" ones who flew by rules and indicators.[20]

Exemplifying the era's changes in flying and piloting was James, "Jimmy" Doolittle who would become a national hero in 1942 for leading the eponymous raid over Tokyo. The SETP honored him at their inaugural banquet in 1957, but not for this feat alone. Before World War II Doolittle had already made a name for himself in scientific aviation.

Originally trained as a mining engineer, Doolittle joined the army during World War I and became a pilot, although he completed his training too late to see combat. He spent the 1920s experimenting with rapidly advancing aircraft, showing what they could do in long-distance flights. During that work, Doolittle observed a split between engineers and pilots. He found that "engineers felt pilots were all a little crazy or else they wouldn't be pilots. The pilots felt . . . that all the engineers did was zip slide rules back and forth and come out with erroneous results and bad aircraft." Doolittle thought it would help "to marry these two capabilities in one person." He went back to school "to be that person" to unify the two professions.[21] He enrolled at MIT to study aeronautical engineering, and in 1925 he earned his doctorate in aeronautics, one of the first in the field. After graduating, Doolittle became the U.S. Army's chief of flight testing, where he could "alternate between the drawing board and the cockpit."[22]

For his MIT dissertation, Doolittle had interviewed pilots and compared their impressions to flight test data. Pilots believed they could intuit the wind direction merely by the feel of their aircraft, thinking that their aircraft flew differently with a headwind as opposed to a tailwind. Doolittle's thesis proved them wrong. Pilots could not detect the direction or strength of the wind from the aircraft itself; only the sight of the aircraft's progress over the ground gave any indication of wind direction or strength.[23]

Doolittle's work began to bring engineering precision to the craft of piloting. A series of experiments sponsored by the Guggenheim Foundation began to look at the problem of "blind landing," when the ground was obscured by clouds or fog. As the project's test pilot, Doolittle found that the instruments pilots used, hardly more than a simple compass and an inclinometer, did not provide enough information for blind landings. He turned to industrial suppliers who were just completing developments of their own. The Sperry Gyroscope Company was then branching out from marine gyroscopes into aviation, and provided an artificial horizon and a directional gyroscope. Startup Kollsman Instrument provided a new kind of altimeter, which would be accurate to twenty feet or less.[24]

In 1929 Doolittle flew an entire flight, from takeoff to landing, using only these instruments and a series of radio aids to keep him on course, his view out the window blocked by a canvas hood. This experiment initiated rapid developments in so-called instrument flying, and Doolittle was awarded the Collier Trophy, the highest award in American aviation.[25] These same instruments—the artificial horizon, the directional gyro, the precision altimeter, combined with radio navigation—remain integral to aircraft cockpits and instrument flying today (though they are rapidly being replaced by virtual indicators displayed on computer screens).

The Guggenheim experiments, and the commercial interest of companies like Sperry and Kollsman, gave rise to a new subfield in aeronautics. The young engineer Charles Stark Draper at MIT, himself a pilot with a background in psychology and physics, began making scientific study of aircraft instruments. Draper called the new field "Instrument Engineering," and his research group the MIT Instrument Laboratory. He used the term "instruments" deliberately, to connote the clean precision of the scientific laboratory and strengthen the association between the aviator and rational modernity.

Until then, the science of aerodynamics and the study of engines had dominated aeronautical engineering. As flying became more precise, Draper argued, "The growth of an industry must be accompanied by a parallel development of suitable instruments." The field had its inaugurating moment in a session at the 1935 meeting of the Institute for Aeronautical Sciences.[26] Working closely with the new, high-tech instrument companies, Draper "instrumented" an MIT research aircraft and continued to develop blind-flying. Twenty-five years later, these same companies, Sperry Gyroscope and Kollsman, would again collaborate with Draper on the Apollo guidance system.

None could question that Doolittle was a pilot's pilot—he accomplished numerous aviation firsts and record flights and won virtually every national air race during the 1930s. Still, his doctorate-level engineering skills portended a change in the craft. With instruments in their cockpits, pilots would rely less on intuition and experience and more on dials and indicators. Instruments could also make flying easier and lessen the burden of training, which proved an enormous boon as flight training expanded during World War II. The Link Trainer, an early flight simulator used to teach instrument flying, enabled students to learn much of their instrument skill in a virtual environment, without actual flight experience.[27]

Furthermore, once engineers had accurate, responsive instruments in the cockpit, the next logical step was to have the instruments fly the aircraft directly, by closing "feedback loops" around the indicators, making flying automatic. Indeed, after decades of experiments, Sperry introduced its automatic pilot apparatus in 1930; airlines adopted it during the 1930s and the military during the war. An insecure pilot of the 1930s might have found these developments disturbing, believing that they would never displace manual skill and visceral feel in an airplane. An optimistic pilot, however, might have seen that instruments and numbers called for different kinds of intuition and skill. They enabled engineers to measure the art of flying, to quantify the sacred stick and rudder skills, and to design more responsive aircraft.

Flying Qualities

Just how did one design an airplane with the right mix of stability and control? Pilots pointed out that mathematics and theory did not necessarily produce the best airplanes. The question of what pilots wanted and what they liked still dogged engineers. In the 1930s, they began surveying pilots on this question, and were surprised to find that the pilots' opinions had little to do with the numbers that defined an airplane's performance.[28] Some aircraft designed by the most exacting engineers proved to be unpopular with pilots, while others that looked less remarkable on paper proved to be favorites. Engineers lacked the tools to describe what became known as "flying qualities"—the varied and subtle characteristics that made an airplane satisfying to fly. If the controls were too "heavy," the airplane felt sluggish. If they were too "light," the airplane was unruly and difficult to control. "Stable, but not too stable"—what did that mean?

Here it is worth taking note of the National Advisory Committee on Aeronautics, or NACA, the predecessor to NASA (unlike its successor, "N-A-C-A" was pronounced as individual letters, and not a single word). One cannot grasp the history of aviation in the

United States without understanding this organization, and one also cannot begin to assess the engineering cultures of Apollo without appreciating their NACA roots.

NACA was among the most innovative and productive research programs in the twentieth century, generating much of the modern science and technology of flight. Founded in 1915 as an effort to organize American research for World War I, it proceeded to operate a set of laboratories that churned out everything from basic airfoil shapes to test data on the latest military bombers.[29]

The agency's flagship laboratory was the Langley Memorial Aeronautical Laboratory in Langley, Virginia, founded in 1917. There engineers and scientists concentrated on aeronautics; there the wind tunnel reigned supreme.[30] NACA Langley also produced the engineers who would lead the nation in manned spaceflight, and who would conceptualize and run the Apollo program.

In the 1920s, NACA Langley engineers began studying the problem of designing stable yet responsive aircraft and brought the subjective notion of an aircraft's flying qualities into the realm of engineering precision. Researchers measured the forces a pilot exerts on the cockpit controls and how they translate into an airplane's motion.

Soon a group grew up around a young Langley engineer named Robert Gilruth. Building on earlier work by engineers like Edward Warner and Hartley Soulé, and working with engineer/test pilot Melvin Gough, Gilruth took fifteen airplanes with different types of controls, installed the latest instruments, and measured a series of flight parameters. Gilruth asked "what measured characteristics were significant in defining satisfactory flying qualities, what characteristics it was reasonable to require of an airplane, and what influence the various design features had on the observed flying qualities." He did not come up with precise specifications, but rather with ranges of control responses that pilots would find agreeable.

It turned out, for example, that a pilot controlling an airplane was more sensitive to the *forces* on a control stick than to its *position*. Gilruth defined a famous parameter, "stick force per g," to measure how much the aircraft accelerated in each axis for a given amount of stick force. He defined similar parameters, and methods for measuring the pilot's opinions, for other control axes as well.[31] In 1941 Gilruth published a set of requirements for the flying qualities of aircraft that the navy and army air force quickly incorporated. Gilruth's paper became a milestone document that guided aircraft designers for decades to come.

Flying qualities research made for interesting work, calling for close collaboration between pilots and engineers. It also required innovations in instruments, for the engineers had to measure and record a variety of parameters in flight—from the forces the pilot was placing on the controls to the airflows over the wings.

Engineer W. Hewitt Phillips exemplified the field. He had studied instrument engineering with Draper at MIT. As a young engineer in 1940 Phillips joined Gilruth's group, the Flight Research Maneuvers section, soon renamed the Stability and Control branch. Phillips would plan the flight tests by providing a series of prescribed maneuvers for a pilot to execute in an instrumented airplane, and then analyze the data. It was the dream job for an engineer who loved aircraft, combining practical technology with data analysis and modeling.[32]

Others in the United States and elsewhere worked on flying qualities during these years (particularly at NACA's Ames Research Center outside of San Francisco), but Gilruth's group was the leader, and has special interest for our topic here. Gilruth later became chief engineer of NASA's manned space efforts, including the Apollo missions. Barely two generations of engineers separated the stability and control debates of the Wright brothers' era from the human-machine issues in Apollo. As we shall see, "flying qualities" and "pilot opinion" became formal terms that would recur throughout the aerospace projects of the 1950s and 1960s, and in engineering discussions of Apollo.

After World War II the arrival of jet engines brought not only higher speeds, but also entirely different aerodynamics and airframe designs. New realms of flying qualities called for new research, and a new kind of piloting. Cultural critic Roland Barthes, writing in 1957, noted "a sudden mutation between the earlier creatures of propeller-mankind and the later ones of jet-mankind." A *Saturday Evening Post* headline simply ran, "Jet Pilots are Different."[33] The jet age required new "flying qualities," for people as well as for aircraft.

Using the Sky as a Laboratory

Flying qualities research, especially in the jet age, emphasized a new kind of professional: the test pilot, both skilled at stick and rudder flying and familiar with engineering.

The Wrights designed, built, and flew their own aircraft, as did many of the early aviators. In effect, the Wrights were their own test pilots, though the term did not exist in their day. In the decades before World War II, test pilots occupied a unique position between those who designed the machines and those who flew them. During the 1930s, the test pilots began to specialize: corporate pilots, who tested aircraft coming off the production line of a manufacturer; service test pilots, who tested aircraft and weapons in the military; and research test pilots, who worked with research engineers developing the fundamental ideas and technologies of flight.

A 1938 Hollywood movie, *Test Pilot,* starred Clark Gable, just one year before he was to portray Rhett Butler in *Gone with the Wind,* as a similar character operating outside the normal social world, always pushing the limits of his profession. In his best selling book *The Right Stuff,* Tom Wolfe portrayed test pilots as reckless risk-takers, cowboys who could not fit into traditional professional molds and who made a living pushing aircraft to their limits, often at the cost of their lives.

Perhaps some of them were, and they did place themselves at risk, but Wolfe's image misses their essential feature: although skilled craftsmen, intimate with the feel of their aircraft, test pilots worked as research engineers. Their goal was to collect data. As the historian Richard Hallion has written, "A research airplane essentially uses the sky itself as a laboratory."[34] Increasingly over the course of the twentieth century, what it meant to be a test pilot was not only one trained in flying airplanes, but also one trained in engineering. Despite his prominence in Wolfe's account, Chuck Yeager belonged to an older breed, neither college-educated nor engineering-trained. "He was very good at flying aircraft and doing aerobatics," recalled Neil Armstrong, "but he seemed to have less interest in precision and getting information and drawing conclusions from that." By contrast, one Ames engineer described test pilot Joe Walker as "the most cautious man I ever met."[35]

Test pilots spent much of their time evaluating flying qualities, always in close touch with engineers on the ground (a feature of flight testing carried to extremes in Apollo). Test pilots understood not only how an airplane flew, but also why it flew. In addition to their cockpit skills, test pilots were also professional storytellers, experts at narrating and recounting their experiences. In the 1950s, as noted at the start of this chapter, they formalized their profession.

The Testy Test Pilots

In 1955, Ray Tenhoff, a test pilot for Northrop Aircraft, brought together a small group of his peers. The seven men met in a restaurant in Lancaster, California, a town in the Mojave desert about halfway between two major test flying centers—the air force's Plant 42 in Palmdale (where the space shuttles were later built), and Edwards Air Force Base, a few miles to the north. The men worked at these facilities or at the numerous aerospace contractors that dotted Southern California—companies like Northrop, Lockheed, Convair, or Douglas Aircraft. Tenhoff wanted to create an organization of pilots and a forum for informal exchange of tips and tricks of the trade. Test piloting was a dangerous profession and anything that one pilot learned could help save the lives of others, he reasoned.

Soon Tenhoff and his peers organized a larger meeting. The group dubbed itself the Testy Test Pilots Society and now attracted seventeen members. All agreed that "the primary purpose [of the organization] should be in the area of personal equipment, escape and survival, and safety of flight"—matters of life and death of interest to all test pilots. Despite this seemingly uncontroversial charter, the members were concerned about "the distinct possibility of misinterpretation of the purpose of the new society by industry." They took pains to assure their employers that this was not an "advocacy group."

Why this concern? The founders wanted to make clear that the new group would not be a labor union but rather a professional group—their model was the prestigious Institute for Aeronautical Sciences (IAS). The group also set its own boundaries, limiting its membership to "engineering type pilots" (i.e., pilots involved in "experimental" or "flight research"). After some heated debate, a few of the production-oriented members withdrew their participation. At the next meeting, the group dispensed with the flippant Testy Test Pilots and adopted a more sober name: the Society of Experimental Test Pilots, or SETP. Soon they had the trappings of a professional society: committees, dues, even stationary, lapel pins, and a logo—a gold X on a blue background (figure 2.2).

In the following years, the friendly group grew steadily and met frequently to incorporate and organize. The society's first roster in 1956 listed more than a hundred members, most representing the variety of aircraft manufacturers working on advanced projects for the U.S. government—North American, Lockheed, Chance-Vought, Curtiss-Wright, and others. Only nine were government employees—six worked for NACA, and one each for the air force, navy, and marines. Curiously, among these early

Figure 2.2
SETP logo (SETP, *History*. Reprinted by permission.)

members, only one would become an astronaut, a young pilot who had just begun flying for NACA at its High Speed Research Station, located at Edwards Air Force Base. His name was Neil Armstrong.

Pilots' Opinions

One way to distinguish the SETP as a professional society and not an "advocacy group" was through publication, and it began publishing a quarterly review in the summer of 1957. The papers published here and in the SETP's newsletter and then-magazine *The Cockpit* (many of them transcribed banquet talks), are a window into the dreams and anxieties of this distinguished group of aviators on the verge of the space age.

If we learn about technology development only from academic journals, then the picture is biased toward the viewpoints of engineers and researchers. If we study only newspapers and magazines, then the public account is all we see. Rarely, however, do the people operating the machines themselves have a voice outside the public, heroic accounts. Crystallizing that voice and making it heard were the goals of the SETP. The opinions in its pages do not all agree, and some weather the years better than others, but all reveal what it meant to be a test pilot in the 1950s and 1960s, and what the prospects for the future seemed in those exciting but uncertain times.

The SETP's first *Quarterly Review* came out just a few months before the inaugural banquet where General Horner questioned the survival of the pilot. It contained four papers. Only one, "The Trend in Escape from High Performance Aircraft," addressed safety, the SETP's primary concern. The other three raised issues that would come to preoccupy the SETP and its members in coming years: the appropriate role of the test pilot, the rise of automation, and how those issues played out in current flight testing programs. For the next decade the pages of SETP publications traced the technology and its social implications.

Consider George Cooper's "Understanding and Interpreting Pilot Opinion." Cooper, an engineer and test pilot at NACA's Ames Research Center in California, expanded on Gilruth's flying qualities work. Cooper noted the importance of pilot opinion, rather than quantitative factors, for determining flying qualities.[36] This situation, he wrote, emphasizes the critical role of the pilot, because it "imposes on the test pilot the responsibility for valid and consistent opinions on which design decisions can be based."

Cooper sought to turn the pilot into a more objective reporter, a kind of human flight instrument.[37] Cooper proposed "a more specific definition of our adjectives," a numerical rating system he devised that broke down flight into regimes of normal operation, emergency operation, and no operation. Pilots would evaluate a characteristic according to these numbers, which corresponded to standard adjectives, varying from "satisfactory" to "unprintable." A rating of 1, for example, represented "excellent, includes optimum," whereas a 6 meant "acceptable for emergency operation only" and a 10 meant "unacceptable under any circumstances." "The use of the rating system," Cooper wrote, "is not intended to discourage pilots from also using other descriptive—even colorful—words in describing their feelings."[38] Cooper's scale, later adopted as the "Cooper-Harper rating scale," became a standard for evaluating aircraft in both NASA and the military.[39]

Cooper's scale sought to standardize the pilot's answers, the raw data. Then the flight test engineers needed to weigh the answers and accommodate for the pilot's background and biases. A test pilot for a fighter plane already on the front lines, for example, would be more critical than a research pilot investigating a new technique. Ideally, a given aircraft would be evaluated by test pilots with a variety of backgrounds. Pilots tended to find characteristics of airplanes objectionable when they were new, but showed a remarkable ability to adapt. Human pilots proved so adaptable, however, that they sometimes actually masked potentially dangerous problems by learning to fly around them. Or perhaps they could master a certain situation, but with a degree of added concentration and distraction from other tasks that created a new problem. Hence, sometimes the pilot's initial impression should be of dominating importance, while other times pilot opinion should be considered only after a period of learning.

Cooper discussed one element genuinely different from the world Gilruth had lived in: the use of "ground simulators in which a human pilot is part of the loop." Simulators created a new role for test pilots: comparing the results of simulations with actual flight. The pilots could fly test flight profiles on the ground before risking their lives in the air and could point out problems for the engineers to help them improve the models and the simulations. Cooper's paper showed that "flying qualities" and "pilot opinion" defined by Gilruth and others in the 1930s were by no means obsolete in the age of supersonic jet aircraft; in fact they were maturing and becoming more sophisticated as the role of the test pilot evolved as well.

Electronic Stability and Supersonic Flight

Accompanying supersonic aircraft was a new technology that helped match the pilots to their machines: electronic flight controls. In the 1950s, the "century series" of jet fighters emerged, designated the F-100, F-101, F-102, F-104, F-105, and F-106 (so called because of their 100-numbers). These sleek, silver machines were designed to take supersonic flight out of the realm of research and into aerial combat. All of them experienced prolonged, difficult flight test periods as the practicalities of supersonic flight ran into novel, potentially fatal problems. "Designers became increasingly discouraged," one engineer recalled, "when the tail end of their latest creation continually competed with the front-end to see which could be first to greet that pure, clean air."[40] Solving these problems brought electronics into the mix.

Early debates over stability had primarily to do with the airframe itself: the size and shape of the wings, their dihedral (i.e., their angle relative to the fuselage), the location of the tail surfaces, and so on. This situation had already begun to change in the 1930s with the introduction of automatic pilots. A device to keep the wings level or hold altitude, for example, can be understood as improving the stability of an aircraft. By the 1950s, stability was no longer an idea limited to the structure of an airplane, but could be improved by active devices that actually "flew" certain parts of the plane for "stability augmentation."

Stability augmentation brought the esoteric, but rapidly growing world of feedback control into the realm of flight. During World War II, electrical engineers had begun to study a wide variety of machines under the category of "feedback systems." These systems compare the "desired" output of a system with its "actual" output, subtracting them to derive an "error signal." The error is then "fed back" into the input, inverted, so it drives the difference between "desired" and "actual" to zero. The system thus seems to pursue a "goal" of bringing the actual and desired states together.

A simple enough idea, but any system has dynamic properties (like inertia) and may overshoot its goal, causing oscillation or instability. An electronic amplifier, just like an airplane, can be stable or unstable, sluggish or responsive. Engineers began to see that electronic amplifiers, speed regulators, automatic pilots, and even early computers could all be understood in these similar terms of feedback, stability, and oscillation. Hence a single set of mathematical tools could be applied to all, and one type of system could be used to simulate another.[41]

After the Second World War, aeronautical engineers began borrowing feedback techniques from electrical engineering. For example, the "frequency response" method introduced by Henrik Bode allowed one to study an amplifier's stability by breaking down its behavior into a series of oscillations of different frequencies. In 1948, engineer Walter Evans, at North American Aviation (the company that would produce the X-15 and the Apollo command and service modules), expanded on Bode's work by introducing a graphical method to plot the roots of complex equations without actually solving them. Evans's technique quickly influenced engineers building flight controls, and his "root locus" plots became a common method for studying the stability of everything from electronic amplifiers to the Saturn V rocket.[42]

Evans developed his technique while working on the Navaho, an unpiloted missile, but the electrical engineers' techniques also applied to traditional aircraft. Much flight testing in the 1950s involved the pilots' introducing "impulses" into the aircraft controls, so engineers could measure the resulting oscillations, all part of the frequency response technique. "One could almost say that a new branch of the engineering profession came suddenly into being," ran an authoritative textbook on flight controls.

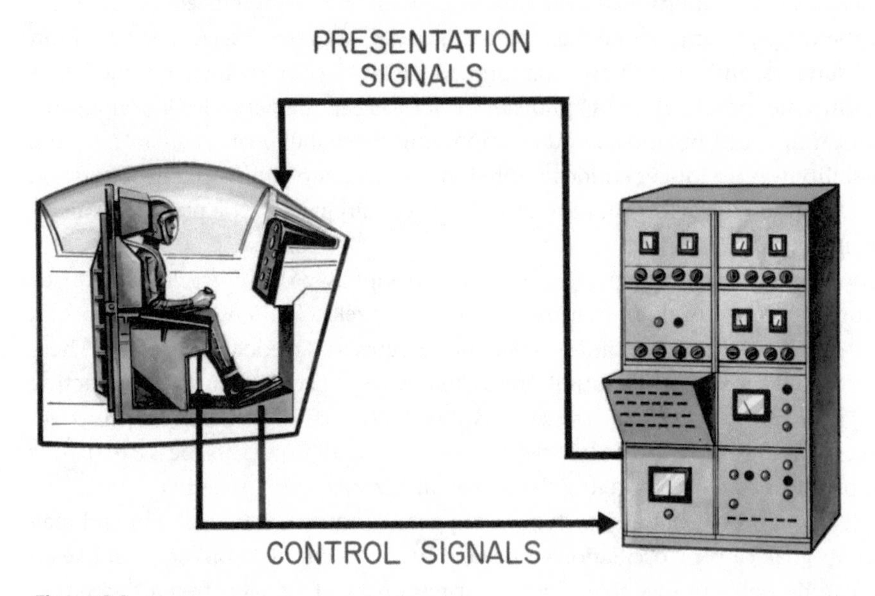

Figure 2.3
A canonical feedback system with a human pilot "in the loop" of a flight simulator. Pilot controls the simulated aircraft using "feedback" from computer-controlled flight displays. Such loops could be stable or unstable. (NASA Dryden photo E-5636.)

"Men were proud to call themselves feedback systems engineers, or 'systems engineers.'" By 1960, engineers began using "pilot-in-the loop" analysis, where the human was included as a component in the feedback system (figure 2.3).[43]

Once the frequency response of a system was understood, these methods also showed how stability could be improved—not only by changing the structure of the aircraft, but also possibly by adding electronics to change the response of the components. Beginning in the late 1940s, engineers began improving the stability of aircraft with stability augmentation devices, colloquially known as "black boxes"—small electromechanical devices that aided the pilot's motions. On one hand, these devices could improve the stability of a given aircraft for less cost, in performance and weight, than adding to or enlarging the wings or the tail. On the other hand, they encroached on the pilot's abilities, for they literally took over some control of the airplane and put it into a circuit or mechanism.[44]

In today's world of semiconductors where millions of transistors exist on a single chip, it is easy to forget the state of electronic devices in the 1950s: they were large, heavy, unreliable, and, worst of all, mysterious. Pilots had trouble accepting them into their craft. Al Blackburn, one of the SETP's founding members and its third president, had a master's degree from MIT and flew flight tests for North American Aviation. Writing in the SETP journal, Blackburn noted the pilots' skepticism of these devices: "Five years ago, should an engineer suggest to a pilot that a given aircraft stability or control deficiency be remedied by the addition of a 'black box' the pilot usually would regard such a designer a complete sadist."

Pilots did not want to entrust flying qualities to a black box. Then as now, black box stood for a mysterious gizmo whose inner workings are obscure to the user. The situation was changing, Blackburn reported, with the increased reliability of electronics, their smaller size, and the increased weight of jet aircraft (which made the electronics a smaller proportion of their overall weight). Most important, the increased performance of jet aircraft meant that sometimes designers could no longer make their aircraft stable. Blackburn noted, "The simple truth is that the range of performance of today's fighter aircraft has exceeded the designer's ability to provide satisfactory inherent stability for all flight conditions of which the airplanes are capable."[45]

During the 1950s, a variety of stability augmentation devices emerged to improve an aircraft's stability characteristics. "Yaw dampers," for example, would subtly jog the aircraft's rudder to keep it from oscillating under certain flight conditions, and are still common on jet aircraft today. Other devices served similar functions for pitch and roll. Some stability augmentation systems improved the pilot's sense of the aircraft, providing an "artificial feel" for hydraulically actuated controls, and some were built into automatic fire control systems to help the pilot keep his weapons on target. In his SETP paper, Blackburn acknowledged that artificial stability augmentation was "changing from a desirable appendage to an absolute necessity."

Not all agreed with Blackburn's assessment. Neil Armstrong remembers, "Some pilots were wary of control surfaces moving without pilot input," because a short circuit or other failure could spin an aircraft out of control.[46] J. O. Roberts, also a test pilot for North American, wrote against the little black boxes that simultaneously aided the pilot and took over some of his control. Roberts argued, in "The Case Against Automation in Manned Fighter Aircraft," that with the current variety of automated systems and instruments, "the pilot becomes excess baggage except for monitoring duties and the question arises as to whether he is paying his way or not."[47] The effort and weight put into automatic systems should go into improving the display of information for the pilot: "Utilizing this improved visual presentation we are able to place the pilot back in the control loop as an active servo mechanism." Up through Apollo, better displays of information became a common pilot request, for the pilots believed that given the proper needles and indicators, they could control anything.

Roberts voiced an extreme view, but he was raising an issue central for the SETP: the role of the pilot vis à vis automated systems. Throughout the society's early history it appears again and again, constantly debated in the formal *Quarterly Review* and the informal newsletter. The question concerned far more than simply new pieces of hardware; automated systems emerged from new groups of people, new types of engineers, and new conceptions of machines. Pilots found themselves enmeshed in the age of systems.

The Age of Systems

Indeed, pilots were not alone in their distrust of black boxes. Some engineers were suspicious of electronic stability augmentation devices as well, arguing that they were simply "band-aids" to make up for the deficiencies of the airframe's designer, who was not skilled enough to make the aircraft inherently stable. The critics updated the old chauffeurs versus airmen debate for an automated age: should the stability be built into the missile's airframe, or into the electronic controls that directed its motion? This question was not purely technical but also social, for stability and control were becoming less and less the domain of an aircraft's structural engineers or aerodynamicists and more the domain of electronics and "systems" experts.

Before going on and considering the test pilots' response to space travel, then, we must examine the rise of "systems," the major intellectual movement in engineering after World War II. The systems experts, the organizations they came from, and the ideas they developed played a major role in Apollo, though their dominance did not go unchallenged.

World War II coalesced systems thinking in several arenas. In response to technical problems like radar and automatic gunfire control, engineers began to see that all components of a system needed to be understood together, rather than as a hodge-podge of components. Engineers learned to conceptualize their machines as integrated systems with feedbacks and dynamics, where the behavior of each part helped determine the behavior of the whole.

By 1950, these ideas and techniques began the self-conscious era of systems thinking. *The Oxford English Dictionary* shows that uses of the term system exploded after 1950, including systems engineering, systems analysis, systems dynamics, general systems theory, and a host of others.[48] Each field had its own innovators, its own emphasis, and its own home institutions and professions, but they shared common concerns with feedback, dynamics, flows, block diagrams, human-machine interaction, signals, simulation, and the exciting new possibilities of computers.[49] Norbert Wiener's *Cybernetics* (1948) exemplified the trend, arguing that feedback control and statistics evoked analogies between computers and organisms, social systems, even the mind itself.[50] The idea of the cyborg, part human, part machine, emerged as Wiener-inspired NACA researchers considered the future mix of mechanical and organic necessary for spaceflight.[51]

The management aspects of systems engineering formalized in the mid-1950s when the air force stretched its resources to quickly build an intercontinental ballistic missile (ICBM). In the Atlas missile project, management began to move beyond the model that had dominated the aviation industry for decades. Aircraft had always been composed of large numbers of components from a variety of subcontractors, coordinated by the prime contractor who built the airframe. In Atlas, airframe companies, with their emphasis on structures and manufacturing, lost their central role. Now dynamics, interconnection, and coordination became the dominant activities. Rather than mechanical designers, engineers with management experience and an understanding of dynamics and control coordinated the project.

The technical change entailed a social shift. New types of engineers, with new backgrounds, educations, and technical values, took charge.[52] "Interface" became the key idea; systems engineers believed that if they could define the connections between elements in the system rigorously and completely, then the internals of the black boxes could be left to the subcontractors.

Many of the innovators in cold war systems engineering had their roots at General Electric and AT&T, via the aviation industry. Simon Ramo had cut his teeth at GE and Hughes Aircraft, and earned a Ph.D. at Caltech. His friend Dean Woolridge came out of Bell Labs. In 1953, the two left Hughes Aircraft to found a systems-engineering contractor, Ramo-Woolridge, to do systems engineering for the air force (it soon became the TRW Corporation). Together with the air force, they coordinated contractors and scheduling and oversaw the project's integration. This systems-oriented ICBM culture trained Apollo leaders like Joe Shea, George Mueller, and Sam Philips.

The navy had a similar project to build a ballistic missile-firing submarine, named Polaris. Here the navy's Special Projects Office performed the systems engineering function.[53] When MIT's Instrumentation Laboratory built the guidance computer for Apollo, they modeled their machine and their organization on the Polaris project.

Ramo became a promoter of systems engineering (even making an appearance on the cover of *Time* magazine), which he defined as "the design of the whole from the design of the parts" (figure 2.4). As he wrote, "Systems engineering is inherently interdisciplinary because its function is to integrate the specialized separate pieces of a complex of apparatus and people—the system—into a harmonious ensemble that optimally achieves the desired end."[54] Atlas included a system of materials, logistics, computers, and ground support, and the missile itself was a system. In Atlas, Polaris, and other large projects of the 1950s, systems engineering meant coordinating and controlling a variety of technical and organizational elements, from contract specifications to control systems, from computer simulations to deployment logistics.

Technical and social aspects of systems coalesced around computers. Both analog and digital computers figured prominently in the image and practices of systems engineers. They could simulate systems and make predictions about performance in an

Figure 2.4
Rise of the systems men: Ramo and Woolridge on the cover of *Time*, with a cybernetic creature in the background. Their mathematical, abstract approach to engineering did not always sit well with the more hands-on aircraft engineers. (Time Magazine ©1957 Time Inc. Reprinted by permission.)

uncertain environment. Both the computer and the systems analysts carried the prestige of science: providing dispassionate, expert advice free of political influence. For the strategy to work, the systems engineers required a certain amount of authority. They sold systems engineering as an authoritative scientific way to transcend "politics" (whether public or military-industrial) with the outside neutrality of the expert. Systems engineering thus elevated the "systems men" to a new level of influence, creating a new niche for engineers as educated managers of large projects and budgets.

Not Just "Fly-Boys:" Pilots in Space

For pilots, the systems men could represent a threat—they had engineered a fleet of air force weapons that had no pilots at all and their abstract, analytical approach to engineering could seem to crowd out the "human factor." These issues came to the fore as the test pilots began to contemplate spaceflight. As one flight test engineer remembered this period: "The 'head shrinkers' and the 'spasmatologists' had decided that man had reached the limits of his capabilities. There followed the great millennium of concentrated effort to design man out of the cockpit to make room for bigger and better 'black boxes.' There was much gnashing of teeth and waving of arms but alas, the day of the 'icy B.M.' was upon us. No one wanted the pilot around."[55] The metaphor "icy B.M." carries a triple entendre: it refers to an ICBM, the computers of IBM, and a pilot's view of a ballistic missile as a frozen bowel movement.

Early SETP publications did not so much as mention space, but in the wake of Sputnik the topic began to appear, and soon it came to dominate. It was first brought up by Al Blackburn, in his second appearance in the *Quarterly Review* in its first three issues. He retained the outlook of an earlier era while looking ahead, explaining the sources of the pilots' early anxieties about space, later obscured by the topic's sheer ubiquity.

Blackburn argued that the ballistic missile, so vaunted by the systems experts, was a transitional technology, ultimately to be replaced by human-operated rockets: "Tomorrow's space explorer will no more yield his place to canines or automatons than would Mallory have been content to plant his flag on Everest with an artillery shell." His analogy elaborates the limitations of ballistic missiles—inflexibility, complexity, and prescribed trajectories, and their incompatibility with heroic exploration (notwithstanding that Mallory never made it home). Blackburn went on to propose a flight test program for manned spaceflight, using as a reference von Braun's book laying out a mission to Mars. Critical to this process, he wrote, would be the pilots' involvement at an early stage: "Let him delay too long in joining the initial design phase and his craft will have been automated sufficiently to relegate its occupant to a human guinea pig, along for the study of cosmic radiation effects or ennui at zero "g," a stunt rider contributing only biological interest."

For Blackburn, the presence of human pilots in space had direct consequences for the pilot breed in particular and "the dignity of man in general": "This pilot must prove his worth not just in terms of the weight of computers, servos, amplifiers, and actuators he can replace. More positively, he must show that no degree of redundancy can replace his reliability, no array of transducers can match his perception, nor can IBM's whole inventory of computers supplant his judgment."

Here one human characteristic begins to appear that will remain throughout Apollo: reliability. The pilot must insist on instrument displays that will show him all the details of the machines' operation so he can monitor automatic systems and take over at a moment's notice in case of malfunction. The pilot is the ultimate backup system. The pilot will ensure the success of the mission.[56]

Blackburn's flight test plan was well conceived and prescient—manned spaceflight programs would soon incorporate the steps he described. But he also moved the debate forward in a critical way: far from representing a threat to the test pilots, the space age offered the possibility of putting their biggest rival out of business, by putting pilots into ballistic missiles. As he put it: "Indeed, before its [the ballistic missile's] perfection, we may see it pass into obsolescence, outmoded by man's conquest of space." Elsewhere Blackburn wrote that "the era of the large intercontinental ballistic missile is merely a phase the duration of which is a matter of speculation but the demise of which is nonetheless certain."[57]

The pilots, of course, recognized that they could not stand still in the face of ballistic missiles, and their creators, the systems men. The answer lay in professional development. "We have also experienced a change in the make up of a test pilot," Blackburn concluded. "Doctors have their medical boards, lawyers have their bar examinations, and there seems to be no reason why professional aviators should not have similarly stringent tests of their qualifications."[58]

General Elwood Quesada expanded on this theme at the third SETP banquet. A decorated military pilot who had recently become the first administrator of the newly formed Federal Aviation Administration (FAA), Quesada told the group that given the recent hype about space travel, "It is not strange, then, that we hear a lot about crewless aircraft, about airpower without manpower ... as pilots, we must not let ourselves be caught in this mass sentiment." Quesada's solution to the dilemma was to improve the professional background of the pilots, emphasizing, *"The day of the throttle jockey is past. He is becoming a true professional, a manager of complex weapons systems"* (italics mine). Some testing of unmanned vehicles was necessary, he argued, but people should fly into space as soon as possible. "The test pilot is a remarkable combination of servo-mechanics and complex computers, integrated with human judgment. The computer element is packaged in a bone-encased box of 75 cubic inches. The servomechanical element of this complicated machine reaches its ultimate efficiency in the operation of the human hand."

The pilots' professional qualifications must improve to maintain their status in this new world, Quesada implored, to the point that "the test pilot, especially the experimental test pilot, is usually a graduate engineer, a superb pilot, a dedicated man, and more importantly, a thoughtful man."[59] Steeped in the imagery of cybernetics, Quesada's language also employs the white-collar keyword of the 1950s: manager.

General Charles Blair, a pioneering air mail and airline pilot who had flown the first solo flights over the North Pole, also addressed the issue head on. Blair speculated on the practicalities of spaceflight, where the astronaut "better have a good autopilot, perhaps a couple of them." He said: "We all realize we can't operate a space ship by the seat of our pants. This contraption will be loaded with automatic equipment, star trackers, horizon scanners and sensors, inertial navigators, accelerometers and gyros, digital computers and numerous other long-haired devices... the further out we venture into space, the more automatic and precise our equipment must be."

"But it's also a fact of life," Blair continued, "that pilots of all shapes and sizes are allergic to complete reliance on automation. Somehow we like to think that our grey matter isn't just going along for the ride." He concluded with a concise statement of the test pilots' professional ambitions at the dawn of the space age: "This space man must be an educated chap, not just a fly-boy."[60]

The SETP publications were not the only or even the most important forum where these debates took place. Its pages doubtlessly reflected numerous informal conversations in hangars, cockpits, and bars as pilots related nightmarish experiences of failing electronics, autopilots going haywire, and nerdy young men with computers who misunderstood their craft. Still, the SETP offered a forum where test pilots could talk freely among themselves while disseminating their ideas to the profession (in effect, creating their profession); its records provide a rare glimpse into the self-image and anxieties of this group of highly trained machine operators. As the decade closed, the experimental test pilots felt keenly the precariousness of their position: automation was making its way into the cockpit and analog computers were changing the basic stick and rudder skills, with digital computers not far behind.

Moreover, systems men were gaining prestige, authority, and experience in laying out the specifications and managing the largest projects. In the military, ballistic and guided missiles were making headway into traditional aviation missions. The systems-oriented mindset of the missileers was threatening to dominate the frontier of spaceflight. For some, the prospects looked gloomy and called for pilots to fight these threats at every new black box. For others, however, the rise of automation would engender a rise in education and professionalism among pilots, the "throttle jockey" to be replaced by the "systems manager." Just at that moment, a new aircraft was emerging that would push these skills to their limits.

3 Flying Reentry: The X-15

There is no necessity to defend an aviator for trying to put things in a perspective from which they've crept and get things back into balance. There is also no challenge on the part of aviators to automatic systems or electronic systems or sophisticated mechanical systems; it's the challenge of putting them into the perspectives that are necessary so that they are the best complement and augmentation and supplementary support you can give an aviator, an astronaut, or a pilot or whatever you want to call him.
—A. Scott Crossfield, "Pilot Contributions to Mission Success," 1963

If something is unmanned, I say, "The hell with it." This is a human endeavor we're living in.... I'm a pilot and an aviator. And I think that if we're going to do anything like go into space, or fly fast, it's pretty much an experience that a man wants to do. We're a human race, not made of machinery.
—A. Scott Crossfield, interview with Merlin, 1998

A Meeting at Edwards

In October 1954, the NACA Committee on Aerodynamics held its biannual meeting. Chairing the group was Preston Bassett, president of the Sperry Corporation, the pioneering control systems company that developed automatic pilots, gyroscopes, and a host of other aircraft instruments. The committee also included leading lights of aeronautics: Clark Millikan, the aerodynamicist from Caltech, Allen Puckett, a guided missile pioneer from Hughes Aircraft, helicopter designer Bartram Kelley, and a variety of other notable academic and industry representatives, including decision makers from the U.S. Air Force and Navy. Their meeting lasted two days, the first at the NACA Ames Research Center in Palo Alto, just south of San Francisco. Here the committee discussed their usual business: the need for more funding in automatic control, engine research, plans for conferences, and proposed projects.

On the second day the group flew south for a tour of Edwards Air Force Base, in the high desert of Southern California, and the NACA facility on the grounds of the base, the High Speed Flight Station. Great things were happening here, making the area second only to Kitty Hawk as sacred historical terrain for American aviation. In the American lore of flight, Edwards is a sort of shrine celebrating the flying machine in the garden of the west.

The NACA facility at Edwards had begun as a satellite of the famed Langley Memorial Aeronautical Laboratory in Virginia, bringing from Langley both its people and its culture of flight testing—without the wind tunnels. Its leader, Walter Williams, was an old Langley hand, had worked on stability and control during World War II, and had been the NACA project engineer for the X-1, in which Chuck Yeager broke the proverbial "sound barrier."

Now, in 1954, the desert station was marking a banner year. What had been barely a collection of huts and barracks during Yeager's flight seven years before now had a new, permanent building and a new institutional identity. It had just become its own center, the NACA High Speed Flight Station (HSFS, later renamed the Flight Research Center or FRC), independent of Langley and free to develop a culture of its own or, more accurately, to build upon the unique culture that had already developed out in the desert (it would later become the NASA Dryden Flight Research Center, and exists in this form today). Some felt that the laboratory-like atmosphere of the wind tunnel still dominated Langley, but the pilot reigned supreme at FRC.[1] Here, engineers and pilots established what center historian Michael Gorn describes as "the ground rules of modern flight research"—careful, graduated experiments, precise and electronic data collection, and, above all, intimate collaboration among engineers, pilots, and technicians.[2] At the same time, Williams established clear directions for the work: higher and faster.

In the afternoon the NACA committee sat down for an executive session to discuss the ultimate successor to these highly successful research programs, following Williams's higher-and-faster imperative. Two years earlier, the committee had recommended that NACA look at the problems of manned and unmanned flight between twelve and fifty miles high, and at unusually high Mach numbers, between four and ten times the speed of sound. Now the NACA group heard a report from a study group headed by Langley's John Becker, an expert on supersonic flow. His team, which included Langley engineer and Gilruth disciple Max Faget, studied the problem of high-altitude hypersonic research and concluded that the critical work concentrated in two problem areas: the heating of an aircraft's structure due to supersonic shockwaves, and the "achievement of stability and control at very high altitudes at very high speeds, and during atmospheric re-entry from ballistic flight paths."[3]

Stability and control: here was the same problem the Wright brothers had encountered, the critical problem of aviation, extended into new worlds. Becker's report noted that while some of these problems could be addressed in the laboratory, much of the work would have to be done in actual flight. The experts pronounced a project feasible to the limits of Mach 7 (seven times the speed of sound) and several hundred thousand feet, the point at which the structure would simply melt or fall apart. They went on to suggest that the best way to approach these flight regimes would be a piloted aircraft, because test flights could begin in relatively benign conditions and proceed gradually to the most extreme. They recommended a new research airplane.[4]

Was it even possible to build an aircraft to fly on the edge of space? What would such a bird look like? Becker's team argued the aircraft should be launched from the air and powered by a rocket and should be "a piloted aircraft capable of being landed in a normal manner." Their arguments were not only technical but also pragmatic: build the simplest conceivable aircraft to accomplish the stated goals and get it flying as quickly as possible. And by all means put a man in it. They considered a man in the loop a high road to simplicity, eliminating the need for fancy electronics and employing the adaptive powers of the pilot to accommodate uncertainty.

The NACA committee reviewed the Becker report with enthusiasm. "It was the general sense of the majority of the members," the minutes reported, "that there are no known limits in flight to which we will or can take human beings, [and] that guided missiles have not eliminated the use of manned aircraft."[5] They recommended that a project to build this new aircraft be initiated immediately. The air force and navy representatives concurred, and the project was born, soon to be known as the X-15.

The Brilliant Dissenter

But there was a problem. The NACA committee was not unanimous. One member asserted, in a confident, expert tone, that a vehicle to explore these flight regimes should be unmanned. With no pilot, it would be less dangerous and could be built more quickly and cheaply. Progress in remote control meant that such a vehicle could be flying within two years.

Who was this lone dissenter, arguing against putting a pilot in the X-15? Some unappreciated engineer resentful of the glamorous test pilots? Some cost-sensitive bureaucrat? One of the steely missile men in love with automatic controls? No, the man who argued against a piloted X-15 was Clarence "Kelly" Johnson, legendary aircraft designer, chief engineer at Lockheed, and founder of its famous "Skunk Works." Johnson had already designed or helped design such pilot favorites as the P-38 Lightning, one of the highest-performance aircraft of World War II; the F-80 Shooting Star, America's first jet fighter; and the F-104 Starfighter, which NACA itself would heavily use in the X-15 program. Johnson would go on to notch his belt with the U-2 spy plane, the SR-71 (designed in 1960 and still the fastest piloted aircraft ever built), and a host of other historic planes, some of which are still in use today. He had won, or would go on to win, nearly every major award in aviation, aeronautics, and engineering, some of them several times. The argument for a human pilot in the X-15 was being questioned by America's leading aerospace engineer.

Taken aback by this heresy, this challenge to their exciting new project, the others in the room vigorously countered Johnson's objections. The whole point of the proposed program, they replied, was to study "the man" operating a hypersonic vehicle in the weightlessness of space. NACA was already doing research into high mach numbers on unmanned vehicles using rockets (Gilruth's Pilotless Aircraft Research Division at Langley), but these tests were not adequate replacements for manned vehicles. There followed a heated discussion.

Finally, the committee overruled Johnson's objections, invoking the cold war imperative of "maintaining supremacy in the air." Its members endorsed the NACA proposal for a new Mach 7 airplane.[6]

But Kelly Johnson would not be silenced. He attached a minority opinion to the meeting minutes. He reiterated his arguments against the manned presence in the X-15, and even questioned the very foundations of high-speed flight testing. If a manned, hypersonic vehicle were to be built, Johnson argued, it should be made useful for military purposes. Johnson simultaneously denigrated the science and impugned human presence, arguing that the FRC's much-lauded high-speed flights "have proven mainly the bravery of the test pilots."[7]

Johnson was an aircraft designer, not a researcher or systems man. He worked for a manufacturing company, Lockheed, not a government research establishment. He often made use of the aerodynamic data that NACA produced when designing his aircraft, but at this moment, in 1954, he did indeed have aircraft on the drawing board that would dwarf the performance gains made by the NACA group. His dissent may have been self-serving—perhaps he was arguing for approaches to the problem that would favor his Skunk Works at Lockheed. Perhaps he was just a headstrong man who liked contrarian arguments to push his colleagues to clarify their positions.

Whatever his motives, Johnson's arguments could not be dismissed as those of a lunatic fringe. The nation's premier aircraft designer, who created aircraft beloved by pilots, was arguing against manned, hypersonic flight. His argument had little effect on the decision to go ahead with the X-15, but it did signal the currency of the opposing arguments, and it helped put the program on the defensive: from its origin, the X-15 had to justify the manned presence in space.

The X-15 Project

The X-15 brought the old chauffeurs versus airmen dichotomy into the world of spaceflight and computerized control. It introduced changes in engineering practice and the role of the pilot that had significant personal, institutional, and technological connections with the Apollo program.

The black airplane had a radical appearance, but it was a logical, incremental step from the Bell X-1, consisting of a straight beam of a fuselage, short, stubby wings, a pilot up front and a rocket engine in the back. No fragile, crunchable airframe this: the entire aircraft was made out of a high-temperature alloy with the space-age moniker "Iconel-X." Far from the sheet aluminum normally found on the skin of an aircraft, Iconel-X was heavy-duty stuff for building pressure vessels and turbine blades, not light, delicate airplanes. The important feature of Iconel-X was its high melting point, so it could withstand the extreme temperatures of hypersonic flight.

NACA administrator Dryden described the flight path of the X-15 as being "like a fish leaping out of the water" because of its unusual parabolic arc into space. A B-52 bomber, converted for research purposes, would fly the aircraft to a launch altitude slung under its wing like a parasite on the belly of a shark. At about forty-five thousand feet, the bomber would drop the vehicle; a few seconds later (and a few hundred feet below) the pilot would start the engine, rapidly leaving the host behind. After accelerating to supersonic speeds, the pilot would pitch up and fly nearly straight up, passing out of the atmosphere into space. After just a minute or two, the rocket would cut off, and the vehicle would zoom up on a parabolic trajectory (figures 3.1 and 3.2).

As the pilot rocketed out of the atmosphere, he could maneuver the vehicle's attitude (i.e., its orientation) using small thrusters or "reaction controls." As it rounded the top of its arc, the pilot would point the X-15 back toward earth. Slowly at first, the craft would descend, and as it approached the atmosphere the pilot would pitch the nose up, to a high "angle of attack" (the angle between the airflow and the wings) for

Figure 3.1
X-15 being dropped from a B-52 bomber. (NASA Dryden photo E-4942.)

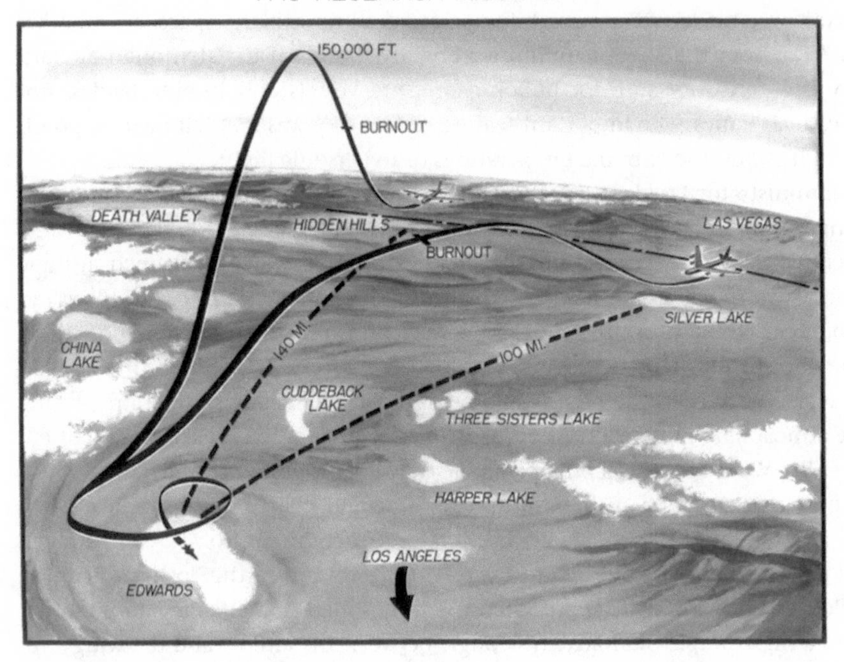

Figure 3.2
X-15 mission profiles, showing launch from a B-52 bomber and parabolic rise into space before reentry and landing at Edwards Air Force Base. Two trajectories are shown, one optimized for speed and the other for altitude. (NASA Dryden photo E-616586.)

reentry. After a series of turns and maneuvers to bleed off speed, the pilot brought the machine in for a "dead stick" (with no power) landing at Edwards (or another of the many lake beds around California and Nevada if something had gone wrong). The whole flight, from the drop of the B-52 to the landing, lasted barely ten minutes. Still, it was expensive, requiring an array of chase planes, rescue craft, and support helicopters.

The whole sequence was tracked and observed from a network of antennas and tracking stations dotted throughout the West. Known as the "high range," the network continuously logged the aircraft's position and speed and downloaded measurements from the vehicle itself. A master control station at Edwards coordinated all the data and plotted it on paper for engineers to monitor in real-time.[8] During the flight, the pilot was constantly in touch with engineers and flight controllers on the ground. A few years later, when NASA built a global network of stations to track the Mercury spacecraft as it orbited the earth, it explicitly modeled it on the X-15's high range, as well as on the X-15's ground-control protocols.

Yeager's original supersonic flights had been shrouded in secrecy; the X-planes of the 1950s, and initially the X-15, were not secret, but were the comparatively esoteric stuff of high-speed flight research. That changed in 1957, when the government and the public panicked in response to Russia's successful Sputnik launch. Suddenly, the X-15 assumed a new prominence as the sole U.S. endeavor to send men into space. For a brief period it became the darling of the press, hailed as a savior, the only American project that even had a hope of putting people in space.

In October 1958 NASA came into being and North American first rolled out the X-15, with Vice President Richard Nixon in attendance at Edwards. In September 1959 it flew under its own power for the first time, achieving Mach 6 and higher than 200,000 feet by the end of 1961. The X-15 flew until 1969, so the program experienced several different phases, the later ones obscured by the glare of Apollo. But when it first flew in 1959 it was the leading manned aerospace project in the country, pushing the envelope for human control.

The X-15 and Its Pilots

While the aircraft was being designed and built, NACA and the military selected a group of test pilots to begin preparation for flying the X-15. Many had flown in Korea. Several would fly in Vietnam. All had bachelor's degrees in science or engineering.

Scott Crossfield, the man more identified with the X-15 program than any other pilot, had been with the program since its beginning in 1955. Crossfield had first flown in an airplane the same year that Lindbergh crossed the Atlantic. He had been the first to fly Mach 2 and had flown virtually all of NACA's early rocket planes, but left NASA to become North American's chief pilot on the project, or as he put it, the "X-15's chief son-of-a-bitch." Though known primarily as a pilot, "I am an engineer, aerodynamicist, and designer by training," Crossfield stated, to explain leaving the exciting atmosphere of Edwards flight testing to manage the X-15 project.[9]

Neil Armstrong, the youngest of the X-15 group by more than five years, joined NACA's High Speed Flight Station as soon as he got his bachelor's degree from Purdue in 1955, at age twenty-five, having already flown nearly eighty missions in Korea. Before joining the X-15 program he had specialized in testing high-speed stability and control and handling qualities on the century series of jet fighters (figure 3.3).[10]

Other pilots who would fly the X-15 during the 1960s included Joe Engle, an air force test pilot; like Armstrong he would move from the X-15 to Apollo (serving as a backup pilot). In 1981, he would become the only person to fly a manual reentry of the space shuttle from Mach 25 back to landing, on Columbia's second flight. One of the later X-15 pilots, Milt Thompson, would spend his career in flight test at NASA Dryden, and would become a prolific and articulate chronicler of the X-15 piloting experience.[11]

Figure 3.3
Neil Armstrong with the X-15. Armstrong was the first to fly the craft with its novel adaptive control system. (NASA Dryden photo E60-6286.)

Breaking the Bronco

On its first flight, the X-15 immediately ran into a control problem. The B-52 dropped the vehicle into an unpowered test, and pilot Scott Crossfield glided to a landing at Edwards. At first, all had worked smoothly for the brand-new, strange, and heavy plane. But as Crossfield brought the X-15 in for a landing, its nose pitched up sharply, catching Crossfield off guard. He pushed forward on the stick, but the vehicle overcorrected downward. Then he pulled up again, but the nose came up too far. Crossfield later recalled, "Now the nose was rising and falling like the bow of a skiff in a heavy sea. I could not subdue the motions."[12]

Crossfield was in a scary situation that would become known as a PIO—"pilot induced oscillation." In effect, the human-machine combination was feeding back on itself and becoming unstable. This had nothing to do with the pilot's skill, but was inherent in the nature of the system. In fact, Crossfield's skill saved the airplane and his life, as he managed to touch down on the runway at the bottom of one of the oscillation cycles. The landing gear was crushed, but otherwise the program was saved.[13] Like the Wright brothers, Crossfield used a horse analogy: "The X-15 had an independent and contrary personality of its own. I was determined to break that bronco before it got me. I won."[14]

Though easily solved, Crossfield's problem with the new plane stemmed from the design of the X-15's control systems. The "systems view" of flight controls that engineers had adopted during the 1950s treated the human operator like a component that could be modeled like any other device.[15] Borrowing from electronics, engineers used the term "gain" to describe one aspect of those components. The term originally refers to an amplifier, and the gain is nothing more than the volume control—how much the amplifier amplifies, and how much output there is for a little bit of input. From a systems view, the airplane itself behaved like an amplifier: the control system amplified a little movement of the stick to a large movement of the aircraft. The pilot has a gain as well—for a given perception of the "error," the pilot will make a movement to correct it, but how much? What is the pilot's "gain"? Put another way, how sensitive should the aircraft's control system be? Should just a slight force on the stick produce a slight change in attitude, or a great excursion? These were questions that Gilruth had defined in his flying qualities research, and the right answer varied for each aircraft.

To answer this question the X-15 designers built a simulator to model the aircraft. A simple idea: put the pilot "in the loop" in the simulator, then adjust the gain until the aircraft becomes easy to fly, until its flying qualities are satisfactory. But there was a problem, which led to Crossfield's PIO: when flying the real aircraft to a real landing, his "gain" was not the same as when he flew the simulator. In fact, it was much higher. Unconsciously, the pilots behaved differently during flights than in simulation. The stress of flying, the concentration, the very real possibility of losing one's life, made the pilot a more sensitive controller in a real airplane than in a simulation, causing the system to oscillate. As with many such feedback problems this one was easily solved by changing the "gain" on the control system—a bit of tuning. But the episode illustrated the importance and the pitfalls of simulation as a tool in development, and the subtle ways that it interacted with the control systems and the role of the pilot.

Analog Simulators at NACA

The techniques and technology of simulation were just beginning in aerospace in the 1950s, matured greatly in the course of the X-15 program, and played a central role in Apollo. Modeling flight was not a particularly radical idea for aeronautical engineers, for they had always worked with wind tunnels, which we might see as highly

Figure 3.4
Joe Walker in an X-15 simulator. Note the cathode-ray tube above the instrument panel for simulating a horizon view out the window in real time. (NASA Dryden photo E-10251.)

specialized simulators, designed to replicate the conditions of flight inside a laboratory. In the early 1950s, NACA engineers began using simulators to model new aircraft as they were testing them, and eventually to model them before they were built. Engineers even sometimes spoke of the airplane itself as an analog computer, solving a host of differential equations in parallel and real-time (figure 3.4).[16]

In an analog computer, the idea basically was to connect a vast array of electronic building blocks into a large model of a system. Because of their complexity, and the parallel nature of their computations, each simulation seemed to develop a personality of its own, being "frequently described as cantankerous, malicious, mulish, or other less friendly terms."[17] Yet this very physicality gave the engineers a deep understanding of the problems they were studying.

While they were designing the X-15, North American Aviation built an analog simulator of the aircraft. Eventually, it grew to six analog computers connected together, and about five hundred patch cords that "mechanized" the equations (the analog computer equivalent of programming).[18]

Simulators quickly became indispensable in the X-15 program. Flight planners "spent their entire working life in the simulator," and hence were experts on the handling of the aircraft. Pilots would spend fifteen to twenty hours in the simulator practicing for an eight- to ten-minute flight.[19] Older NACA pilots tended to be reluctant to accept the devices, and they avoided practicing maneuvers, or even entire flights, in the simulators before flying them in the air. Younger pilots, by contrast, like Armstrong, embraced the simulators more quickly, and became active participants in their design and improvement.[20]

Pilot Milt Thompson believed he could not have flown the aircraft successfully without practicing in the simulator, and he was not alone.[21] Pilots understood the benefits of the simulators—it was their lives, after all, that were at stake—but were also keenly aware of their limitations. Most of the simulators used for research were "fixed base," which meant they did not move, depriving the pilots of a significant source of sensation. Early simulators also lacked the "feel" of the control stick that, in a real airplane, allowed a pilot to sense the airflow and resistance forces around a plane. Until these features were added in the 1960s, one pilot commented, "it was just a pinball game," and pilots would sometimes correct the engineers when they found the simulators did not "feel" right.[22]

Thompson noted how the X-15 simulator never quite prepared him for the experience of flight. "In all my simulation practice for the flight, I had been very relaxed, smoking cigarettes, drinking coffee, and sitting in a slumped, head forward position," he recalled. On his first real mission, the drop from the B-52 jolted him to the top of the cockpit. Said Thompson: "Now all of a sudden I was viewing the instrument panel from a completely different perspective and suffering from tunnel vision."[23] In addition to their lack of acceleration or motion cues, the simulations did not include all physical phenomena, but rather only those the engineers already understood, such as the basic aerodynamics, the aircraft's major structural features and their bending modes. During one flight, feedback through the control system excited a high-frequency oscillation in the X-15's structure. Because the simulator model did not include subtleties of the structure, it did not predict the oscillation. "The success of a mission," Thompson concluded, "is still up to a pilot."[24]

The most glaring, and the most troubling, limitation of the simulators was their lack of visuals. During the 1930s, pilots had learned how to fly on instruments alone, especially through fog or bad weather, and for much of the high-altitude research flying of the 1950s and 1960s, the view outside the window was less important than the dials and gauges in the cockpit. But at low altitudes, especially during the critical operations of landing, the scene outside the window was key for the pilot. "The biggest problem I had was with the displays," wrote Thompson. "They didn't seem real. It wasn't a true-life thing. It was hard to correlate between real-life and looking at a meter."[25] Problems in simulating the visual environment of the moon would be endemic to Apollo, and an erroneous model of the lunar environment would in fact cause the Apollo 15 astronauts to lose their bearings in critical moments.

The other critical effect missing from a simulator experience was anxiety. As Thompson pointed out, the pilots could work in simulators with coffee and cigarettes, and could easily take bathroom breaks. They could start or stop the simulation with a switch installed in the cockpit. Yet when they flew the real airplane, particularly for the first time, the drama and danger of the experience changed the pilots' characteristics as black boxes in the control systems—frequently operating with higher "gains" than they did in the laboratory. "It is simple to evaluate a flight condition on a simulator, rate it subjectively, and reset when you lose control." But Thompson concluded, "Until a reset capability is provided in the airplanes, the success of a mission is still up to the pilot."[26] No matter how good the virtual world, no computer was going to keep pilots from flying. As one contemporary account dryly put it, the X-15 simulator "lacked the in-flight realism afforded by the rapid approach of the ground."[27]

The X-15 Control Systems

The X-15 was an unusual aircraft to fly. The pilot could not see the nose, and he could not see the wings. His full-pressure suit wrapped him up tight and isolated him from the outside world. He could not feel or touch anything directly other than the suit and gloves. He could smell nothing other than the pure oxygen he was breathing. In Thompson's words, "I was in my own little world. I was comfortable and secure and protected from harm. I had complete control over the environment in the cockpit." He remembered the wonderful feeling of being able to breathe and smell fresh air when the canopy was raised after the flight, but also a sense of intrusion, "that someone was invading my privacy" when he had to step out of the womb-like vehicle.[28] Many of these characteristics, of course, would reappear in subsequent manned spaceflights.

As if symbolizing the air force's preference for bombers over missiles, the X-15 didn't take off on its own, nor was it boosted from a rocket, but was dropped from an aircraft like the B-52's other, more deadly payloads. The mother ship, in effect, served as the first stage of rocket flight; then the pilot took over and flew the vehicle out of the atmosphere into space. In fact, the X-15 was intended to investigate the other end of the flight—what was then called "entry" but was soon rephrased "reentry" into the atmosphere. How would a vehicle negotiate the dynamic and uncertain transition from the vacuum of space to the flow and friction of the atmosphere? From spacecraft to airplane: this problem called for the latest in control systems, and, of course, for the pilots' skill.

The X-15 actually had *two* control systems, one with small rocket thrusters to control its attitude in space ("reaction controls"), and another with the traditional aerodynamic control surfaces (i.e., ailerons and elevators) for use in the atmosphere ("aerodynamic controls"). Spacecraft, of course, require reaction controls because the usual flapping control surfaces on an aircraft require air flowing over them to be effective. The real trick, when flying reentry, would be to make a smooth transition from reaction controls to aerodynamic controls. Could a pilot handle this complex task, especially the difficult period when both sets might be necessary (figure 3.5)?

To investigate this question, the X-15 actually had *three* control sticks. It had a traditional stick between the pilot's legs for aerodynamic controls. A smaller stick on the left side of the cockpit controlled the reaction thrusters. A third stick on the right-hand side also controlled the aeronautical control surfaces, but it was intended for high-acceleration flight when high g-force loads might preclude the pilot from handling the center stick. Thompson remembered that it became "a macho thing" not to use the center stick, but to fly the whole mission in the atmosphere with the side stick. Thompson felt he could fly better with the traditional center stick, but "my ego would not let me use the center stick, even in an emergency."[29]

Reentry Control

Ordinary aircraft operate in a kind of viscous soup, the air tending to damp down any small motions. In space, with no air and no such damping, a vehicle would continue any small motion until counteracted by an opposite motion. Hence maneuvering the X-15 in space required the lightest touch on the controls, as too much thrust would cause the motion to overshoot. The three axes of yaw, pitch, and roll simultaneously could be too much for the pilot to handle. This became a real problem during reentry, when the X-15 needed to stay within a tight range of attitudes (orientations or pointing angles) to avoid overheating. Also during that critical period, the response of the vehicle, its flying qualities, changed rather quickly in the transition to atmospheric flight airspeed and dynamic pressure changed. At first, when flying the simulator, the pilots found that they could not fly the reentry because even the slightest pilot input produced oscillations. The X-15 was unstable on reentry, even with the pilot in the loop. Only an automatic system could make it stable (figure 3.6).[30]

Fortunately, a solution to this problem was built in to the X-15: a stability augmentation system, or SAS. This device would automatically damp the natural oscillations of the airframe at high speeds by adding subtle control forces more quickly than the pilot could react. The SAS had the effect of electronically putting the aircraft into an air-like viscous soup, which would tame its unruliness in all three axes.

Figure 3.5
The X-15 cockpit. Note the three control sticks: one in the center for traditional aerodynamic controls, one on the left for reaction thrusters while in the vacuum of space, and one on the right for aerodynamic controls while under high-g conditions. The stick on the right also controlled the adaptive controller connected to an analog computer. Note the controls for the adaptive controller in the black box just above the center stick. (NASA Dryden photo E63-9834.)

But with the SAS required for reentry, what if it failed? Designed as an aid to the pilot, it now became a life-critical system. Yet the SAS was plagued by troubles, and often the X-15 pilots had to turn it off. The SAS problems put the pilots on notice: the automatic system could help you, or it could kill you. No wonder they distrusted black boxes.

Another solution proved simpler. Unlike typical aircraft, which have a rudder and vertical stabilizer only on the top of the fuselage, the original X-15 also had a ventral fin below the fuselage to keep it stable at low speeds and high angles of attack (on reentry, the angle of attack is high, as the vehicle "surfs" its way in). Engineers hypothe-

Figure 3.6
Comical view of the pilot's role during X-15 reentry. (NASA Dryden photo E-13794.)

sized that the vehicle should be more stable in reentry without that fin. They removed the fin, and a test flight showed that indeed this measure improved the stability of reentry. The solution supported the arguments of those who disliked artificial stability augmentation, believing the black boxes were mere bandaids to poor aerodynamic design. Nonetheless, this was the research process—a little pilot skill here, a little electronics there, some structural modifications. Similar trade-offs would come to characterize systems engineering on Apollo.

Eventually, because of these and similar efforts, one pilot recalled, "The entry piloting task is not nearly as formidable as once envisioned."[31] His comments point to an interesting paradox within the program. Did the pilots learn this difficult task? Or did the engineers redesign the aircraft and its systems to make the task easy? Both types of skills were required for reentry—those of the operators and those of the designers. Neil Armstrong remains uncertain that an X-15 reentry required a great deal of skill, but "it did require a good deal of practice on the simulator."[32] Electronics and computers tamed an unknown flight regime into a training task in a laboratory.

Adaptive Control on the X-15

Nothing pressed issues of control more than the innovative adaptive control system installed in the third X-15. In 1958, engineers at Minneapolis Honeywell, a leading manufacturer of control systems, built an analog computer that ran a model of a desired aircraft's response and handling qualities, and then programmed the computer to adapt as necessary to match the model. The analog computer literally ran a simulation of the ideal aircraft while installed in the real one. For the X-15, Honeywell's system could use both the reaction controls and the aerodynamic surfaces, thus combining the two separate flight regimes into one, and allowing the pilot to fly the reentry with a single control stick. The adaptive system could fly "outer loop" functions like holding a particular pitch angle or angle of attack (useful for reentry) or moving at particular angular rates in pitch or roll. The computer could theoretically give the X-15 ideal handling qualities under a wide variety of flight regimes.

In 1960, before it flew, the third X-15 blew up during an engine test on the ground, propelling pilot Scott Crossfield out of the aircraft but leaving him uninjured. NASA sent the vehicle back to North American for repair, and took advantage of the overhaul to install the Honeywell adaptive controller, dubbed the MH-96.

The MH-96 contained within it an analog computer running a simulation of the X-15, the very airplane it was supposed to be flying. This model represented an ideal X-15 under ideal conditions. When the pilot commanded the stick to move the airplane, he actually commanded the electronic model. Then a feedback control system, "the fighting heart of the adaptive system," gave whatever commands were necessary to the airplane itself to make it respond like the idealized model.[33]

The MH-96 could cause the "real" X-15 to fly like an "ideal" one, which would make it behave exactly the same under all flight conditions, from the vacuum of space right down to the ground. It would automatically mix the reaction controls and aerodynamic controls, so the pilot only needed one control stick, whether flying in the atmosphere or in space, or during reentry. Also included were several autopilot modes: the pilot could command the MH-96 to hold the X-15 constant in roll, pitch, or yaw, as well as to keep a constant angle of attack, which was most useful during reentry. Rather than fighting with three different control sticks, the pilot could usually fly by twiddling a few knobs. It seemed the ideal solution to extend the airmen's manual skills to the vacuum of space.

In 1961, with this system installed aboard the third and reconstructed X-15, Neil Armstrong, who among the active test pilots had come to be something of a specialist in control systems, flew the first three flights. On the first MH-96 flight (the forty-sixth X-15 flight overall), on December 20, 1961, at a conservative Mach 3.76 and 81,000 feet, the system had a variety of problems as the axes switched in and out, and Armstrong landed with the adaptive system engaged only in roll. Still, he flew the entire flight with a single side-stick controller. By the third test, he got the vehicle up to 180,000 feet, and noticed that the controller was holding the aircraft unusually stable. "At this time, I took a time out to look around and see out the window," said Armstrong, which he had barely been able to do before.[34] The adaptive control system relieved him of enough workload that he could enjoy his environment. He then flew the entire approach and landing with the side stick.

By the fourth flight, things were well tuned up and Armstrong was reporting to the ground: "The reaction control damping is exceptionally good. It flies as good as the airplane does on aerodynamic controls at low altitudes." He took the X-15 to Mach 5.3 and above 200,000 feet. While Armstrong was concentrating on the controls and evaluating their various modes, however, "I did not properly appreciate the altitude I was at... which caused me to go sailing merrily by the [landing] field." Upon reentry, he bounced off the atmosphere and reentered far down range. Armstrong banked to turn, but he was still high enough that the wings were ineffective, so the vehicle continued going straight; gradually he dropped the nose to gain speed, and the aircraft banked. Finally, the wings bit the air and he entered a supersonic turn, far south of Edwards, nearly to Pasadena. After the overshot turn, Armstrong just made it back to the Edwards lake bed. "Handling qualities during the flare were considered to be less desirable than on previous similar approaches," he reported. At more than twelve minutes, this was the longest ever (in endurance and distance) X-15 flight. It would be Armstrong's second-to-last before he was selected to join the astronaut corps.[35]

Despite this unusual close call, the pilots built up confidence in the MH-96 and preferred it to control for the high-altitude flights. "In theory, the pilots should have been completely satisfied. In real life, however, this was not necessarily true," Armstrong wrote, for the pilots had reservations.[36] For them, the MH-96 presented an unusual situation. Thompson found it "somewhat unnerving" to fly, because "some electrons were now moving the control system without any inputs from the pilot.... As a pilot, you hope the guy who designed this electronic control system knew what he was doing. In fact, you would like him to be in the airplane with you to be exposed to any adverse results."[37] Some pilots didn't like to use the pitch or angle-of-attack hold modes during reentry, because in case of an emergency, they preferred to be actively in the loop and ready to respond.[38] Said Thompson: "Such [automatic] modes can greatly reduce the pilot's concentration and workload, but this can boomerang."

By automatically making the aircraft's response invariant, the adaptive system could mask the effect of changes. Normally, when the nose goes up, the airspeed decreases, for example, but not on the X-15 with adaptive control. Thus "the pilot has the impression that the aircraft does not want to land.... The aircraft feels completely solid to the pilot right up to the point at which loss of control occurs."[39]

On one flight, Thompson felt the adaptive autopilot go "berserk." The stabilizers began limit-cycling (a kind of oscillation), at Mach 5.5. "The aircraft was essentially out of control" in at least two of its three axes, he reported dryly. "It was quite a ride."

The problem would come back to bite them in a fatal crash.

Killed by Adaptive Control

In 1967, pilot Mike Adams was flying the X-15 with the adaptive control installed, zooming past 250,000 feet. Once out of the atmosphere, instead of flying upward like an arrow, the nose began to fall off to the side. This change did not affect the trajectory; through the vacuum of space, the aircraft flew sideways as well as it flew straight. But Adams corrected for the change not by yawing to the left, but to the right—which exacerbated the error, and the vehicle continued to drift off to the right. He may have been distracted by an electrical arcing in one of the instruments the aircraft was carrying, or he may have been experiencing a fit of vertigo, which he had been prone to on previous flights. Unbeknownst to Adams, noise from the disturbance on the instrument forced the adaptive systems gains way down, essentially disabling the reaction controls. Flight controllers on the ground, commented, "A little bit high. . . . real good shape" was their comment.[40] They had no idea that the aircraft had turned completely around.

But as the X-15 arced over the top of its trajectory and began to reenter the atmosphere the attitude mattered a great deal. Becoming an airplane again, it needed to fly, and the X-15 wouldn't fly sideways or backwards. Soon Adams recognized these troubling effects. "I'm in a spin, Pete," he radioed to the ground. Spins are dangerous at any altitude and airspeed, although recoverable and not inherently fatal. Nobody had ever entered a spin, however, at Mach 5 and 200,000 feet. Supersonic shockwaves could impose any kind of damage on the aircraft.

Miraculously, at Mach 4.7 and 120,000 feet, the aircraft recovered from the spin—due to the SAS, the pilot's inputs, the X-15's inherent stability, or some combination of all three. Yet here Adams lost control to the computers. During the spin, the adaptive controller drove the gain to the highest extreme in pitch and did not reduce it as the vehicle entered the atmosphere. With hypersensitive controls, the aircraft now became unstable and began oscillating while the vehicle was falling 160,000 feet per minute. The nose pitched up and down like a porpoise, forty to sixty degrees every few seconds. The pilot could have manually reduced the system gain, but he did not—an understandable omission given the stress he was experiencing. Within a few of these violent cycles, after about a minute, the X-15 broke apart at about 60,000 feet and fell to the desert floor in several pieces, killing the pilot. It was the only fatality of the nine-year X-15 program.

The accident report blamed a combination of pilot error and the MH-96 oscillation. "The system functioned as designed," Thompson wrote, "but the design did not consider this particular and unique combination of conditions and pilot response."[41] An automatic pilot with complex, adaptive loops could behave in unpredictable, even dangerous ways as it encountered unforeseen circumstances. The X-15 stopped flying soon after.

The Adams tragedy highlighted the discomfort with which the pilots adapted to the new world of computers and automatic controls. "When first exposed, the pilots tended to distrust the systems and occasionally have the feeling that the system, rather than the pilot, is controlling the aircraft." Some pilots gradually learned to appreciate the benefits offered by the automatic control—the hold modes, the improved damping on reentry. Others had been "bitten" by it and came to distrust adaptive control.[42] Similar problems would arise in Apollo, whose control systems depended on digital computers and software with similar complexity, adaptability, and potentially unpredictable behavior.

Black Boxes and Gray Matter

In 1962 President Kennedy presented the Collier Trophy, the highest award in aviation, to X-15 pilots Crossfield, Peterson, Walker, and White. The award marked the end of the first phase of the program, when a number of the original research questions had largely been answered. This official pronouncement of success, from the highest levels, credited the project's pilots more than its engineers. The initial doubts about the feasibility of humans in the loop for spaceflight research, raised in private by Kelly Johnson, seemed to have been put to rest. But they persisted, for we find supporters of the X-15 constantly justifying themselves.

X-15 publications, from technical papers to statistical analyses and breathless PR, consistently emphasized the role of the human. Neil Armstrong was typical when he published his conclusion that "electronic equipment figures prominently in the X-15 flight and ground systems, but this hypersonic vehicle is an instrument of the pilot, depending on him for control and flight success.... The vehicle is controlled exclusively by the pilot in the conventional fashion."[43]

"How well do black boxes replace grey matter?" asked a speaker at an air force science and engineering symposium in 1962. The speaker was reporting on a comprehensive study of all X-15 flights up through January of that year to quantify what impact the human pilots had on the program. The study considered what would have happened were a "relatively simple automatic guidance and control system [installed] in place of the pilot." It found that numerous incidents during X-15 flights required the flexibility of the human in the cockpit. The study's conclusion: "the program would not have been very successful without both the pilot and redundant/emergency systems."

If the X-15 lacked either pilots or emergency backups, the report stated, then all three research aircraft would have been lost after an average of five flights. The study compared the X-15 to the BOMARC missile, which failed on more than half of its missions. If a man were in the loop of the BOMARC, the study argued, then it would achieve 97 percent mission success—about equivalent to the X-15's 96 percent. An extensive statistical study of the X-15 missions showed that the rate of success would have been considerably lower were it not for the human presence in the cockpit, about equal to the BOMARC's 43 percent success rate.[44]

Written with a statistical bent and rigor that seemed immune from the passions surrounding humans in the cockpit, the X-15 report explicitly addressed the ongoing debates over manned spaceflight. The air force's director of flight testing coauthored the foreword to the report, noting that "The X-15 program, because of its currency and similarities to the next generation of aerospace projects, provides a quantitative insight into the relative merits of piloted versus unmanned space vehicles."[45]

Yet the study proved deeply conservative about technology. It contained no deep consideration of digital computing or any discussion of remote controls from the ground, two technologies that would help make Apollo's pilots successful. Rather, the X-15 report compared human actors only to "a relatively simple and reliable 1958–59 state-of-the-art guidance and control system in place of the pilot." The justification was that any more complex automation would have its own failure rates that would have to be considered, complicating the problem. Improvements in reliability for automated systems, of the kind that would occur in the following decade, were not considered. Nor were the implications of applying to remote systems the meticulous, high-quality engineering routinely devoted to manned systems. The debate was framed as one between skilled pilots and mindless, unreliable automata. Despite its limitations, the X-15 report was frequently cited in later years as quantitative support for the human pilot in space.

Not only technical publications supported the human role in the X-15. A 1962 Hollywood movie, titled simply *X-15*, starred Charles Bronson and Mary Tyler Moore and repeated the official story. The air force contributed wonderful color footage to the film, reviewed the script, and asked the producers to ensure they gave adequate credit to the skill of the pilots.[46]

The film opens with a dramatic launch sequence narrated by actor and pilot (and Air Force Reserve General) Jimmy Stewart. "Some believe that the guided missile and electronically controlled missiles," Stewart begins, "are the ultimate answers to space flight." Released after John Glenn's orbital Mercury flight, the film promoted the X-15 as an advance over Mercury. "Man will never be satisfied sitting in a nosecone, acting as a biological specimen," Stewart continued. "Now the X-15 is ready, manned by a pilot who will make all the decisions for accurate control in flight, and reentry, and recovery." The script continually touts the technical skill and emotional stability of the pilots and the X-15's superiority over the Mercury capsule. "The X-15 pilot will be able to choose his angle of reentry, and control his speed and altitude and glide to his landing area," the chief engineer tells the press, "always under the pilot control. He has a choice."[47]

In 1964, on the tenth anniversary of the inception of the X-15 program, NASA compiled a summary of its research results. Where Mercury was a test of a human's ability to function well in space, "the X-15 was demonstrating man's ability to control a high-performance vehicle in a near-space environment." By this time Mercury was flying, Gemini was well on its way, Apollo was well defined, and "certainly the problem of launching the lunar-excursion module from the surface of the Moon through the sole efforts of its two-man crew must appear more practical and feasible in the light of repeated launchings of the X-15," with just the pilot and the B-52 crew to prepare for launch.[48]

An entire chapter of the research results volume was devoted to "man-machine integration." Here, the authors tackled the human versus machine question head on:

The X-15 program alone cannot disprove the merits of unmanned vehicles, since it contributes to only one side of the argument. Nor, on the other hand, does it glorify the role of the pilot, for it was only through the use of automatic controls for some operations that the full potential of the X-15 was utilized. Rather, the real significance of its excellent mission reliability is that it has shown that the basic philosophy of classical, piloted aircraft operation is just as applicable to the realm of hypersonic and space flight as it is to supersonic flight. That philosophy decrees that the pilot is indispensable, and that he must be able to override any automatic control, bringing his skill and training to bear upon deficiencies of machinery.[49]

"A fantastic variety of skills"

In 1969, NASA donated one of the X-15s to the Smithsonian Air and Space Museum; it hangs today in the main hall. The official press release for the donation read, "The capability of the human pilot for sensing, judging, coping with the unexpected, and employing a fantastic variety of acquired skills remains undiminished in all of the key problem areas of aerospace flight."[50]

In addition to contributions in hypersonics and related sciences, the X-15 put human pilots in space, ensuring them a place in the cockpit in future space missions. While the skill of reentry was easily handled by automated systems, the pilot's primary function evolved to be a redundant system, a systems manager, coordinating a variety of controls as much as directly controlling the vehicle.

The X-15 represented the state of the art in research into aerospace control systems and pilot interfaces when Apollo began. It also had had several significant connections to Apollo. Most obvious is Armstrong himself: the man who piloted the first lunar landing was an X-15 pilot who specialized in flight controls. The Apollo spacecraft was designed and built by a team at North American Aviation derived from the one that built the X-15, with Scott Crossfield, the original X-15 test pilot, as the program manager (by contrast, the Mercury and Gemini capsules were built by McDonnell Douglas). X-15 ground control and tracking protocols (including using a test pilot on the ground to communicate with the pilot flying the mission) set the stage for NASA's Houston-controlled approach to manned missions.[51]

Through these informal channels, and more explicit efforts discussed in the next chapter, the pilot-centered, "airmen's" approach to spaceflight, originated at NACA Langley and thrust into the space age by the X-15, began finding its way to the moon. It would not be unchanged by the journey.

4 Airmen in Space

The development of manned space flight is not just a matter of replacing a warhead by a manned cabin. Suddenly, a switch is thrown between two parallel tracks, those of missile technology and those of aviation technology, and an attempt is made to move the precious human payload from one track to the other. As in all last-minute switchings, one has to be careful to assure that no derailment takes place.

—Joachim Kuettner, Huntsville engineer

Last-Minute Switchings

In May 1961, a year and a half after the X-15's first flight, Alan Shepard became the first American to fly in space. He flew not in a rocket plane but atop a Redstone ballistic missile. Public response to Mercury quickly overshadowed the X-15, and missiles superseded aircraft as the first stage of human delivery into space. But the human role was far from settled. As Joachim Kuettner, a former test pilot and member of Wernher von Braun's engineering team, described above, human spaceflight in the United States required merging two technologies.[1] Each had developed an engineering culture of shared assumptions, approaches, and techniques. Pilots and flight test engineers envisioned astronauts as active controllers rather than passive cargo, guiding powerful rockets into free spaceflight. Rocket and missile engineers, by contrast, guided their vehicles with automatic feedback controls to transport astronauts from launch to orbit. Both Project Mercury (six manned flights, May 1961–May 1963) and its successor Gemini (ten manned flights, March 1965–November 1966) brought these cultures together, advancing but not resolving their divergent notions of human control in space. The merger, neither smooth nor preordained, discarded promising ideas, closely held beliefs, and envisioned identities.

Human spaceflight evolved in the 1960s into something quite different from how it was initially imagined. Rather than piloting rockets and hypersonic spaceplanes, astronauts found themselves enclosed in capsules aboard automated rockets. Nevertheless, NASA found roles for human operators that allowed them to "fly" their craft in new, unexpected ways. Pilots took on new roles as systems monitors, backup systems, and computer operators in novel operations like rendezvous and docking. As in the X-15, enabling these new tasks were electronic controls and computers. As engineers outlined early ideas for lunar flights, they drew on these earlier experiences and added still newer computer technologies. Apollo systems engineer Joe Shea called Apollo a "balanced" program: "balancing the history of missile technology and the history of aircraft technology. You'll note that the system is neither all man nor all machine. Man is in effect a sub system."[2] Whether human beings would be subsystems, commanders, or pilots to the moon depended on how Apollo engineers designed their systems.

The Two-Headed Monster

In August 1959, in the "Satellite Room" of the Miramar Hotel in Santa Monica, California, von Braun addressed a special dinner meeting of the SETP. The audience of more than fifty all-star pilots included Scott Crossfield, fresh from his first X-15 flight, and Harrison Storms, North American's star engineer-manager.

The renowned German rocket engineer was a former Nazi and S.S. member, but perhaps worse for the U.S. Air Force and NASA test pilots, he was working for the *army*. Where the X-15 was providing challenging tasks for pilots on reentry, von Braun's group focused on the critical problem of boosting payloads into space. As director of development for the Army Ballistic Missile Agency, von Braun spoke about new conflicts between missile engineers and pilots. He would articulate a powerful vision of the human role in future spaceflights that the pilots would find impossible to defeat.

Von Braun began his speech by acknowledging that the pilots saw him and the army as the "two-headed monster who . . . is trying to throw airplanes into the trash can." He did not deny the role.

Von Braun, himself a private pilot, professed to hold test pilots in high esteem. He found them invaluable as monitors and observers during test flights. "Just see how many firings of unmanned experimental missiles are still necessary these days to make a new prototype acceptable," he pointed out, versus the small number of aircraft required to certify a new design. Still, missile engineering was a completely new realm, distinct from aircraft development, and in the language of cybernetics, human senses must yield to machines. "We have to substitute telemetry for the human mind, eyes and ears. We have to substitute automatically controlled guidance for your hands and muscular systems." Missiles were not simply new machines; they also engendered new techniques and new engineering cultures.

Then came the German engineer's disturbing thesis: "When you consider the velocities and the forces involved in missile launchings, you come to realize that human intervention is not only impossible from the physical standpoint: it is actually undesirable." Not only would human payloads cause problems of reliability and stability for a missile, they would not even be able to help get a rocket off the ground. "There is little time for intelligent reaction during the powered phase of flight," von Braun told the assembled group of skilled pilots, men who liked to describe themselves as finely tuned machines with human judgment. "We like to think of man as an amazingly versatile computer. But in missile terms, he is outrageously slow and cumbersome."

In case he wasn't yet clear, von Braun answered the question of whether human pilots could fly rockets by hand: "In our more advanced existing rocket systems, human operation of controls, or human observation to follow a predetermined flight path during the high acceleration ascent, is simply out of the question." Guidance, abort sensing, even emergency ejection of the pilots from a failing rocket must all be under the control of automatic systems. The pilots, whom von Braun called "missile riders," would be unable to determine if anything were going wrong. Potential problems would be both subtle and fast. In such circumstances, von Braun said of the failing rocket's pilot, "He will just have to push the ejection button if the abort signal flashes and the automatic system fails."

Von Braun's picture was not all doom for the pilots, for he added that they must exercise their flying skills once the spacecraft and its human payload are comfortably in orbit and separated from the booster. Pilots would find their new home in the vacuum of space, and the unknown, new realm would bridge two cultures. "There is a common meeting ground—it is between us missileers and you experimental test pilots: in outer space."

Von Braun concluded his missive with what must have been a faint palliative to the assembled pilots, echoing their kickoff talk two years before when the issue was "survival": "Let us not tell you that your profession is dying: its greatest challenges lie ahead."[3]

The speech landed like an "icy B.M." on the group of space-hopeful pilots. The X-15 had first flown under its own power one month before. The Mercury spacecraft was under development; its silver-suited new astronauts debuted to the world earlier that year. They were already in training and would appear on the cover of *Life* magazine a month after von Braun's speech. Yet here was the nation's top rocket engineer barring these brave explorers from a major phase of future spaceflight: the exciting, fiery launch, forcing them to redefine their notion of "control." Now, being in command would mean having a finger on the abort button as a backup for an automatic system and doing something, as yet undetermined, when reaching orbit.

If they had read von Braun's 1953 book, *Conquest of the Moon*, compiled from his articulate, beautifully illustrated articles in *Colliers* magazine, the pilots might have grown even more concerned. Von Braun proposed sending dozens of men to the moon, but the "expedition leader" would be a scientist. The remaining crew was to include navigators, engineers, physicians, astronomers, photographers, even a mineralogical team, but no pilots. It would not need them, even for the lunar landing, which von Braun pictured as fully automated. Von Braun's grandiose vision of the future of spaceflight had enormous impact, but it did not include hands-on control (figure 4.1).[4]

After the talk, Al Blackburn, the SETP president, stood up to rebut, taking "sharp exception" to von Braun's presentation. Blackburn recounted his own experiences with brain-dead autopilots, broken fire control systems, and failed cockpit computers. Blackburn remembered the evening in his memoir, recalling that von Braun called for anesthetizing the astronauts during launch.[5] X-15 pilot Milt Thompson reported intense discussion following, continuing well into the after hours. As the official SETP history put it, "The generous ministrations of the local hostelry supplied more than ample quantities of unneeded lubricity in the management of which the German rocket pioneer showed astonishing prowess."[6]

The SETP published von Braun's speech in its monthly newsletter with a comment that they were glad to hear that such a leader as von Braun still considered the pilot an essential element in outer space. The society also took on a renewed mission, "to convince individuals such as Dr. von Braun that the pilot can offer even greater aid to the space craft designers during the high acceleration launch period."[7] The pilots were not easily dissuaded. After half a century of wonderful, hard-earned progress, flying the shiny new rockets off the pad seemed their destiny. They would hold onto their dream, at least until the prospect of true control in space required comparable skills, and offered comparable rewards. But that was years away.

Dyna-Soar and the Johnsville Tests

Von Braun's speech eliminating the pilots from the boost phase actually countered an active research program. While the X-15 program dealt primarily with the transition from the vacuum of space back to the atmosphere, pilots and researchers were beginning to realize that there would be more to human spaceflight than reentry. In fact, the most dramatic and dangerous part of the flight might well be the launch, and here pilots wanted a role.

An air force program called Dyna-Soar inspired their position (figure 4.2). The U.S. Air Force began the program in 1957 to develop a manned bomber or reconnaissance plane to be launched into orbit. Boeing got a contract for a one-man test vehicle dubbed X-20 or Dyna-Soar, for its ability to glide hypersonically across the upper atmosphere. Dyna-Soar would take off like a rocket, fly through space, reenter on wings, and land on wheels like an airplane.

Numerous changes in direction, a lack of overall coherence, and its eventual cancellation in 1963 showed Dyna-Soar was aptly named. It eventually consumed more than $400 million, and is frequently hailed as a predecessor to the space shuttle.[8]

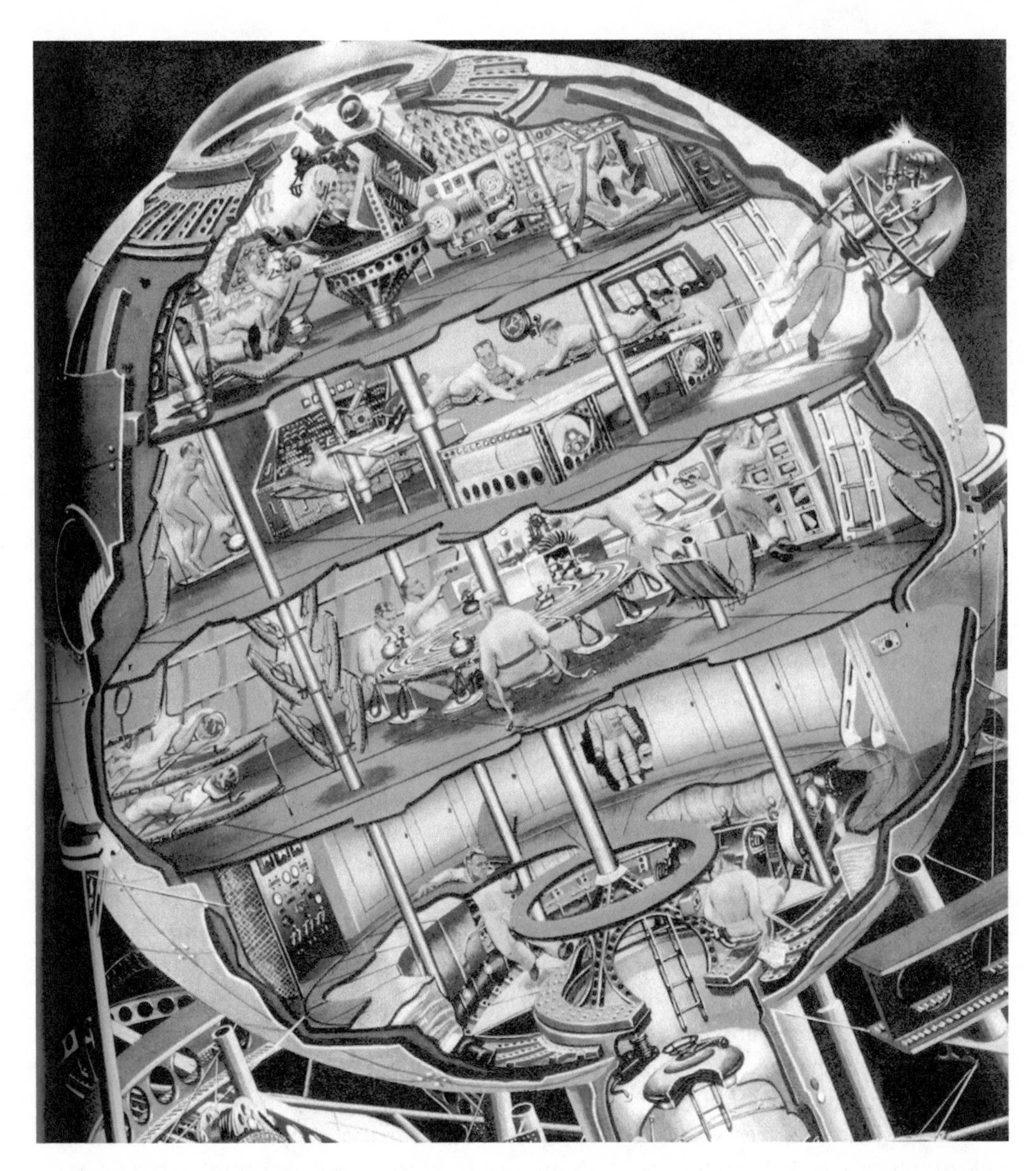

Figure 4.1
Wernher von Braun's early vision of spaceflight included scientists, engineers, physicians, and photographers, but not pilots. In this illustration by Fred Freeman the ship is on automatic pilot, monitored by the crew on the top deck. The commander sits in a cybernetic chair in the center of the top deck, reminiscent of a World War II fire-control system. Note the navigator on the top right taking star sightings, a prescient analog to the Apollo optical system. The figure on the top deck on the right is a caricature of von Braun, and the figure floating between the bottom decks is Willy Ley. (Von Braun and Ryan, *Conquest of the Moon*, 63. Reprinted by permission.)

Figure 4.2
Dyna-Soar reentry. (U.S. Air Force.)

Dyna-Soar consisted of a manned, unpowered glider launched atop a large booster like a Titan II ICBM, or even on von Braun's new Saturn stages then in development. Proponents saw the X-20 as "a combination of airplane and guided missile, and the duties of the pilot will be comparable with the duties of the engineer in a railroad locomotive bound to a given track."[9] An air force promotional film left no doubt as to the level of automation to be deployed in the reentry: "This Dyna-Soar project puts an emphasis on the pilot, on the *man*. By putting *man* at the controls, the Air Force has carried forward into space that journey started by the Wright Brothers a little more than a half a century ago."[10]

A look at advanced studies for Dyna-Soar recovers a critical, lost piece of pilots' initial picture of human spaceflight: flying the rockets off the pad. To demonstrate this ability, a group of test pilots used a novel simulator to conceive and advocate a role for themselves in manned rocket flight.

In 1959, just a few months before von Braun's speech, Neil Armstrong and a team of X-15 pilots and engineers went to the Naval Aviation Medical Laboratory in Johnsville, Pennsylvania, which had the country's most capable centrifuge. By spinning the pilots around at prodigious rates, the centrifuge could impress enormous g-forces on the human body. A "capsule" at the end of a massive arm could carry a person, instrumented for experiments. Moreover, a simulated cockpit in the capsule allowed pilots to perform tasks under the g-forces they would experience in flight.[11] The X-15 pilots went to Johnsville to practice controlling a vehicle under high g-force conditions; all agreed

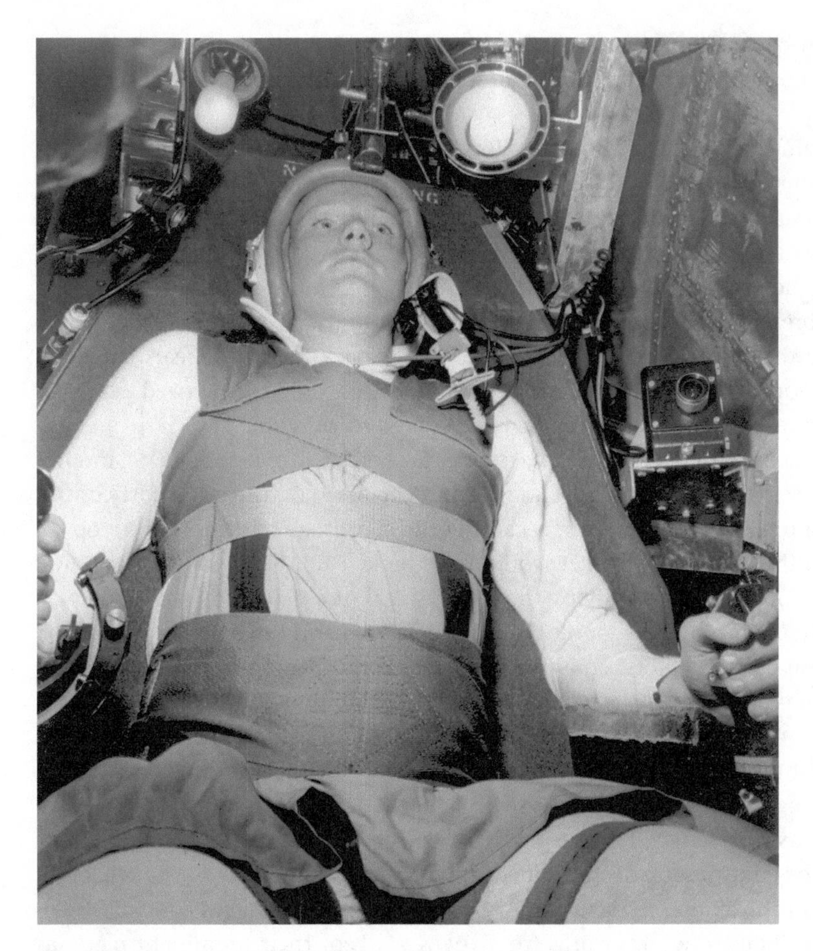

Figure 4.3
Neil Armstrong in the Johnsville centrifuge for testing piloted boost control, 1958. (NASA Dryden photo E-5040.)

the experience improved their confidence about going into the X-15's extreme environment (figure 4.3).[12]

While at Johnsville they also experimented with controlling a booster rocket. Engineers programmed an analog computer to simulate the equations of motion for a multistage launch vehicle like the proposed Saturn V. Reclining on custom-fit couches designed to support their weight under a high g load, the pilots stared at a control panel with an array of dials and needles. They held a side-arm controller in their hands to "fly" the simulated rocket. A set of circuits and switches operated by an engineer in a control room could mimic the transitions between rocket stages, aborts, and other anomalies. Seven pilots, including Armstrong, tried flying the simulated rocket under these conditions.

As the centrifuge whirled around, the pilots in the capsule watched a needle on an indicator and moved their stick to keep the needle centered. They performed as human servomechanisms, using feedback to keep the "error" indicated by the needle at zero. High g forces tend to narrow the vision, and at 9g, the pilots could still see a few instruments. At 12g they could still operate the hand controllers but experienced chest pain and difficulty breathing (although the pilots still believed they were in control). Forces fourteen times the force of gravity pushed the pilots back into their seats, and narrowed their vision so the only instrument they could see would be the needle right in front of their faces providing the error signal necessary to guide the vehicle.

Staging proved particularly difficult, when one stage of the rocket cut off and another kicked in with a powerful jolt after a short delay. Several times, the pilots lost control and the virtual rocket veered off course, causing the centrifuge to trip off and stop. In the real world, this meant the rocket would have broken up. Numerous other times, however, the pilots succeeded at flying the rocket stack into predetermined orbits.

Armstrong and his colleagues found the Johnsville tests and the data they produced to be critical support for pilots' envisioned roles in rocket launches. They published their work after von Braun's speech, countering his prohibition. "The control task was well within the capability of the human pilot," the pilots and engineers concluded, contending that a human should serve as "the primary controller of a vertical launch vehicle"—as long as they were provided with adequate information displays to guide their control. "As a passenger, he [the pilot] can be very expensive cargo; but as an integral part of the control loop of the vehicle, he might add materially to the reliability and flexibility of the launch maneuver," Armstrong and his colleagues concluded. Reliability and flexibility—two keywords from the X-15, kept their prominent place, as did pilot preferences for data displays over automation.[13]

The role of the pilot in space missions was on the table. Pilots had already lost a battle with the advent of the ballistic missile, in their view little better than a dangerous, unpiloted drone. Would the giant space rockets then under construction be mindless machines, taking mere "missile rider" payloads up for a ride? Or would they be human-guided craft, directed by keen eyes and hands aiming them into orbit, like trains on virtual tracks? Would the X-15's skilled reentries be the way of the future, or a forgotten sidetrack on a ballistic path? Chauffeurs or airmen?

SETP publications took up the airmen's case. Boeing test pilot Arthur Murray blamed the need for automated boosters on the primitive state of rocket technology. Older boosters, he believed, from the V-2 to the Atlas missile, required complex automation because they were crude and risky.

"Mission success afforded by piloted control of a booster," he told his colleagues, "may well be one of the significant contributions made by Dyna-Soar." Murray advocated "an X-20 design which not only tolerates pilot control, but which insists that he be the primary element wherever a better overall operation will result."[14]

Indeed, the early 1960s were replete with experiments and arguments for piloted boost. X-15 pilot Milt Thompson participated in a series of trials designed to show that pilots could manually fly the Titan booster into orbit with the Dyna-Soar vehicle on top. "This was a very controversial issue," Thompson recalled, as "the booster designers had been using automatic control and guidance systems from day one. In their minds it was the way to go."[15] As late as 1963 another SETP member wrote that "piloted boost is almost certain to be a reality in the future."[16] Scott Crossfield believed that the reason for the low reliability of rockets as opposed to aircraft was directly related to the former's lack of a human pilot.[17]

When provided with proper visual indicators and displays, the pilots believed they could execute nearly any task. X-15 pilot Joe Walker argued to the SETP that displays of real-time information would enable pilots to control any part of a system. "The history of flying shows that pilots want to be in the control loop during critical phases of operations, for instance, take-off, approach, and landing."[18] Walker believed advanced space vehicles would have similar requirements, his conclusion "backed up by proof from electronic static simulation and closed-loop dynamic simulation." The X-15 had taught pilots not only to embrace simulations, but also that properly deployed, earthbound virtual worlds could help them defend their roles in outer realms.

Despite the arguments, experiments, data, and simulations, the pilots lost their case. No American would ever manually fly a rocket off the launch pad. Human space vehicles, rather than bold, powerful airplanes, would be oddly shaped cocoons, enclosing their inhabitants, leaving them barely enough room to control their fates, and asking for virtually no input on the way up. Aircraft might elegantly obey and express their pilots' wills, but rockets fly as mindless automata. Von Braun's V-2 and the Redstone booster that derived from it followed a few stable feedback loops under the control of rigid circuits. The precious skills of takeoff and landing would succumb to automated sequencers and parachutes.

Project Mercury and the Pilot's Role

Project Mercury evolved during these debates and solidified their outcomes. By 1958, it had become clear that NASA, and not the air force, would run the first American program to put people into orbit, in a project that would come to be known as Mercury. NASA introduced its first set of astronauts in April 1959, followed by a glossy exclusive spread in *Life* magazine that fall (and more than seventy articles on the astronauts and their wives in the following four years). Public response, fueled by *Life*'s exclusive, heroic coverage, set a celebratory tone that only intensified in response to successful flights.[19] While the astronauts explained their role as the ultimate backup system, ready to take over if the machines were to fail, the press and the public imagined them as American heroes in full control of their fates. As Mercury took center stage, what had been a professional debate among pilots and engineers now became a matter of public interest and cold war politics: just what should the human do in space?

Writer Tom Wolfe's book *The Right Stuff* and the popular (somewhat fictionalized) movie of the same name captured these tensions surrounding Mercury. Wolfe cast the debate between traditional test pilots like Chuck Yeager who compared riding atop a Redstone rocket to "spam in a can" and the hopeful, though uncertain astronauts. "The difference between pilot and passenger in any flying craft," Wolfe recounted, "came down to one point: control."[20] Wolfe wrote about the macho conflict between the Edwards test-pilot crowd (the men who founded the SETP) and the relative upstarts that would form the team known as the Mercury Seven (none of them SETP members), and how the test pilots gradually saw their glory eclipsed by the huge public response to Mercury. Whatever its limitations, the resonance and public impact of Wolfe's writing testifies to the issue's central place in the cultural image of the program.

Yet as we have seen, Wolfe overemphasized the "cowboy" nature of the test pilots and missed the scientific-technical dimension of their professional identities. He saw friction between the pilots, who wanted to fly, and engineers, who focused on automation and technology. But the conflict was not simply between pilots and engineers, but also between groups of pilots and groups of engineers, a clash of cultures with different visions of the human role. Manned spaceflight in America would be a synthesis of these visions, led by one group with a particular background: the Space Task Group.

The Langley Group

In 1958 NASA put together an upstart team at its Langley Research Center called the Space Task Group (STG) under the leadership of Robert Gilruth, the pioneer of flying qualities research. Gilruth assembled twenty-five young Langley engineers willing to stake their careers on a risky endeavor.[21] They focused on the capsule itself; the Redstone rocket to loft it into space would be supplied by von Braun's group in Alabama. The Germans had spent more than twenty years building and flying rockets without human occupants. Similarly, the Atlas booster that would fly the orbital Mercury flights was built by Convair for the air force—founders of the West Coast ICBM culture focused on systems engineering and automatic control. In contrast to these newfangled systems types, many of the young engineers in Gilruth's Space Task Group had technical backgrounds in flight control and had worked closely with test pilots for much of their careers. The STG formed the core of what would soon become NASA's Manned Spacecraft Center in Houston, with Gilruth as its first director.

Since 1945 Gilruth had headed the Pilotless Aircraft Research Division (PARD) at Wallops Island, Virginia. The group's name sounded ominous for pilots, but these were hands-on engineers who flew things. PARD engineers began firing rockets at hypersonic speeds to create aerodynamic conditions that the day's wind tunnels could not mimic. Fourteen of the original STG engineers came from this group, including Max Faget, who designed the shapes for the Mercury, Gemini, and Apollo capsules, and even for the space shuttle.[22]

The remainder of the original STG came from NACA groups in instrumentation, flight testing, and stability and control (along with a group of Canadian aircraft engineers laid off from the Avro company, in part because their government chose to rely on missiles rather than aircraft for air defense). These research and design engineers had roots in flight control and piloting and included a number of men, in addition to Faget, who would go on to become household names in the American program. Caldwell Johnson, who worked closely with Faget, recalled incorporating pilots' opinions into design decisions, "as a result of the background I had in aircraft work, I brought into play the pilot and aircraft aspects of the system."[23] Chris Kraft had cut his teeth under Hewitt Phillips at NACA Langley in flight test, instrumentation, and telemetry.[24] The shrewd, technical Kraft remembered that when the STG began working on Mercury, they were "still thinking in airplane terms"—that is, giving the astronaut everything he needed to know to fly the rocket. "But it quickly became obvious from Air Force input that things happened so fast in rocketry that an astronaut couldn't do anything anyway."[25]

Gilruth's group also included the man who would drive the inclusion of a digital computer in Apollo. Robert Chilton had flown B-17 bombers during World War II and then earned bachelor's and master's degrees at MIT, studying under Stark Draper and Robert Seamans. He spent ten years at Langley in the Stability and Control branch working with Hewitt Phillips and his assistant Charles Matthews (who went on to become Gilruth's assistant on Mercury). Among other projects, Chilton used control theory to make mathematical models of the responses of a human pilot.

Chilton helped specify the Mercury control system and selected and oversaw the contractor to incorporate it into the spacecraft. For Mercury, Chilton conceived that "the pilot would perform a balanced role," of a variety of control, monitoring, and abort functions. Yet the pilot's primary job, in Chilton's view, was "being the captain of the ship and not just the pilot."[26]

A philosophy began to emerge: during normal operation, the control loops would be automatic, stabilized by feedback controls. If the primary systems failed the pilot would take over. The consequence was that in an emergency, the pilot's workload went up, because then he would rely on simple, less-automated backup systems that were presumably more reliable than the automatic systems.[27] "If the pilot had to take over [in Mercury], which we didn't think he would . . . he would just simply use his controls to zero, null these needles and help him maintain attitude, and look through the periscope just like centering the bubble." Nevertheless, Chilton recalled, the astronauts didn't like the idea—"they wanted to fly the thing."[28] As John Glenn told a press conference soon after his selection to the Mercury team, "We don't want to just sit there and be just like a passenger aboard this thing. We will be working the controls."[29] Nevertheless, the automatic-as-primary approach would survive through Apollo.

One other man joined the STG with a unique point of view. Robert Voas was trained as a psychologist, and he worked for the navy studying "human engineering" or "ergonomics" for pilot selection, training, and operations. The navy assigned him to the STG at its founding, and he immediately began working on the role of the pilots.[30]

Voas realized that the question of pilot involvement bore heavily on the selection and training of the astronauts. Technical decisions had human implications: what the operator needed to do affected who would be chosen. "Was this just someone along for the ride," he stated, "or were they expected to be a very significant part of the operation?"[31] Voas deemed that individuals with physical fitness and experience operating technical systems would fit the bill (a decision he later regretted, because he found the Mercury Seven deficient in engineering skills).[32] Nothing in these original qualifications necessitated experience in aviation. On Voas's recommendation, NASA approved a set of selection criteria that might include "test pilot, crew member of experimental submarine, or arctic explorer." NASA was ready to go public with an announcement, but at the last minute President Eisenhower decreed the astronaut candidates should be chosen from the ranks of military test pilots—for both security and secrecy reasons.[33] (Such limitations alone, however, should not have excluded submariners.)

The Mercury astronauts distrusted Voas, as they did human factors experts in general, whom they classified with an old pilots' enemy—flight surgeons. One NACA engineer referred to "industrial psychology and human factors" as "dirty words around here." Chris Kraft thought Voas "wasn't nearly as good as his advance notices."[34]

But Voas did strongly advocate for the human role, "to have a man in space and not just send up empty spacecraft that were completely automatic." Because a Mercury flight "begins and ends with periods during which the astronaut does not control the vehicle's attitude or flight path," wrote Voas, the human role in Mercury had been "underestimated."[35] Voas called the astronaut "not a mere passenger, but an active controller of the vehicle" and argued that human intervention in control loops (along with ample displays of information) improved the flexibility and robustness of the system. "The astronaut operates as an integral part of the Mercury system," Voas concluded, and pointed out that Mercury would show "man's proficiency as a controller for space vehicles."[36]

Voas listed seven roles for the Mercury astronauts, in order of priority: (1) "systems management," which mainly involved monitoring the automatic systems; (2) "programming" or "sequence monitoring," which involved monitoring critical events during launch and reentry; (3) controlling vehicle attitude; (4) navigation; (5) communications; (6) research observations; (7) keeping attentive and fit under flight stresses; and (8) ground preparation. The pilot functioned either as a backup system himself, or as a switcher who enabled backup systems to come into play. For example, if the sequencers were to go awry, "he must substitute himself for the malfunctioning automatic programmer," a task more subtle and demanding than simply responding to green and red lights like a laboratory mouse (although that is how the task might have appeared to an observer).[37] Voas mapped the manual and automatic functions and the numerous ways that a given task could be divided between the two (figure 4.4).

Voas participated not only in the selection process, but also in shaping the innovative training regimen for Mercury. Again, training depended on what would be expected of the astronauts while in orbit; both changed over the course of the program. While Voas and his colleagues had deep experience in flight training, for Mercury the goal became as much to condition the astronauts not to do things (in panic or erroneous response), as it was to train them for positive actions. It also sought to condition their responses to the physical and psychological stresses of spaceflight. Because the astronauts were seen as redundant backup systems, knowing the spacecraft systems and being able to detect and diagnose failure became primary objectives.[38]

The Mercury Control System

In Mercury, as with the X-15, questions about the human role were built into the machine. McDonnell-Douglas built the Mercury capsule itself, but its control system was designed by the Minneapolis-Honeywell Regulator Company (then also building the MH-96 adaptive controller for the X-15). Company experts initially wanted a fully automatic spacecraft, with little human operation, but after consulting a "human engineering" expert who favored a more active pilot, McDonnell engineers believed more human control could improve the reliability of their spacecraft. Nevertheless, all agreed with Robert Chilton's view that unlike in an aircraft, the automatic modes would be primary, with the human-actuated controls reserved for off-nominal conditions or emergencies. Indeed, humans were not essential for flight: four Mercury capsules launched on Redstone rockets and three on Atlases under fully automatic control before being qualified for human flight.[39] The conception gradually shifted from complete automation to "monitored automatic." The human function was "something more than secondary, if still less than primary."[40]

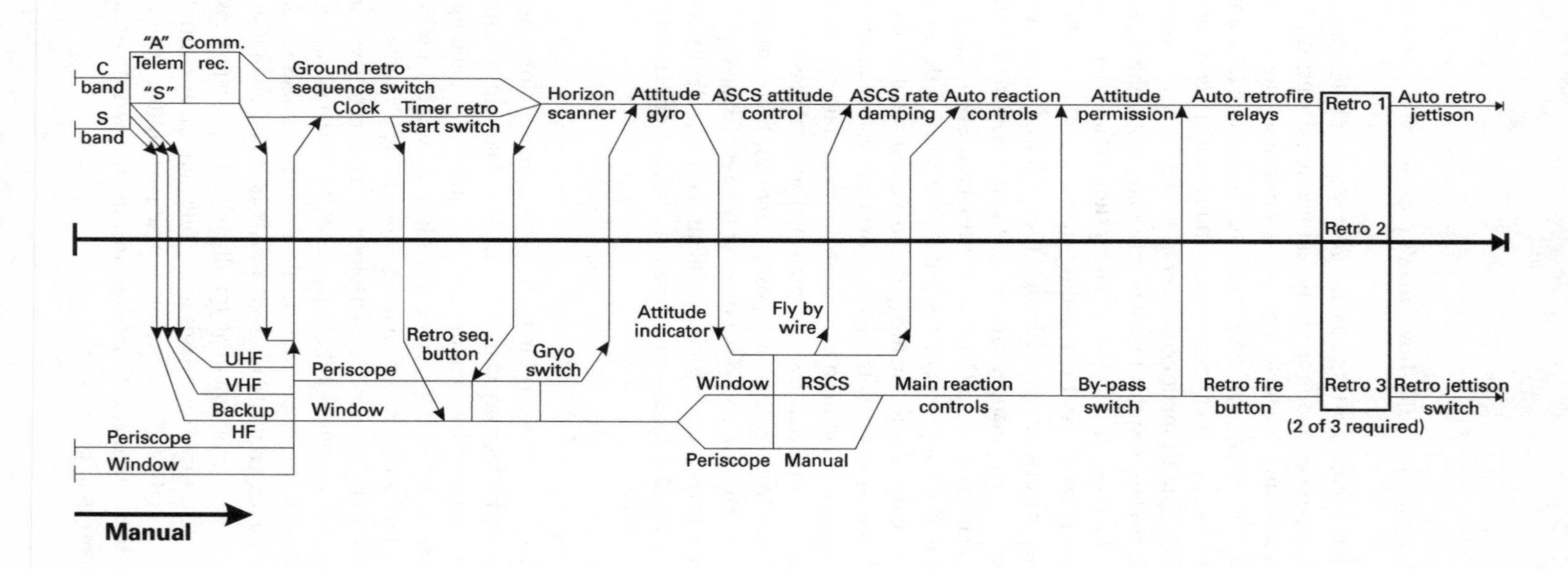

Figure 4.4
Automatic versus manual procedure for Mercury Retro Maneuver, showing the numerous pathways by which the pilot can switch between automatic and manual modes. (Redrawn by the author from Voas, "A Description of the Astronaut's Task in Project Mercury.")

The Mercury spacecraft had two redundant sets of control systems, one automatic and one manual, each with their own separate reaction jets and fuel supplies. The pilot had no control over the spacecraft's position; in other words, he could not change its orbit. He could, however, control its attitude, that is, the orientation around the center of gravity. Prior to firing the retro rockets for reentry, the attitude was normally set automatically, but if the primary failed the pilot could point the spacecraft and fire the retro rockets under his own control. During reentry itself, Faget's capsule design meant that the vehicle would passively orient itself with the heatshield down, though it might oscillate a bit. The pilot could use either automatic or manual control to damp the oscillations.

Some pilots, particularly Deke Slayton, wanted rudder pedals in the cockpit, but in the end astronauts used one hand on a three-axis stick. The pilot could push fore and back for pitch, side to side for roll, and twist the stick (instead of pressing on pedals) for the yaw control. In automatic mode, the system would assume the proper attitudes and hold them via feedback control with an autopilot. On the other extreme, in fully manual, the pilot could control the attitude of the spacecraft by commanding the thrusters directly, opening valves for fuel in proportion to the deflection of the stick.

This fully manual mode took a great deal of thrusting to hold attitude, so it was hopelessly inefficient with fuel. A more efficient, hybrid solution allowed "rate control," where the pilot commanded a servo that stabilized the vehicle's rate of motion about each axis. As with the automatic control on the X-15, when the pilot moved the stick, the spacecraft would move, and when he took his hand off, the spacecraft would stop and hold where it was. This simple idea required a finely tuned servo loop to be responsive and stable. In a mode known as "fly by wire" (although rather different from the modern usage of that term for aircraft), the pilot could issue "on-off" control commands to the solenoid valves in the automatic control system. The pilot could use more than one of these modes at one time—for example, if the autopilot was holding in pitch and yaw, the pilot could still control roll with the fly-by-wire mode. He also had display of rate information in three axes, a periscope, and the window, added to the second capsule after a request by the astronauts.[41] The newly prominent astronauts, of course, had opinions about these engineering decisions, and were still fighting the battle of whether they would fly the rocket off the pad.

Defusing Spam Bullshit

Design decisions about Mercury and its control systems had largely been made by the time von Braun spoke to the SETP in August 1959. Yet persuasive and powerful as he was, he did not close the debate on piloted boost; in fact he fueled the controversy about whether a Mercury astronaut would have a meaningful role. Questions lingered about the pilot's tasks, especially from the test pilots themselves.

Just a few weeks after von Braun's speech, the SETP held its 1959 annual meeting. The conference opened with a session chaired by Neil Armstrong on the human operator in spaceflight. A paper by Scott Crossfield, as well as the other presentations, argued for the importance of the pilot.

Irritated by criticism of the imagined passivity of the astronaut aboard Mercury, two young leaders of the program made the case personally. Chris Kraft and Deke Slayton went to the SETP meeting hoping, in Slayton's words, "to defuse some of this spam bullshit."[42] Yet Kraft did have to break the bad news: "These guys were used to having their hands on the throttle of whatever they were flying, and to making it turn or climb or dive according to their input. So I went into some detail about what happens during a rocket launch and in orbit. Most of it was automatic, I had to admit. An astronaut was strictly a passenger when that rocket ignited and he began the ride into space."[43] Even the abort system was going to be automatic—the really scary things that could go wrong would happen too quickly for the human pilot to react. For Mercury, von Braun had won.

Slayton defended his own role (this before he was grounded by a medical condition in 1962). The case for piloted boost seemed lost, but what a pilot might do in orbit remained an open question. His presentation aimed to show the "pilot's point of view," and to "establish the requirement for the pilot, or astronaut, in Project Mercury." Slayton contrasted his view with that of the engineer, who "semi-seriously notes that all problems of Mercury would be tremendously simplified if we didn't have to worry about the bloody astronaut," and with those in the military who believed that "a college trained chimpanzee or the village idiot" could replace a pilot. Slayton's thesis: "if you eliminate the astronaut, you concede man has no place in space." Not only a pilot, Slayton argued, but an *experimental test pilot* was required (an ironic position, given that several of the Mercury astronauts would not have qualified for membership in the SETP before their selection). In summary, Slayton argued that for Mercury "if everything works perfectly the pilot's task will be quite simple." But in the case of a failure or emergency, "the task would become quite complicated but not beyond the pilot's capabilities."[44]

The 1959 SETP meeting articulated the pilot's role for Mercury—passive passenger on the way up, active controller in space. But Slayton argued that even this situation was only temporary. He understood that the rocket experts believed the boost phase should be handled automatically, but added a hedge that "we agree, at the present time." Slayton believed that in a future era of space station and orbital rendezvous, "it will be desirable to give the pilot complete control over booster firing and powered flight trajectory."

"Pilots get some good news," ran a *New York Times* headline about the October meeting, glossing over the bad news about piloted boost. "Group is assured machines won't supplant humans on probes into space." The *Times* noted "this was music to the assembled test pilots' ears, because their major theme long has been that they are not obsolete."[45]

Pilots in Space

When Mercury actually started flying, its pilots exercised their numerous options for control in response to a series of contingencies and failures. In May 1961, on Alan Shepard's landmark, suborbital flight, a leaky thruster gave the Mercury capsule a slight but constant rotation about its roll axis. Shepard used the fly-by-wire feature to compensate, and during reentry he used it to damp the oscillations.[46] As a test, and also from necessity, Shepard exercised all three of Mercury's modes of control—automatic, manual, and fly-by-wire. Shepard's flight did not include the rate-control feature, but Grissom's suborbital flight (July 1961) did, and Grissom found rate control to be the most responsive of the various options.

Grissom's flight also pressed the issue of pilot control in another way when his hatch prematurely popped off during recovery. The spacecraft sank to the bottom of the ocean and the astronaut almost drowned. Ensuing debate over whether the cause was pilot error or mechanical failure raised the question: if the pilots are to get credit for successful flights, how much blame do they share when something goes wrong?

There followed two automated flights, one including "Enos" the chimp, to verify the orbital system for Mercury. Automating the capsule simply required adding a sequencer device in the pilot's seat, with minimal modification to the spacecraft. When John Glenn finally flew his orbital flight in February 1962, he switched between modes to determine the most effective and economical way to control the spacecraft. Glenn's capsule developed a malfunctioning thruster, and the manual modes allowed him to compensate. In the words of the official program history, he decided to "become a full-time pilot responsible for his own well-being," manually orienting the spacecraft and commanding the retro rockets to fire.[47]

On the second orbital flight in May 1962, Scott Carpenter had a more equivocal performance, having used up all of his maneuvering fuel, but he too exercised all of the control modes. Though several of the control systems failed, Flight Director Chris Kraft was furious with Carpenter for not following the flight plan and suitably obeying directions from the ground. Voas felt that Kraft's anger deprived NASA of the "public relations feature that the man had performed and brought back a damaged craft, or a partially nonfunctioning spacecraft," which he believed would have built support for the human role.[48]

Not all pilots performed equally, but NASA was loath to compare them in public. Whatever they thought of their own skills, even before the astronauts tightened their harnesses in the tiny capsules they were harnessed to a social structure on the ground that reached through radio channels and ballistic trajectories, a social structure with its own agendas of power, authority, and control, a theme that would continue throughout Apollo.

Like the X-15, the Mercury program constantly found itself justifying the human presence. In the words of Mercury's excellent official history, "Mercury saw the evolution of the astronaut from little more than a passenger in a fully automatic system to an integral and fully integrated element in the entire spaceflight organism. By the end of the project, the Mercury capsule, instead of simply being a machine with a man in it, had truly become a manned space vehicle."

That evolution took place in public presentation more than in the actual spacecraft. In early articles in *Life*, for example, astronauts sometimes referred to their upcoming flights as "rides" and to themselves as "passengers." *Life* compiled those articles, largely verbatim, in the 1962 book *We Seven*, but rewrote certain sections to emphasize human control. For the book, *Life*'s ghostwriters systematically replaced the word "capsule" from the *Life* articles with the word "spacecraft," a substitution NASA made more generally between October 1961 and January 1962. After then, the astronauts used the word "passenger" only in a negative sense.[49] As Michael Collins later wrote, "Capsules are swallowed. One flies a spacecraft."[50]

Public statements surrounding the flights also emphasized the pilot's control. John Glenn proclaimed that his flight would have failed without a man "aboard to assume control and bring the capsule back" (he did not mention that without a man aboard there would have been no need to bring the capsule back). Wally Schirra related how he turned off all automatic sequences after liftoff—"the capsule was all mine now"—and he declared that before his flight, no one had "flown a capsule before, much less under pilot control." Deke Slayton's wife professed that she feared the prospect of her husband going into space, until she learned that the "pilot, and not the automatic system, was the most important thing."[51]

Press responses to the Mercury missions echoed NASA's selling of pilot control. Journalist and science fiction writer Robert Heinlein typified their conclusions when he wrote, "The Mercury shots proved that an astronaut can actually control his ship." The *New York Times* published an editorial after Glenn's flight titled with his words, "Let Man Take Over," and hailed the flight as a symbolic victory in a battle between man and machine. "We need not be ruled by machines," ran the piece, and extending the lesson to broader social currents, it continued, "people should reject a belief in an automatic stream of history over which we have no control." Historian James Kauffman, in an exhaustive survey of press responses, notes that "every article on Glenn's flight refers to Glenn's active control."[52] Independent of the technical facts, the notion of the pilot in active control of the spacecraft engendered narratives of human spaceflight well suited to public presentation.

"The first true pilot's spacecraft"

Notwithstanding Mercury's triumphs of human control, rhetorical or practical, Project Gemini began to fulfill von Braun's vision: pilots as monitors during launch, but in the loop while in orbit. Gemini planners sought to explore longer-duration spaceflight, learn how to rendezvous and dock, achieve precision landings, establish the capability for "extra vehicular activity" or EVA (colloquially referred to as "space walking"), and provide additional experience for flight and ground crews in manned spaceflight. Woven throughout these objectives—from orbital maneuvering to accurate reentries—was the desire to expand human control.

The Mercury capsule had been heavily dependent on sequencing devices, even with the human in the loop. Gemini simplified the system by putting much of the sequencing responsibility onto the human operator and providing numerous new thrusters and tools for control. Unlike Mercury, the Gemini spacecraft could not be controlled from the ground without significant additional equipment.[53] Inside, the human operator stood at the nexus of an inertial platform, an optical tracker, and a digital computer—an assemblage that set the stage for Apollo.

In late 1962, before Gemini flew, Gus Grissom visited the SETP to explain the new program and to contrast it with Mercury. He now characterized the earlier project as containing a passive human passenger. "The most important difference," Grissom told his colleagues, "is the amount of control the pilot exercises over all functions." In Mercury, pilots could contribute to the mission, Grissom continued, only by manually overriding automatic systems. "Until now, man has been a self-experimenting guinea pig." By contrast, "Gemini is the first true pilot's spacecraft," he claimed, allowing the astronaut to step into his new role as "the explorer of space." Grissom argued that Gemini would require the pilot's responses at all times, from his finger on the abort button during boost to his judgment about the time and location of reentry. "Gemini will be a pilot controlled operational spacecraft, not just a research and development vehicle." Grissom pointed forward to a change in professional identity: "the test pilot will have stepped into his proper role—the explorer of space."[54] Given the controversy that surrounded his botched recovery on Mercury, Grissom had some interest in portraying the role of the Mercury astronaut as something less than totally in control.

Unlike Mercury, where the craft reentered the atmosphere in an open-loop, ballistic fashion, Gemini would be steered by the pilot right down to the point of landing—yes, landing, a proper return worthy of a pilot's dignity. The Gemini capsule would spring a special parachute shaped like a hang glider. Under this "paraglider" control, the pilot would "fly" down from the last few thousand feet, deploy a few skids from the underside of the capsule, and land on a runway. "The paraglider will be controlled by either pilot," Grissom explained, "with the same hand controller he used in orbit...the pilots will have to locate the landing area on their own." The paraglider's sluggish handling qualities would call for the greatest skill.[55] Guidance and control would assume new importance, both for getting the pilot to the point of rendezvous, and for selecting and navigating to a landing site.

Reporting on the SETP meeting, the *New York Times* echoed Grissom's words with a front-page story in 1962 under the headline "Pilots will control Gemini spacecraft." The story described how after Mercury, NASA would be "reverting to the concept of a pilot-controlled aircraft" for Gemini. "The Gemini capsule will be under the control of the astronauts rather than automatic instruments."[56] Where Mercury had been fully automated out of concern for possible degradation of pilot abilities in space, the program had proved the pilots' abilities and Gemini would take full advantage of them. Advancing technology would mean not more automation, but more human control.

As the tasks and technology changed, the crews evolved as well. Grissom was in the minority: of the seventeen men who flew on Gemini, only three came from the original Mercury Seven. The remainder came from subsequent sets of astronauts in the second ("the new nine") and third (fourteen) groups of astronauts. These men tended to have higher levels of education and greater technical skills in addition to their test-pilot backgrounds, reflecting the changing nature of their tasks and a new NASA emphasis on professional training over physical characteristics.

The ten Gemini missions did indeed fulfill Grissom's hopes and von Braun's vision of astronauts as active, on-orbit controllers. The first Gemini flights, unmanned in April 1964 and January 1965, flew suborbital and required a special sequencer installed in the astronaut's couch to put the craft through its paces. On the first manned Gemini mission (Gemini III, March 1965), Grissom maneuvered the spacecraft with its new set of thrusters, changing not only its orientation (as on Mercury) but also its velocity, dropping the orbit down about thirty miles, which Mercury could not do.

During rendezvous, the pilots flew a "final approach," in an analogy to landing an aircraft, and carefully maneuvered into close proximity with another vehicle. Astronauts practiced rendezvous with spent booster stages on several flights, with varying degrees of success. Then, Gemini VI (piloted by Wally Schirra and Thomas Stafford) and Gemini VII (Frank Borman and Jim Lovell) achieved a dramatic first rendezvous between two manned craft, four pilots in all. From separate launches more than ten days apart, the pilots maneuvered the two craft within inches of each other, "window to window and nose to nose." Schirra, commander of Gemini VI, emphasized in his memoir his ability to control the spacecraft: "Using what I called my 'eyeball ranging system,' I did an in-plane flyaround of Gemini 7, like a crew chief inspecting an aircraft...I was amazed at my ability to maneuver, controlling attitude with my right hand and translating in every direction by igniting the big thrusters with my left hand mechanism."[57] Dramatic photographs of manned spacecraft from the outside highlighted the new, active role of people in space.

Subsequent Gemini missions accomplished rendezvous on different orbits (including the first one after launch, simulating a launch from the lunar surface), from orbits equal, below, or above the target vehicle, and using several different techniques, including with an unmanned target vehicle called Agena. Some of these exercises were designed to simulate maneuvers and abort modes anticipated for lunar missions, some with failures of various parts of the system.

Gemini's project summary concluded that "the extensive participation of the flight crew in rendezvous operations is feasible."[58] The pilots liked docking; it allowed them to fly.

Technical and budgetary problems had forced the cancellation of the paraglider and with it hopes for controlled, terrestrial landing, so Gemini craft came down in the water as Mercury had. Still, Gemini pilots could maneuver during reentry. Because of its offset center of gravity, the spacecraft's roll determined the direction of lift during a critical few minutes in the atmosphere. By rolling the spacecraft to cancel or reinforce lift, the pilots could effectively control the splashdown point by up to three hundred miles down range and more than twenty-five miles side to side. Continuously rotating the spacecraft would null the lift altogether, and make for the shortest landing. Gemini missions came down within a wide range of distances from their aiming points; some were off by nearly one hundred miles (Gemini V) but a majority were within ten miles of their targets. Ten of the twelve Gemini reentries were flown with a pilot's hand on the stick.[59]

Gemini also produced its share of anomalies and human responses hailed as critical or heroic. On Gemini VI, with Wally Schirra and Tom Stafford in the cockpit, the launch sequence initiated but shut down prematurely. Rather than issuing an abort, in which the pilots would have dangerously ejected sideways from the spacecraft, ruining it, against established procedure Schirra kept his finger off the abort handle. He sensed that the vehicle had not moved, even though the mission clock had started, and he did not fire the ejection, avoiding the risky bailout and saving the capsule and the rocket for another day. His quick thinking was hailed as a triumph of man in the loop as he spontaneously violated an erroneous mission rule.[60]

On Gemini VIII, when docked with the Agena target vehicle, the spacecraft started spinning out of control. Astronauts Neil Armstrong and David Scott immediately undocked from the Agena, only to find that the problem was with their own spacecraft and not the target, and they began spinning faster. They isolated a stuck thruster while in a life-threatening tumble, and immediately aborted the mission and landed in the Pacific. Questions arose about whether the immediate undocking was the right decision, but Houston praised the astronauts and explained the situation to validate their decision. On Gemini XII, the rendezvous radar failed and Buzz Aldrin called on manual techniques to perform the rendezvous. By coincidence, he had written his MIT dissertation on pilot-oriented techniques for rendezvous in the absence of automation.[61] Numerous other systems failures led to successful recoveries or real-time mission replanning.

Automation in the Gemini Program

Less heralded than the human controls but equally important for the technical learning of spaceflight and its implications for Apollo was a new type of human-machine interaction, subtly importing the chauffeurs versus airmen dichotomy into the orbital realm. Most of Gemini's rendezvous and reentries relied on programs running on a digital computer.

Intuitive piloting alone proved inadequate for rendezvous. Following Grissom and Young's successful demonstration of manual maneuvering on Gemini III, on Gemini IV astronaut Jim McDivitt attempted to rendezvous with a spent booster. He envisioned the task as "flying formation essentially in space," but quickly found that his aviation skills would not serve him in this situation. Schedule and some other unexpected parameters constrained the first attempt at rendezvous, so McDivitt employed "a brute force technique where instead of assuming that we were in orbit we just assumed that we were flying across the earth like in an airplane."[62] McDivitt turned around to attempt a practice rendezvous with his spent booster stage, but found that as he thrusted toward the target, it inevitably moved away. After McDivitt spent a significant amount of maneuvering fuel, flight director Chris Kraft (working for the first time from the new mission control center in Houston) called off the exercise, in favor of Ed White's historic space walk (the first for an American). The human image of White floating through space overshadowed the disappointing experiment in orbital flying.

Afterward McDivitt attributed the problem to poor lighting, but he was facing something new. Orbital dynamics created a strange brew of velocity, speed, and range between two objects and called for a new kind of piloting. Catching up to a spacecraft ahead, for example, might actually require flying slower, to change orbit. "It's a hard thing to learn," observed Deke Slayton, "since it's kind of backward from anything you know as a pilot." A NASA investigation concluded that training had been inadequate, and that the entire program had underestimated the subtlety of the rendezvous task.[63] Gemini IV taught an important lesson: pilots could not rendezvous by eye.

A successful rendezvous would require more than keen eyes, stable hands, and cool heads. It would require numbers, equations, and calculations. It would require simulators, training devices, and electronics. Running them would be another Gemini novelty: an onboard digital computer, first flown on Gemini III. This machine, manufactured by IBM, had 4,096 words of 39-bit memory. It collected data from a variety of sources, including the spacecraft's inertial measuring unit and could display calculations and quantities on a variety of displays.

For every phase of flight, the pilots shared control with their new companion. On ascent, the computer served as backup for guidance of the launch vehicle. If the primary system in the booster failed, the astronaut could switch over the control to the cockpit, or the Gemini computer could take command automatically. During rendezvous, the computer collected data from the ground, from the astronauts, and from sensors on board the spacecraft. It calculated commands and trajectories and displayed thrust and orientation that the astronauts should fly. Astronauts interacted with it by means of a mode switch that selected a variety of programs, a numerical keyboard and display, and an IVI or "incremental velocity indicator" that displayed "velocity to be gained" for thrusting maneuvers. The astronaut fired the thrusters with his stick, keeping his eye on the IVI, which measured accelerations and counted down in velocity, indicating zero when it was time to stop. The astronauts used the computer in a simulator on the ground to practice and to develop a variety of procedures for rendezvous.[64]

The computer had seven operating modes, corresponding to seven programs and seven functions. During pre-launch, it performed some self-diagnostics. In "ascent" mode, it provided backup guidance to the Titan booster and could take over if the rocket's computer failed.[65] In "catch-up" mode, it provided pointing commands and velocity increments to the crew to begin the rendezvous. In "rendezvous," it took data from radar and the inertial platform to calculate the velocity changes required for rendezvous, and drove the IVI display to zero as the astronaut gave thruster commands. In "reentry" mode, the computer solved the reentry equations and displayed data for the crew to follow when flying manually.

Beginning with Gemini VIII, the computer included a magnetic tape storage system that allowed the crew to read in programs and orbital parameters, a necessary feature since the computer programs had grown in size with the increasing sophistication of the missions and could no longer fit into the memory. Neil Armstrong and David Scott first used this procedure while their spacecraft was spinning, reading in the reentry program for their emergency landing.[66] Heroic action now involved not only controlling the spacecraft but also loading code into a digital machine.

For rendezvous, the crew used data from their digital computer and checked them against a printed chart, which served as an analog computer like a slide rule (a practice continued on Apollo). Usually, the ground staff computed the necessary mid-course or catch-up maneuver numbers and called them up by voice. On Gemini X Michael Collins computed the maneuvers on board, using a combination of the onboard computer and star sightings from the optics, a difficult and frustrating process that did not produce accurate results.

Once the accurate data was calculated and displayed in the cockpit, the actual maneuvering became an ordinary, mechanical task, the pilot serving as a human servo-mechanism and following the computer's instructions. In Apollo, these maneuvers would require no manual thrusting, but would be controlled by the computer from beginning to end based on a series of keypunches.

The computer also commanded the reentry maneuvers, indirectly, through the crew, or directly, in an automatic mode. When coming in from orbit, at about 400,000 feet, the astronaut manually kept the spacecraft aligned with the horizon, when the computer began indicating roll commands for the astronaut to follow. Or, in automatic mode, the computer commanded the attitude system directly.

By Gemini XI, the computer automatically steered the spacecraft and landed within three miles of the target with no crew input, a feat repeated on the final Gemini XII flight. "Reentry of the Gemini spacecraft was successfully controlled both manually and automatically," concluded the project's final report. Which mode was desirable depended on the complexity of the task, the number of control commands, and the desired accuracy of the landing spot.[67]

Jack Funk, a member of the Space Task Group who specialized in trajectories, recalled that the astronauts slowly realized that the dynamics of flying in space were so complex, as were the numerous systems aboard the craft, that anything that helped them manage the chaos would extend their ability: "Instead of having a battleship with 6,000 crew on board, you just have a small craft with a computer which is your big crew that does all the work for you."[68] Aaron Cohen, a control engineer who would play a central role in developing the Apollo system, believed that "the piloting, the man interface with the computer, the display, and the techniques of how he used it," stood out as the major contribution of the Gemini program.[69]

Automation in the Soviet Space Program

Astronauts liked Gemini's pilot-centered philosophy. For them, it was not only solid engineering but also an expression of national character. "During a mission we were independent thinkers and decision makers," wrote Walter Cunningham, who flew on Apollo 7. "It was the American way of life carried into one more challenging environment." By contrast, Cunningham believed, the Soviets would over-automate their spacecraft, and "never fully trusted the individual as opposed to the 'collective' on the ground." This difference of approach, Cunningham and many others felt, was responsible for the Soviets losing their lead in space in the 1960s.[70]

The Soviet approach to spacecraft automation did differ from the American approach, but not in the ways most people thought. Like their American counterparts, cosmonauts were public symbols and served political goals. When crafting their roles, the Soviet regime drew on twentieth-century images of machine-men and scientific organization of labor, including Stalinist iconography of aviation and piloting. In the 1960s, cosmonauts became the ideal of "New Soviet Man," a cog in the larger machine of state, taking individual initiative within heavily proscribed constraints of ideology and authority.[71]

But the ideal did not translate into technical systems in any straightforward way. Lt. Gen. Nikolai Kamanin, who headed cosmonaut selection and training, described his approach as "the domination of automata." Early cosmonauts were not accomplished test pilots but junior fighter pilots—Yuri Gagarin had a mere 230 hours of flight time when he was selected. The first manned Soviet spacecraft, Vostok, was, like Mercury, heavily automated and first flown unmanned (most Soviet rockets and spacecraft flew two unmanned missions before being human-qualified). But Vostok did not offer the range of backup manual modes that Mercury offered. Yuri Gagarin, in fact, was blocked from using the manual reentry function on his spacecraft by a combination lock. In the event of an emergency, the ground controllers planned to read him the combination over radio (although after some debate it was placed in a sealed envelope on board).

Why were Soviet spacecraft automated to this degree? Fully automated spacecraft had "dual use," and could be used for remote as well as manned missions. Additionally, the heavier payload of Soviet rockets could afford the weight of extra electronics. Engineering culture also played a part: Soviet engineers tended to have rocketry backgrounds, fewer of them had roots in aviation. Historian Slava Gerovitch, one of the few who has examined the question from primary sources, points out that "heated debates over the division of function between human and machine often broke out within the space engineering community." As in the United States, professional interests, organizational politics, and national ideology all influenced the debate. Spacecraft design was organizationally separated from cosmonaut selection and training, so the cosmonauts had little input into development and the engineers were not familiar with their points of view.[72]

Soviet engineers discussed humans in cybernetic terms, evaluating their abilities for logical switching, amplification, integration, and computing. Indeed they avoided the term "pilot" in favor of "spacecraft guidance operator." Tensions arose between military and civilian astronauts, between those trained as pilots and engineers or scientists (once two astronauts debating the subject nearly came to blows while in orbit). Legendary designer Sergei Korolev saw the humans as integral parts of larger technological systems, with which they would trade control and authority.[73]

As in the American program, the cosmonauts fought for active roles in their flights, arguing they would increase reliability and the likelihood of mission success. The Soviets began developing automatic rendezvous and docking systems in the mid-1960s, with good results, but when they failed the cosmonauts had trouble intervening at the last minute. As in the American case, the planned Soviet lunar spacecraft included a digital computer with the operators typing in commands to be executed by its programs. "The Soviet approach to automation was never fixed," Gerovitch concludes, and "it evolved over time, from the fully automated equipment of Vostok to the semi-automatic analog control loops of Soyuz," to later digital controls. "The role of the cosmonauts also changed, from the equipment monitor and back-up on Vostok to the versatile technician on Soyuz to a systems integrator on later missions."[74]

American engineers and astronauts, while well aware of their competition with the Soviets during the 1960s, had little detailed knowledge of their technology, much the less the designs of their control systems. Nonetheless, the counterpoint served as an enabling myth. Imagining Soviet technology as fully automated in a totalitarian mode supported the American association of active pilot control with the democratic individual, and the astronauts as "New American Men," full of initiative, freedom, and skill at the controls of their free-flying spacecraft. Ironically, on Apollo only *advances* in electronics and computers would allow American engineers to build control systems to suit this ideal human role.

From the High Desert

Meanwhile, amid flight programs and international competition, lunar plans germinated. As von Braun had realized, opposing views aligned not simply by professions (engineers versus pilots), but by institutions and engineering cultures. Among the earliest to play their cards were researchers at NASA's high-speed flight station (also known as the Flight Research Center, or FRC) at Edwards, the high temple of flight test, who sought to ensure that their pilot-centered culture had a place in the budding lunar program for every phase of flight. In 1960, when the program was just taking shape, FRC Director Paul Bikle wrote to NASA headquarters of the "spectacular competence" of the FRC in flight research that could be brought to bear on a lunar program. Citing their work on piloted boost, he proposed to study the role of the pilot in a Saturn launch and to examine the pilot's role in stabilizing a spacecraft in orbit. Bikle argued that automation should be included only when "the particular task is not within the pilot's capabilities." Similarly, for reentry, "the role of the pilot as an active participant for control of the vehicle to a precise landing point should be studied, to determine the degree of automaticity (if any) required."[75]

Milt Thompson and colleagues argued that the pilots' alert senses—feeling vibrations, hearing tanks pressurizing, smelling smoke—could not be easily replaced by machines.[76]

Test pilots Joe Walker and John McKay argued that the problem with Mercury was that the rockets had been designed for automatic operation and then modified to include a pilot as a monitor. This afterthought, they believed, "compromises both the automatic system and the pilot." Rather, they advocated "inclusion of the pilot in the control loop at all times" so the pilot could swiftly take corrective action, using his feel for the system obtained in normal flight.[77]

FRC engineer Hubert Drake framed the pilots' role in rendezvous while articulating the FRC's position on some general principles. He began with a series of axioms:

- Design vehicle and mission to rely on pilot
- All flights piloted
- No unmanned mission capability
- Avoid excessive safety emphasis
- Concentrate development on few systems

Furthermore, rendezvous would be no big deal: "the pilot comments indicated that the rendezvous maneuver was comparable to docking a boat or parking a car," he stated (a view that contrasted with that of the Gemini astronauts, who emphasized the difficulty and the skill required). In sum, Drake reported the FRC's position: "He [the pilot] should be in complete, direct control and able to use this control in the most flexible manner possible by the use of his own intelligence, senses, and manual skills. Any artificial aids provided should be designed solely to assist him and should not have primary control during any part of the operation."[78]

Even at this early date, the FRC's approach of "complete, direct control" by the pilot represented an extreme view, and had already been superseded for Mercury by Robert Chilton's philosophy of including the pilot as a backup system.

Crafting a Human Role in Lunar Flight

Before Gemini or even Mercury missions flew, and before President Kennedy's decision to send Americans to the moon, Robert Chilton began looking at the human role in a lunar flight. Meeting the requirements for a moon mission would be hugely more difficult and complex than for the simple orbital strolls: hitting a lunar entry corridor required accurate navigation; maps of the moon had crude precision compared to those of earth; lunar reentry speeds far exceeded those from orbit; and people had to withstand much longer missions and perform at peak capacity up to the end.

"One of the biggest, toughest problems we had," Chilton reckoned, "was to put the astronaut in the loop"—not just in the loop of a little ground-based tracking like Mercury, but in the loop of the complex guidance system that would be required for lunar flight.[79] Because a spacecraft returning from the moon would reenter the atmosphere from a trans-earth trajectory at a much higher speed than Mercury or Gemini spacecraft had, it would need to target a narrow reentry corridor, which called for accurate guidance operating autonomously on board.

In the summer of 1960 NASA held an industry conference to define the moon landing project and let potential contractors know what to expect. They also announced a new name: Project Apollo. Chilton presented a "command and control" configuration for the pie-in-the-sky project. The crew for Apollo, Chilton told the audience, would assume a role "comparable to the crew of a transcontinental jet airliner." This was not the mundane comparison it would be today, for 707s had only started flying these routes in 1958. Each airliner had a pilot, a copilot, and a flight engineer. Apollo, too, would have a three-man crew. Chilton did not mention a lunar landing.

Chilton predicted that Apollo would rely heavily on automation. Computers would relieve the pilots of routine tasks and assist them for those that required high speed, accuracy, or computational complexity. That did not mean, Chilton hastened to add, that the mission could be conducted with full automation. Unmanned flights might provide data for the development program, but "specially instrumented vehicles" should fly only suborbital routes as tests.

Chilton defined the primary crew tasks as making command decisions, monitoring the systems, and supervising navigation and control. Tasks also included "maintain[ing] system performance by accomplishing the necessary maintenance and repair, or by engaging appropriate backup systems" and emergency manual modes. If, as was argued for Mercury, the crew were the ultimate backup system, then it would seem only reasonable to ask them to fix the machines when they broke.[80]

Under ideal circumstances the flights would be relatively automated, but humans would get "into the loop" in an emergency—much as John Glenn had taken over control of his spacecraft when the thruster failed. Only later did Chilton recognize the problem with this approach: the astronauts had to train for numerous failure scenarios. That problem "rose up and bit us," because the training of the crew became a major bottleneck in meeting the flight schedules. "The training load on the crew was just so horrendous," Chilton recalled, "that it paced the [Apollo] program after a while," because the crew were so busy training not only on the primary procedures, but on all the emergency procedures as well.[81] Such problems lay far in the future when Chilton laid out the requirements for Apollo in 1960.

NASA then requested proposals from industry for several large-scale feasibility studies for the lunar mission. Goodyear Aircraft proposed to study the social structure of the crew and the "man-machine variables that affected optimal balance between guaranteed mechanical reliability and need for operator activity and challenge." Only a thorough "man-machine analysis" would define the requirements for the crews, who might be younger than the Mercury astronauts. A Convair feasibility study identified the crew members as "commander," "subsystem operator," and "scientific operator." The Boeing study argued that "man's role must transcend that of a monitor of an automatic program," and proposed "to enter man in the system loop," to allow the commander to "look ahead" and select optimum solutions before committing to a course of action. A McDonnell study divided the three crew member's tasks into systems management, flight control, and navigation. Vought aircraft divided them into "Pilot-Vehicle commander, Navigator-Second Pilot, and Flight Engineer-Maintenance-Scientific Observer," and articulated a philosophy of the crew in direct command of all vehicle control loops during all phases of flight.[82]

The Apollo feasibility studies implicitly questioned the assertion that all astronauts had to be trained pilots, positing roles for flight engineers, systems managers, and scientists. Reflecting on this early, formative period, Michael Collins and others believed that Apollo did not adequately incorporate the lessons of Gemini. Stated Collins, "When Gemini methods were suggested to Apollo engineers, there was no eagerness to accept or even to listen."[83] A simple explanation attributes this disconnect to the parallel schedules of the two programs. A subtler analysis notes that, despite the common role of Robert Chilton, from its beginning Apollo developed its own vision of the human role, one augmented by and dependent on machines. By the time Gemini flew for the first time, Apollo engineers had already designed the astronauts into a rich, flexible system for flying to the moon, one organized around a digital computer.

From Piloted Boosters to Digital Aids

Wernher von Braun began the space age by excluding human pilots from the boost phase of flight. What had seemed the ultimate pilot's task, skillfully guiding a powerful rocket along a precise trajectory, became relegated to servos, automatic control loops, and eventually computers. Conflicts over human roles in spaceflight reflected the last-minute switchings between the cultures of missile and aircraft engineering; one eliminated and one celebrated the human pilot. Such tensions between engineering cultures, and between engineers and operators, continued to arise as Mercury astronauts vied with automatic systems and ground systems for control of their machines. Mercury's machinery remained in primary control, but technical failures (which were numerous) called for the pilots' skills. These "ultimate backup systems" proved their worth by getting themselves home. NASA publicity, *Life*'s hagiography, and general press coverage continually told stories of human triumph in the face of technological failure.

Gemini put Mercury's publicity into engineering action, as rendezvous became the new realm of human piloting in space. Gemini astronauts recast Mercury as mere trials to demonstrate its abilities for control in space. Orbital changes and rendezvous and docking maneuvers were now the frontiers of human performance. Yet a pilot's intuition proved inadequate for rendezvous, and the problem was only solved with a variety of mechanical aids, from graphical charts to digital computers. Nonetheless, graceful meetings of two human-occupied spacecraft in orbit extended the beauty and control of flight into the orbital realm. All seemed to contrast with the archrival Soviet program that supposedly overautomated its spacecraft to the point of disallowing human intuition and skill.

As Project Apollo germinated, it sought to extend the Gemini experience to another planetary body. But the steep learning curve also extended the utility of automation. People seemed to be necessary for the flights, but what kinds of people? Pilots, engineers, or scientists? More realistically, what mix of the three? Early studies based on terrestrial models suggested crew structures in which technical experts would chauffeur a scientist-observer on a lunar expedition.

5 "Braincase on the tip of a firecracker": Apollo Guidance

It was a curious ship, a braincase on the tip of a firecracker.... Without fire it could not move; without electricity it could not think.

—Norman Mailer, *Of a Fire on the Moon*

In this guidance and navigation business, we're kind of one down in the dramatic art. We have to compete with people who build engines and make lots of smoke and flame... in the computer area, we don't even have any moving parts. We have some small flashing lights... and that's about as dramatic as we get.

—Ralph Ragan, Raytheon Company, Apollo Project Operations Manager

When John F. Kennedy took office in January 1961, human spaceflight did not appear on his political radar screen. Kennedy's science advisors believed that emphasizing human flights would overshadow the country's strengths in space science. They recommended the president distance his administration from Mercury, an expensive and potentially dangerous holdover from the Eisenhower administration that had received a great deal of attention but had not yet flown with a human aboard. NASA had been working on a lunar program for two years, as a technical extension of earth-orbital work, holding industry conferences and letting study contracts, but the political support and massive funding required for such an endeavor remained uncertain at best.

Other members of the new administration, however, understood that human spaceflight could only be justified within a larger context than scientific or technical imperatives alone. NASA administrator James Webb and Secretary of Defense Robert McNamara argued that "these [space policy] decisions can and should not be made purely on the basis of technical matters," and rather pointed to the "social objectives" of human spaceflight, arguing that "it is man, not merely machines, in space that captures the imagination of the world."[1]

When the USSR put Cosmonaut Yuri Gagarin into orbit that April, apparently securing second-place status for the United States in space for years to come, the president's mind began to change. "Is there any other space program," Kennedy asked Vice President Lyndon Johnson, "which offers dramatic results in which we could win?"[2] After consulting Webb, McNamara, von Braun, and other technical advisors, Johnson recommended a program to achieve a manned lunar landing by 1966 or 1967, citing the "margin of control over space and over men's minds through space accomplishments." Alan Shepard flew on May 5, 1961, and the overflowing public response persuaded Kennedy to make the bold move of committing the nation to landing a man on the moon "before this decade is out." Congress soon granted his request for financial support.[3] Project Apollo was underway, its political (i.e., financial) support secured by the public impact of a human in space.

NASA engineers had been studying and planning, but now had money to spend and systems to build. Just two months after Kennedy's speech, NASA let its first contract for Apollo. It would not go McDonnell Douglas, which had built the Mercury spacecraft, or to Minneapolis Honeywell, which had built Mercury's control system. Neither would the Apollo guidance system be made by North American Aviation, which would manufacture the spacecraft itself, nor by Grumman, which would make the lunar lander. These contracts were still in the future, the companies still jostling for selection, when Robert Chilton decided to get started.

In fact, the contract for Apollo's navigation and control system, the first in the entire program, would not go to a company at all but to a university: the Massachusetts Institute of Technology. MIT's Instrumentation Laboratory (IL) had developed the "instruments" that helped make flying "scientific" during the twentieth century and had trained the NACA engineers in flight controls. This lab had built a national reputation with innovative inertial guidance systems for missiles, and its engineers had recently begun envisioning an unmanned probe for navigating to Mars. Following Kennedy's imperative, IL would now incorporate human users into its precise guidance schemes.

Autonomy and Accuracy

During the 1930s, when Charles Stark Draper became interested in aircraft instruments and developed the field he called "instrument engineering" to advance the technology, he also developed an institution. Its very name, the Instrumentation Laboratory, emphasized the scientific nature of the work. NASA officials always referred to the IL simply as MIT, ignoring the complex relationship between the lab and the university. Soon after Apollo, the IL was no longer even part of MIT, having been spun off in the early 1970s in response to student protests' against its weapons work. Today the lab is an independent center known as Charles Stark Draper Laboratories.

Working in close contact with the armed services and companies like Sperry Gyroscope, the engineers in Draper's lab fostered the "scientific" attitude toward flying pioneered by MIT graduate Jimmy Doolittle in his early instrument flight of 1929. The IL developed a specialty of building instruments based on spinning gyroscopes, and became a center of what sociologist Donald MacKenzie called the "gyro culture"—a group of individuals, techniques, and institutions experienced in the art of applying gyroscopes to control vehicles.[4]

During World War II, Draper developed a gunsight, based on his gyroscopic turn indicators for aircraft, which could be deployed cheaply and quickly to help defend navy ships against close-proximity attacks. Sperry Gyroscope manufactured what became known as the "Sperry-Draper Mark 14 Sight" and produced 85,000 of them during the course of the war. Draper's industrial contact at Sperry Gyroscope was a young vice president, a rising aviation lawyer named James Webb. Twenty years later, as NASA administrator, Webb again relied on his old friend from MIT to build him gyroscopes for Apollo.[5] Robert Seamans, Jr., who would become NASA's top technical manager during much of the 1960s, worked on these projects as a young engineer in Draper's group. Seamans titled his memoirs *Aiming at Targets* because with his background, hitting the moon with a spacecraft seemed a similar problem to hitting an aircraft with a shell.[6]

During the 1950s Draper and his group continued to work with gyroscopes and control systems, while developing the critical technology of inertial guidance. An inertial navigation system used highly precise gyroscopes and accelerometers to keep track of all the forces acting on a moving body; it could navigate ballistic missiles through space to precise targets in the Soviet Union, with no outside references and hence no possibility of jamming. By the end of the 1950s, IL-designed inertial systems were in the Polaris and Thor missiles, in addition to a variety of ships and aircraft.[7]

A question arose, leading to what one of the IL's principals called "the beginning of a philosophic controversy"—should all navigation in space be self-contained, or aided with radio communications from earth?[8] Again, the lines formed around professional cultures: advocates for external navigation tended to be electrical-radio engineers, whereas the inertial advocates stemmed from mechanical engineers piqued with the interest of making super-precise gyroscopes.[9] The air force's first ICBM, the Atlas missile, was guided by a combination of inertial systems and radio beams.

As Donald MacKenzie pointed out, gyro culture developed two technological values by which IL engineers measured their success: *autonomy*, or the ability to navigate without external reference, and *accuracy*, the ability to navigate to precise points. Both derived from military requirements for ballistic missiles, by what was technologically interesting and achievable, and by the local culture in the IL. Both would dominate the engineers' thinking on Apollo, and both would come into question during the course of the program.

Polaris

The Polaris project gave the IL team members the experience they needed for Apollo. The Polaris submarine was to launch nuclear missiles while submerged, requiring boosters with solid instead of liquid propellants to make the system simpler and safer. It also had to be extremely small (less than half the diameter of other missiles of the day). In 1956 designing this miniature, precise inertial guidance system for Polaris became an IL project.[10]

The navy ran Polaris out of the Special Projects Office (SPO), which played a role of systems engineer similar to that played by TRW in the Atlas program. To manage the complexity of the Polaris program, SPO engineers developed the Program Evaluation and Review Technique, or PERT, a method of plotting and tracking complex schedules that survives to this day as a staple of project management, and which NASA and the IL team would see again as part of Apollo.[11] Indeed SPO's role on Polaris became one model for NASA's management of Apollo.

On the Polaris project, Eldon Hall, a quiet, lanky Idaho farm boy with an intense interest in electronics, led the IL's Digital Development Group. Although the IL had previously built guidance systems with analog computers, Hall persuaded the navy to bet on a digital machine for Polaris, based on the accuracy of computations required.

The Polaris computer was digital, but not a general-purpose machine that could run any program from software. Rather, Polaris used a "digital differential analyzer" architecture that drew on the early mechanical computing technologies developed at MIT in the 1930s by Vannevar Bush. The original ENIAC computer built at the University of Pennsylvania used this architecture as well. It was in effect a digital implementation of an analog computer, where the equations were wired in between computing elements (in ENIAC, these routings could be changed and the computer reprogrammed by rerouting cables, whereas in Polaris routings were fixed by the wiring of the machine). The Polaris computer integrated the accelerations measured by the inertial system through a series of difference equations, repeatedly adding and multiplying (and combining with the submarine's position and velocity) to get the missile's position. The computer used these data to issue commands to steer the missile along the desired trajectory. A new idea developed at the IL, called "Q-guidance," allowed much of the computational load to be undertaken by ground-based digital computers ahead of time, enabling use of simpler machines in the missile itself.[12]

The Polaris electronics were not particularly complex, but making them work with the required reliability, robustness, and light weight challenged Hall and his IL engineers. They recognized the effective use of electronics to be as much about mechanical packaging as about circuit design.[13] Hall explored new areas in construction, stacking modules like "welded cordwood," and wire-wrapping the interconnections. Hall also designed the computer circuits so they all shared a single type of germanium transistor—effectively implementing the interchangeable-parts philosophy that had characterized mechanical manufacturing for more than a century. Transistors, barely a decade old at this point, could still be suspect in reliability, requiring rigorous qualification to be included in military hardware. In Hall's design, only one type of transistor would have to meet the intense testing and support criteria for the missile.

In the course of his work on Polaris, Hall visited the production line at Texas Instruments and met engineer Jack Kilby, who showed him a new invention: the integrated circuit, which incorporated several transistors on one semiconductor. Hall also visited the Fairchild Semiconductor company in California and met there with Robert Noyce; the company was working on making their integrated circuits scalable and robust.[14] Hall then convinced the navy to incorporate the novel devices into the second-generation Polaris II computer. Again, Hall would build it out of a single component: a two-transistor NOR gate integrated circuit, sixty-four of them in all, costing $1,000 each.[15]

The first guided Polaris flight took place in July 1960. When the early Apollo studies were beginning at NASA, then, the MIT IL was achieving success in this high-profile, high-risk missile project.

Going to Mars on Paper

Not all of the IL's engineers were absorbed in Polaris. After Sputnik, some sought to expand their horizons beyond earthbound missiles. Hal Laning, a mathematician and control engineer, had been at the IL since 1947, and was head of the small but crucial mathematics group. Laning recognized the intrigue and potential of the computer when he worked on Whirlwind, MIT's first computer, specially built for research in real-time control systems. He wrote a program called George, a small compiler for Whirlwind that allowed engineers to write equations for the computer to solve rather than obscure assembly language (features soon incorporated into the high-level language FORTRAN). Another IL engineer, Milt Trageser was a physicist who concentrated in optics. The launch of Sputnik in 1957 convinced Laning and Trageser that there could be a future in engineering for spaceflight. After years of figuring how to guide nuclear missiles to their targets, perhaps they were attracted to the prospect of a scientific mission.

They began a small design study of a Mars probe, a small spacecraft that would fly by the red planet, snap a single picture, and return the film to earth. Today, the idea of a small, free-flying space probe is familiar, and we are used to seeing stunning images from a variety of such devices, several of which have traveled even further than Pluto. At the time, however, the idea seemed innovative, even fanciful.

The design of the Mars probe reflected Laning and Trageser's expertise and also set the basic configuration for the Apollo guidance system several years later. It incorporated the strength of the IL: a set of gyroscopes to keep the probe oriented. In addition, Laning added his own interest: a digital computer. Trageser incorporated an optical telescope with which the probe would orient itself relative to the moon and the stars, and a camera to capture the image of the planet as it flew by. As Laning put it: "I think we had our own concept of the project. Milt Trageser was a physicist. He, I think, viewed it as a giant eye. I was a computer scientist, and viewed it as an opportunity to put a totally self-dependent machine in the air for a period of time."[16]

"Self-dependent" is a significant phrase—it reflects the IL philosophy of self-contained navigation, inherited from the ballistic missile world. One other group in the country was working on deep-space probes, the Jet Propulsion Laboratory (JPL) in Pasadena; it emphasized a technique, still used today, of tracking space probes through a network of ground-based antennas. These approaches would evolve and complement each other on Apollo.[17]

In Laning and Trageser's design, the onboard computer allowed the probe to do more than follow a simple fixed series of events, but to choose "alternate courses of action" that would increase the chances of a mission's success. A central computer, they wrote, would serve this function better than controls that were "distributed among a variety of servo-control systems."[18] The Mars probe navigated by automatically measuring four angles, between the sun, stars, and planets, to determine the spacecraft's position. Unknowingly, Laning and Trageser were providing a counter to the X-15 and Mercury philosophies that human pilots added reliability. They argued instead for an onboard computer that could use complex logic to respond to unforeseen problems, just the type of control system ignored by the air force X-15 study comparing human controls to automatic systems.

The air force continued to fund the Mars probe study, and by 1959 as many as fourteen members of the IL were working on the project. These included Ramon Alonso, who joined the IL from Harvard's Computation Laboratory, led by the pioneering computer scientist Howard Aiken. Also joining the Mars probe group was Richard Battin, an electrical engineer with a Ph.D. in applied mathematics, who became a pioneer in the design of interplanetary trajectories. Battin had worked in the lab from 1951 until 1956, and then left for a career in business. The Sputnik launch convinced Battin the future was in space and he immediately returned to the lab, whence he developed the critical Q-guidance technique. Battin and his colleagues developed the general, abstract methods to bring new computing power to bear on guidance problems in real-time.[19]

Today, when interplanetary probes accelerate around planets on their way to distant journeys, they are executing Battin's idea for "a kind of celestial game of billiards." Battin's book *Astronautical Guidance* (1964) laid out the basic ideas for a generation of students.[20] Beginning with basic celestial mechanics, he showed how to calculate transfers from one orbit to another, and how to approach a target planet, either for a roundtrip "reconnaissance" trajectory or to enter an orbit. Battin developed techniques for taking navigation fixes by measuring the angle between the sun and a planet, between a planet and a star, between a star and a landmark on a planet, and a variety of other observations.

Battin's book culminated in a case study of a lunar reconnaissance mission. In 1960 he began teaching a course called Astronautical Guidance that became foundational to the field (he still teaches at MIT today, after nearly sixty years in the classroom). Nearly a third of the men who walked on the moon took this course. A genial man highly skilled in mathematics, Battin often opens his lectures with lighthearted numerological analyses of the Apollo program. (For example, he notes that 1961, the year NASA awarded the Apollo computer contract, is an invertible number, meaning it reads the same upside down and backward, like 1881, 1691, and 1111. The next year that would be an invertible number after 1961 is 6009, some indicator of providence for Apollo). Behind the levity lies a serious, creative mathematical mind; one colleague called Battin "the guiding light in this whole midcourse guidance and lunar navigation theory."[21]

With this added brainpower, the Mars probe group produced a more extensive report published in five volumes in 1959. They drew on the initial design but fleshed out a number of other subsystems and added a more complete technical analysis. "The project was not intended, on anything other than a very superficial level, to be a scientific mission," writes Alexander Brown, but rather a demonstration of the IL's skills applied to the new arena of spaceflight. Brown argues that the Mars probe study was not only the design for a piece of flight hardware, but also a "probe" enabling MIT/IL engineers to foray into the potential new market of space exploration (figure 5.1).[22]

Figure 5.1
Milt Trageser (left), Hal Laning (center), and Richard Battin with a mockup of their Mars probe. The capsule-shaped component Battin is holding is to protect the camera film with images of Mars during the return into Earth's atmosphere. (Draper Laboratories/MIT Museum.)

Battin's Recursive Estimation

By the time the study was finished, however, the world had changed. The air force was no longer the lead service on space exploration. That role had shifted to NASA. An MIT colleague arranged for a meeting between the IL group and Dr. Hugh Dryden, now serving as deputy administrator of NASA. Unfortunately, the day the MIT group chose to go to Washington, D.C., in September 1959, to present their work, Nikita Khrushchev was visiting the capitol, and the high-level people at NASA were preoccupied.

NASA did not fund the IL's proposed Mars probe, noting lack of funds and lack of interest in unmanned missions. Also, NASA had its own group that would rival the IL: the Jet Propulsion Laboratory (JPL) in Pasadena was working on unmanned probes for lunar exploration, and they dismissed the MIT approach as naive.[23] Several similar attempts to attract funding failed to find support for the ambitious project.

Still, NASA issued small grants to Battin to continue studies of interplanetary navigation, and to Laning and Alonso to continue development work on the computer, asking them to work with JPL. At least IL's foot was in the door. By April 1960, Laning and Alonso had a more complete design for their interplanetary computer, featuring a variety of low-power electronics and real-time interrupt techniques, making the machine suitable for control of a spacecraft.

As NASA's interest began turning to the moon in 1960, Battin began thinking through the practicalities of navigating to the moon and their mathematical implications. He asked these basic questions: given a spacecraft on a trajectory to the moon, where is it? How fast is it going? Assuming that it is not perfectly on course, some mid-course corrections must be made. When?

To fix the position, an optical instrument (or an astronaut with a telescope) could take sightings of celestial objects and send that data to a computer for calculation. Three simultaneous measurements would fix a point in space. Measurements could include the angle between the sun and a planet, the diameter of a planet, the time at which a star passes behind a planet ("occultation" in astronomical terms), the angle from a landmark on a planet to a star, or radar measurements between the spacecraft and a known point (such as another spacecraft). But given the numerous objects and angles available, which ones should be chosen at any particular point?

Battin studied the problems of identifying the best sources of data for navigation, defining the optimum processing for the measurements, minimizing the amount of navigational data needed, and the amount of corrections required for a mission with a given accuracy requirement.[24] For example, measurements with very acute angles would produce relatively uncertain estimates, whereas those with wider angles would provide better data and should be weighted accordingly.

For a lunar trip, say a computer is responsible for maintaining a "state vector," a list of numbers specifying the spacecraft's position and velocity. Battin broke the problem down into "decision points" on a lunar flight: an observation is made, the state vector is updated, and then either a correction is made (i.e., the rocket is fired) or no action is taken. A set of decision rules control the number and frequency of these observations. Each measurement would update the state vector in some way, depending on the statistical level of uncertainty in the measurement. Battin showed how to derive the optimal estimate for a given set of observations, given that each of those observations had some level of uncertainty. Furthermore, his scheme was *recursive*, which meant that it constantly evaluated the uncertainties of each observation as it updated the state vector. He imagined a computer with a catalog of stars and planets; at each decision point it would suggest which combinations would provide the best new information. In a typical earth-moon trajectory, taking about sixty-two hours, Battin developed a plan for forty-one observations, resulting in four velocity corrections. He calculated that the position uncertainty when arriving at the moon would be 1.2 miles, easily tight enough to enter a precise orbit.

NASA, of course, inherited NACA's experience in flight control, and had experts of its own looking at problems of space guidance. At the industry conference in 1960, Stanley Schmidt, the NASA guidance guru from the Ames Research Center in Palo Alto, presented the analytical work he had done to support the Space Task Group on lunar navigation. He made a critical connection: the previous year, Rudolf Kalman had published a paper that detailed statistical methods for optimizing the control of a vehicle, and Schmidt had been among the new techniques' first users. He explained to Battin the similarities between his statistical techniques and Kalman's. Battin realized that "we were really doing the same thing as Kalman's paper."[25] What became known as the "Kalman filter" remains a fundamental technique in navigation and a host of other estimation problems. Among its first mission-critical applications were the flights to the moon.

Guidance to the Moon

Based on Battin's recent work, the IL thought they could develop a system that provided onboard guidance with periodic updates supplied by the crew through star sightings, and they could also provide a programmed return mode if the crew were "decimated" by an accident. NASA told MIT to expect a contract for approximately $100,000 for about six months, which would fund a study but not hardware design, including topics like midcourse guidance, instrument design, computer requirements, and reentry guidance. MIT would agree to submit a work statement for the basis of a contract, and would report to the Space Task Group (the study was to complement the feasibility studies then underway). In sum, NASA and the MIT group agreed that the project would study "a *manned spaceflight system capability*, not simply to circumnavigate the moon with an encapsulated man." As in Mercury, NASA made the distinction between a manned craft with a passive passenger and a piloted spacecraft. Lunar landing was not mentioned in the discussion.[26]

A few months later, in April 1961, Robert Chilton visited MIT and toured the facilities. Though he had worked with Draper as a student, he didn't know anyone else in the group. Chilton now met the core talent like Trageser, Battin, and Laning. He found them "very impressive people." He was also amazed to learn about the role MIT had played in Polaris—performing "overall system management" for the guidance system, and then integrating it into the missile itself.

Chilton also saw that the Polaris system, although radically different in form, shared some key human-machine features with Apollo. "What they needed to do was have a guidance system on the Polaris missile so that as it was stored in the submarine, wherever they went, they'd know where they are and how to reach the target. When they got to the point where they could shoot off the missile, it would know where to go. That was obviously, a very close, happy marriage for us [in Apollo]."[27] The Apollo study contract with NASA had not yet been signed, but the IL was already gearing up its staff in preparation for the work. Polaris was just winding down, freeing up considerable personnel.

As Chilton met with the IL group, he began to understand their unique position as a university-based laboratory that did more than basic research and actually developed operational hardware. IL engineers pointed out that they could not compete with industry for contracts, but were rather directed to undertake projects "on the frontier of industrial capability."[28] Chilton worked with Trageser on some basic specifications, and determined that an Apollo design would consist of

- A general-purpose digital computer
- A space sextant
- An inertial guidance unit
- A console for the astronauts
- Other electronics for support

It would be, in effect, an expanded version of the Mars computer—but with a human interface. They chose as well to keep the in-flight autonomy of navigation that characterized the IL's work and factored in their earlier studies, so Apollo navigation would not be vulnerable to jamming from the Soviets. Chilton emphasized the importance NASA would attach to astronaut participation, "to utilize the man in carrying out his complex mission rather than merely to bring him along for the ride."[29]

Apollo's First Contract

Chilton's visit to the IL was fortuitous, for one month later, in May 1961, President Kennedy made his famous speech. By igniting the race to the moon, Kennedy unknowingly raised the stakes for what had been an obscure, preliminary IL study. The IL engineers completed their analysis a month later. It would be feasible, they showed, to take the existing Polaris system, modify it for a moon flight, and put it quickly into production. Yet the new system had one major difference: it would have to be operated by a human being. Chilton recalled that integrating an astronaut into these automated control systems presented "one of the biggest, toughest problems we had."[30] What would the person in the loop do?

Now the IL engineers had to face the astronaut's role for real, and actually build a system. Things really got going in the summer of 1961. Ralph Ragan, who had been director of the Polaris program, was called back from his summer vacation to work on a proposal for Apollo. How would you do this? How much would it cost? The IL proposed an ambitious project, ranging from basic research in guidance methods to actually building flight hardware.

David "Davey" Hoag, another of the IL's guidance gurus, had been technical director on Polaris and became program director on Apollo. Hoag recalled NASA's clear instructions: "Anything that the crew could do, you should have the crew do it and not have this system automated to the extent that it would do something that the crew could do. So if the crew could do it, then you weren't supposed to. This was the early philosophy and that did color the way we've made decisions."[31]

IL engineers interpreted the requirement in their scientific mode, linking it to the question of accuracy. Just to put a probe on the moon, in any random location, would not require a human operator. Indeed, an unmanned lunar lander, called Surveyor, was already in development and would precede Apollo to the moon. Landing with pinpoint accuracy, however, in some predetermined spot, was a difficult problem that challenged the engineers and called for human intervention. The IL translated NASA's requirement for active human involvement into a specification for their accurate guidance. "The observation of star occultations by the moon and earth," ran the IL's first progress report in 1961, "is an extremely accurate measurement which can best be made by a human."[32]

The engineers saw the astronauts as calibrators of their delicate inertial equipment; were such calibration not required, a fully automated trip might be possible. The gyroscopes that held the inertial platform stable tended to drift and required periodic recalibration (inertial navigation is essentially a very fancy form of dead reckoning). It was basically the Mars navigation scheme, which had been completely automatic, adapted to a human operator. For a lunar landing, the proposal indicated the astronauts would be taking landmark sightings for navigation, but that the computer would control the precise landing.

To this scheme, IL engineers added the characteristics of their Mars computer: high reliability, low power consumption, reasonably high speed, versatility, and direct control of the guidance system through hardware interfaces. They proposed a computer that varied the speed of its processing to save power, another idea from the Mars probe project. In addition to controlling midcourse guidance, the computer would direct most of the flight including lunar landing, lunar takeoff, orbiting, reentry, and rendezvous. Interestingly, they also proposed that the Apollo computer would control "launch vehicle guidance"—it would fly the rocket off the pad, the very task that von Braun had rejected for human control (indeed the final configuration did allow some manual or automatic steering during the later boost phase if the main computer in the Saturn V failed).

IL called the system to be built "AGE" for Apollo guidance equipment, and proposed building two different generations. The first would be quickly put together from off-the-shelf components and would fly on test missions and help study performance for later optimizing. These AGE I models would be prototypes built in parallel with a research program intended to advance the state of the art in components and techniques.[33] AGE II would then be built for lunar missions, optimized for weight, power, and reliability (later these became known as Block I and Block II).

The first functional AGE I was to be delivered to NASA by July 1963 for a flight that October. The final model II hardware would be delivered in 1964 for flights that year. NASA was hoping to fly within two or three years, in hindsight an impossibly optimistic schedule. Still the schedule's basic form did outline how Apollo would proceed. The IL would seek industrial support for any production runs beyond the prototype stage. The projected cost of the IL project was $4.375 million, estimated to support two hundred people by the end of six months, and three hundred steady-state after one year.[34] From 1962 on, Apollo constituted one-third to one-half of all IL funding. By 1969 the lab would spend upwards of $100 million; Apollo would be twice as large as any other IL project.[35]

Years later, this vague, hurried proposal would be blamed as the program exceeded its budget and had trouble meeting schedules. There was never any clear performance requirement for the inertial system (how accurate did the system *need* to be?), it simply evolved as the project developed (violating a basic tenet of systems engineering). "Nobody ever said you have got to be so accurate on entry or so accurate on launch into earth orbit," John Miller said, "A great deal of it was judgment and much of this judgment ended up with me."[36] Trageser similarly felt the system probably was more accurate than necessary, but in the absence of a compelling specification, even that was difficult to confirm.[37] What did accuracy even mean when landing on the moon—hitting a set of coordinates, or landing close to a feature of interest? Such questions would dog the program.

The IL submitted its proposal on August 4, 1961. Within days it was awarded the contract to develop the guidance and navigation system for Apollo, the first of the entire project. Chilton modeled the Apollo contract—wherein MIT had the design responsibility and the separate pieces were parceled out to companies for production—

on the way the navy had managed the Polaris project.[38] Of the prime contracts on Apollo, it was the only one given to a university, awarded with no competitive bidding. The proposals for the Apollo spacecraft themselves were not even submitted to NASA until months later, in October 1961.[39] No technical sideshow, NASA recognized that the guidance system would be central to Apollo's success.

Such an early award of the guidance contract gave the IL engineers a head start, by as much as a year in comparison with some of the other major subsystems. "I think Bob Chilton was really the one who saw that the contract was let early," remembered Aaron Cohen, who would later manage the program for NASA. "He had the foresight" to recognize that the guidance system was the most complex of the Apollo subsystems.[40] Chilton himself pointed upward for the early impetus: "Bob Gilruth said, 'Go ahead,' and I started writing sole-source justifications and stuff like that, and MIT got this overall job as an extension [of the earlier studies]... we expanded the contract, but without any competition. So it was a natural controversy."[41]

Indeed, a clamor arose from industrial contractors. If a competition had been held, it might have attracted a dozen or so bidders. One engineer from G.E. expressed his "keen disappointment and surprise" that the contract was given to MIT with no competition.[42] Guidance contractors said they had been led to expect a call for proposals by NASA, and were "seriously disturbed" by the choice of a nonprofit organization in a sole-source contract with no competitive bidding. Company presidents wrote to their representatives in Congress who wrote to President Kennedy of this "economically unsound decision of NASA." Yet the award was legal, so the challenges went nowhere. NASA simply responded that it was a "very difficult problem and one that required and extremely high level of competence and imagination" on the part of the contractor, on a tight schedule.[43]

A famous anecdote captures some of the factors that led to the sole-source contract, and the Apollo computer's implications for the professional identity of its users. Doc Draper proposed to NASA Administrator James Webb that at least one of the Apollo astronauts be scientifically trained, arguing it would be easier to train a scientist to perform a pilot's function than vice versa. Draper volunteered himself to go along and ensure the system would work. "I fully realize my limitations as a test pilot," he wrote to Robert Seamans, his protégé and now NASA's top technical manager, "but I feel my qualifications in science and engineering fields should be considered as worthy background for a crew member" to run the equipment. Seamans gracefully wrote back to his former mentor that he would forward the request to the appropriate authorities.[44]

Participants often retell the story, in various forms, of Draper volunteering to go to the moon; it serves as a kind of origin story for the Apollo guidance project. They argue that the IL was simply the obvious place to build the Apollo computer. But others, particularly in industry, were equally or more familiar with inertial guidance, digital computing, and flight controls. Draper had many personal connections within NASA—

Seamans and an old relationship with Webb. Seamans wanted to see some Apollo contracts in New England and announced the guidance contract early, so when they went out for bids on the prime contracts for the spacecraft itself, it would be clear that they did not include guidance.[45] And, of course, the president was from the same state as the IL.

The "Command Center"

With the momentum from Kennedy's speech, their head start, and the enthusiasm of an exciting project, NASA and the IL refined an initial statement of work, laying out their vision for the Apollo guidance system. Written during the early Mercury flights in August 1961, the statement built on and refined the Mercury approach toward automation and the pilots' roles. Again, in keeping with the IL engineering culture, they stipulated that navigation be autonomous, that "the primary command and decision making responsibility shall be on board the spacecraft." That is, the astronauts should be able to complete the mission with no guidance or information from the ground, although they might wish to incorporate such data "to increase reliability, accuracy, and performance." Second, the crew was to control or direct the spacecraft during all flight modes. Despite this manual control, the statement noted, automatic systems should still be employed, "to obtain precision, speed of response, or to relieve the crew of tedious tasks; but crew monitoring of these systems with provisions for crew override or mode selection is required."

Emphasizing human control of the mission, NASA described the capsule for the astronauts as a "command center" (later officially the "command module"). It was to include "features which allow effective crew participation," such as windows, equipment that could be maintained in flight, and "simple, manually operated functions in lieu of complex automation." Means were to be included to allow the "pilot" to control the attitude of the spacecraft by hand. The work statement also called for "direct electrical control of the valves"—that is, the astronauts would be able to control the thrusters without mediation by the computer (several astronauts expressed desires for even more direct mechanical—as opposed to electrical—connection to the thrusters. They lost: unlike aircraft, Apollo would not have control cables). In an attempt to use the latest media technology to document the project, the contractor was to provide quarterly reports and submit color 16-mm motion picture films of the highlights of the program.[46]

This original idea also included two elements that would not survive into the final version; both reflect changing ideas of the astronauts' roles. A "map and data viewer" was to be a database, like a microfilm-based server, allowing the astronaut to call up "thousands of different frames of information, very high information density," including lunar features, star charts, and procedures for operating and repairing the equipment, a "combination road-map, almanac, and manual of special instructions."[47] Were the spacecraft navigated in a truly autonomous mode, such information would have been necessary and the map and data viewer an efficient way of reducing weight and bulk.

But as Apollo evolved, controllers on the ground assumed greater real-time input into the mission. Close cooperation with the ground eroded the need for autonomy in orbit, and the vast amount of reference materials moved to the other side of the radio link. Indeed, complementing mission control in Houston were large numbers of engineers poring over charts, manuals, specifications, and calculations. In general, the pilots did not have huge data books cluttering up the spacecraft (unlike aircraft pilots who shared their cockpits with navigational charts and books of procedures). The reference data for troubleshooting was to come from the ground, and the astronauts survived on a rather small number of checklists, timelines, and emergency trajectory data.

The onboard teleprinter also would not survive the design process. It was to allow commands and coordinates from the ground to be uplinked and printed out in the cockpit. The idea was to save time, particularly at critical moments such as when the LM emerged from the far side of the moon, and needed a quick navigation update before descending to the lunar surface. Flight Director Chris Kraft strongly advocated for the printer as an error-free way to get data and commands up to the pilots, but Deke Slayton, NASA director of Flight Crew Operations (informally the "Astronaut Office") killed the idea.[48]

Throughout Apollo, instead of reading numbers off the teleprinter, the astronauts would listen to numbers from the ground, write them down, read them back for verification, and then enter them into the computer (a process that took several minutes for a good navigational update). Human hearing replaced the technical printer (the state vector could also be digitally updated from a ground telemetry link). These techniques were more time consuming and error-prone than a simple printer, but they did save a few pounds of weight. One result of this decision meant that the Apollo voice links provide a particularly rich record of the flights. Even in these small technical choices engineering trade-offs between human and machine embedded in the very early stages of the project.

Designing the Project

IL engineers quickly set about specifying the requirements. During the first year engineers tried to understand the scope of their commitment in order to define the details of a full contract.[49] Initially, it seemed that the key components of the system would be the space sextant, to allow the astronauts to align the inertial platform during midcourse navigation, and a series of gyroscopes to stabilize the accelerometers. Cost estimates came in at about $150 million.[50]

The original proposal was simply for an inertial guidance unit, to take navigation fixes, but before the end of the year, the IL staff proposed that the Apollo system contain a digital computer to keep track of the spacecraft's progress and perform navigational calculations. By then, NASA's original statement of work had been significantly expanded to include guidance during all phases of flight, not just the navigation to the moon.[51] It also defined the crew of three: the commander would "control the spacecraft in manual or automatic mode" and select and monitor the various guidance modes; the co-pilot would support the commander and serve as an alternate pilot. The third crewmember, the "systems engineer," would monitor the other associated systems like propulsion, electric power, and life support.[52]

As with Polaris, the IL could not provide production hardware to NASA—Draper's lab was no factory, but rather a design and prototyping shop (actually, it was a research laboratory, but departed from that role in Apollo from the beginning). By early 1962 MIT issued requests for contracts for the production of the major systems, including the inertial unit, computer, accelerometers, and optical system.[53] Twenty-six companies submitted proposals, representing a cross-section of the controls and computing industry in the country at the time: the Autonetics division of North American Aviation, Burroughs, Control Data, Minneapolis-Honeywell, and Sperry Gyroscope, to name but a few. A NASA board evaluated the contractors. Administrator James Webb, in consultation with Dryden and Seamans, made the final decision in May 1962.[54]

AC Spark Plug (a division of General Motors, soon renamed AC Electronics) would build the inertial platform, as well as ground support and checkout equipment, and would assemble and test the system. AC Sparkplug of course had emerged from the automobile industry, and on the surface seemed an unlikely choice to build a precision space system (spacecraft don't have sparkplugs). Yet the company was actually a leader in inertial guidance: they had done the guidance systems for the THOR and TITAN missile programs.[55] For Apollo they narrowly edged out Hughes Aircraft, which had worked with the MIT/IL on Polaris. Minneapolis-Honeywell was among the top control systems companies in the country and achieved high technical ratings, but the company was busy with other projects and NASA felt the Apollo contract might exceed their capacity.[56]

Raytheon, the Massachusetts-based military electronics company, would manufacture the digital computer. The company was then producing the computers for Polaris and had worked closely with the IL engineers. Kollsman Instrument would build the optical systems, which included a space sextant.[57] Kollsman had been started in Brooklyn, New York, by Paul Kollsman to market his invention: a precision altimeter for aircraft, still a standard component of nearly all aircraft today (the calibration dial in an aviation altimeter is called a "Kollsman window"). Kollsman had worked on Doolittle's early instrument flights in the 1920s, and during the 1950s built precision optical instruments for sun- and star-tracking on air force bombers.

The LOR Decision

All of these events need to be understood in the context of the overall Apollo program and the engineering issue that preoccupied it during the year after Kennedy's speech: the famous "mode decision." How would people fly to the moon? The obvious answer seemed to be to launch a spaceship from earth and have it land on the moon, then take off again and return to earth, which became known as "direct ascent." Or one could launch several rockets into earth orbit and assemble them for the flight to the moon. These ideas drew on the famous Chesley Bonestell images published along with von Braun's articles in *Colliers* magazine in 1952. The landings featured large, sleek rockets landing fins-first on lunar terrain. This meant the astronauts would be at the top of a large stack, and would somehow have to control the rocket several stories below during descent. The initial MIT study even included a large periscope so the astronaut could see the surface as he landed from an awkward position at the top of a twenty-five-meter-tall vehicle.[58]

But as spaceflight matured and engineers became more familiar with the possibilities, other options emerged. A mathematician at NASA named John Houbolt began looking at an idea called "lunar orbit rendezvous," or LOR. In this mode, NASA would send two spacecraft to the moon, and one would stay in orbit while the other descended to the surface. This scheme introduced a new kind of vehicle into the equation, and brought several advantages: the actual lunar landing craft could be a small, specialized machine, and all of the equipment and fuel required to return to earth could stay in orbit and need not be landed on the moon. It also gave the astronauts some leeway, for the critical lunar descent could be initiated from a relatively safe lunar orbit, rather than from a high-speed entry from earth. It provided some measure of redundancy, and had a better chance of meeting the end-of-the-decade schedule.

LOR also had costs, two of which became paramount. First, it required two spacecraft instead of one, doubling the complexity of the machinery and the project to get it off the ground (and possibly more than doubling the cost). Second, in order to return home, this new spacecraft would have to launch itself from the moon and rendezvous with the first craft in lunar orbit. This was a risky and unknown maneuver; launching a rocket from earth required weeks of preparation, numerous support personnel, and tons of specialized equipment. How would two astronauts perform this maneuver far from home, and find their way to a precise orbital rendezvous? Their lives would depend on getting it right.

The LOR decision and the controversy leading up to it is a rich and complex story and is still being examined by students of engineering decision making.[59] We need not go into it in detail here, other than to explore its effects on the guidance system, which were significant.

LOR had two major, immediate implications. In November 1962 NASA selected Grumman Aircraft Corporation to build the Lunar Excursion Module or LEM (later shortened to "LM" but still pronounced "lem"). The LOR decision gave the Gemini project its vision and relevance as a way to develop rendezvous practice and experience for the lunar mission. Needless to say, the astronauts had no problem with LOR—it put heavy emphasis on human flying skills.

Initially, it was not clear who would be responsible for the guidance system on the LM, but NASA immediately began thinking about sharing components of the guidance system between the two Apollo vehicles. LOR also required two additional radars: a landing radar device, which would measure altitude and velocity of the LM to the lunar surface, and a rendezvous radar device to track the LM back to the CSM (command and service module) for rendezvous. IL engineers wrote the specs for these because the Apollo computer would need to read their signals, and Grumman subcontracted the radars to RCA, which subcontracted the landing radar to Ryan Aeronautical.

MIT as a System Integrator

For the first couple of years the Apollo project was largely undefined, the money flowed freely, and the nerve-wracking deadlines seemed far in the future. IL engineers remembered a loose, creative time. "NASA was very thin," one IL member recalled, "they were busy with other things, so we didn't have a lot of direction. They gave us a lot of rope." This was the period when Mercury, Gemini, and Apollo were all running in parallel, and the Space Task Group was in the process of relocating to Houston. A young engineer named Aaron Cohen began to manage the guidance and navigation project for NASA; he led a guidance and control panel that had weekly meetings with representatives from Ames, Langley, and the spacecraft contractors. At these conferences, remembered IL engineer Dan Lickly, "You told what you were doing, you got a lot of criticism, you had comments. And then you went back and looked at it, studied it, put together something, came back in another month."[60]

Nevertheless, Apollo was growing into a large project and changing fast. The MIT engineers soon had to contend with other groups who were also beginning to work on the system. In November 1961 NASA granted a contract for the command and service modules to North American Aviation—the same group that had built the X-15. The following summer, after the LOR issue was settled, they contracted with Grumman to build the LM. The IL role was initially limited to a *guidance* system—to answer the question "where am I?" rather than a *control system* to allow the pilots to move or orient the spacecraft. Under the North American contract, Minneapolis-Honeywell Regulator Company would build a stabilization and control system for the spacecraft (analogous to the one in Gemini), which would actually allow the astronauts to fly. The IL engineers would have to cooperate, if not to compete, with these other systems and contractors.

In addition to LOR, the Apollo guidance system continued to expand throughout 1962 as the technical nature of the system came into focus. The Saturn rocket had its own guidance, but at one point NASA briefly added a requirement that the IL system should be able to back it up (it was eventually able to back up the Saturn guidance, but only after the first Saturn stage separated). Also because of LOR, the IL system had to work within two separate spacecraft—the CM and the LM, and the IL team now had to contend with two separate contractors—North American and Grumman. North American was generally unhappy with LOR, as it meant that their machine would not land on the moon at all. IL engineers visited North American to begin talking about integrating the computer into the craft, and they were given a nice, round number for the volume it could occupy—one cubic foot.

Furthermore, North American and Grumman were both reluctant to let go of the responsibility for guidance of their vehicles. "They [the two prime contractors] were really trying to understand how to design a guidance system," Aaron Cohen recalled, "which we really didn't need. We really needed for them to understand what were their requirements for interface"—classic systems engineering language, which focuses on relationships between organizations and components.

Both companies wanted to define a guidance system that MIT would then build for them. The contractors were used to the air force or the navy which specified the system and let them do their thing. NASA was much more involved in the details, treating the contractors more like hired fabricators than design houses, a position hard for the contractors to accept. "They felt they should have overall cognizance of the guidance system," Cohen said.[61]

The IL, too, saw itself as more than a designer of somebody else's machines. Trageser wanted complete responsibility for the guidance system, right down to installing it in the vehicle and ensuring it would work. Grumman engineers, by contrast, considered themselves not simply the designer of the "airframe," but the systems integrator. If they were going to put the Grumman name on the spacecraft, then they'd do all the testing and ensure the guidance system was compatible and deliver to NASA a complete, working unit.

Joe Gavin, an MIT graduate, managed the project as a Grumman vice president. For years, the genial, soft-spoken Gavin had built fighter jets at Grumman under contracts traditional in the aviation industry. "Now, we had the feeling," Gavin remembered about the early days of Apollo, "that NASA thought they were hiring a job shop to carry out their design of a LM. We very clearly understood my god, we were going to design and build a LM that we had our name on, and we would argue every inch of the way, and we did argue every inch of the way... we quickly became known as difficult."[62]

Nor were the two contractors the same—IL engineers remembered the "arrogance" of Grumman—a very tough, technically competent outfit that gave them a difficult time—versus the "pussycat" of North American, a less experienced group (with many recently hired young engineers) more easily intimidated by the technical firepower of MIT. In Davey Hoag's words, "it seemed to me that I could walk all over them [North

American]. I had to tell them the questions they should be asking me, and what did they really need to know so they could design their side to the guidance side." Whereas at Grumman, he said, "they had a sharp team. I met my mettle there... I enjoyed that part of the stuff, but it was a struggle."[63]

Eldon Hall remembered "there were these two giants [North American and Grumman] beating us over the head from both sides.... It was just completely impossible to work with the two." Finally, in January 1963 NASA decreed there would be a single, common guidance computer, and it would be copied, nearly identically, in both spacecraft. This decision kept the work at the IL to a reasonable level, and consciously or not, added a level of redundancy to the system. Indeed, having a second identical guidance computer in the LM would do much to save the lives of the astronauts on Apollo 13.

The Deplorable Optics

The guidance and navigation system contained two similar optical instruments, the "scanning telescope" and the "space sextant." Each had a similar function, but the scanning telescope had no magnification, whereas the sextant had 28 power magnification. Thus the astronaut could use the scanning telescope to sight a star or a landmark, and then track it precisely with the sextant. It was also the key human interface for the system before the digital computer became central. The precise angles of the optical systems were fed into the computer, which could include them in computations. Motors connected to the computer could drive the optics to point in a particular direction (the radar and communication antennas had similar interfaces, so the computer could keep them pointed as the spacecraft moved) (See figure 1.1 on page 4 and figure 5.2).

An inertial navigation system (INS) needs initial calibration and alignment to know where it is and which way it is pointing. To align the INS, the computer would use its previous calibration and point the telescope toward a selected star. The astronaut would then see the star in his eyepiece and center on the crosshairs, which ideally would be a small correction. The offset between the star's estimated position (predicted by the computer) and the actual position (indicated by the astronaut) would allow the computer to update its alignment.

The space sextant was a variant on the maritime instrument used since the eighteenth century for measuring the angles of celestial objects above the horizon. Indeed Philip Bowditch, an IL engineer who did major design work on the optical system, was descended from Nathaniel Bowditch whose maritime navigation books are still used at sea. The space sextant had two lines of sight—one fixed, called the "landmark line," and one moveable, called the "star line." The astronaut would measure angles by aiming the landmark line at a reference point (such as the horizon of the earth, moon, or sun), aiming the star line at another object, and adjusting the instrument to bring the

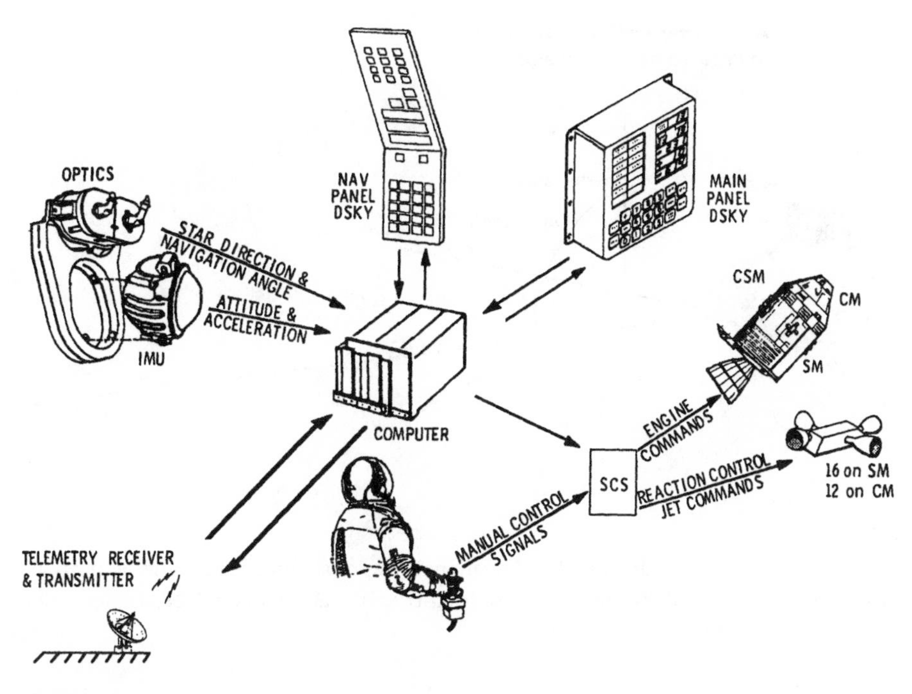

Figure 5.2
Diagram of Block I computer, showing relationship among optics, keyboard controls, and spacecraft control. Note that in the Block I configuration the astronaut does not control the spacecraft through the computer, but through an independent analog "stabilization and control system," or SCS. This was changed with the addition of the digital autopilot in Block II. (Hand, "MIT's Role in Project Apollo, Vol. I," 52A.)

two images into coincidence. To do this the astronaut had two hand controllers—a two-axis stick in the left hand that pointed the optics and a right-hand controller that moved the spacecraft itself. To move the landmark line he oriented the spacecraft, and when the appropriate image appeared, the astronaut moved the optics with the left hand controller to superimpose the star image. Pushing a "mark" button signaled the computer to record precise angles between the two images. The computer gave the astronaut a chance to approve the data, which it would then use to update the state vector (figures 5.3 and 5.4).[64]

Early on, a question arose about how to mount the optics in the spacecraft. It made sense to have the eyepieces mounted so the crew could look through them while on their "couches," their primary position in the command module in front of the instrument panel. But there was more than simply the eyepiece to accommodate. The angular relationship of the telescope to the inertial sensors had a significant impact on accuracy—any bending or slip in the link between them would degrade the quality of

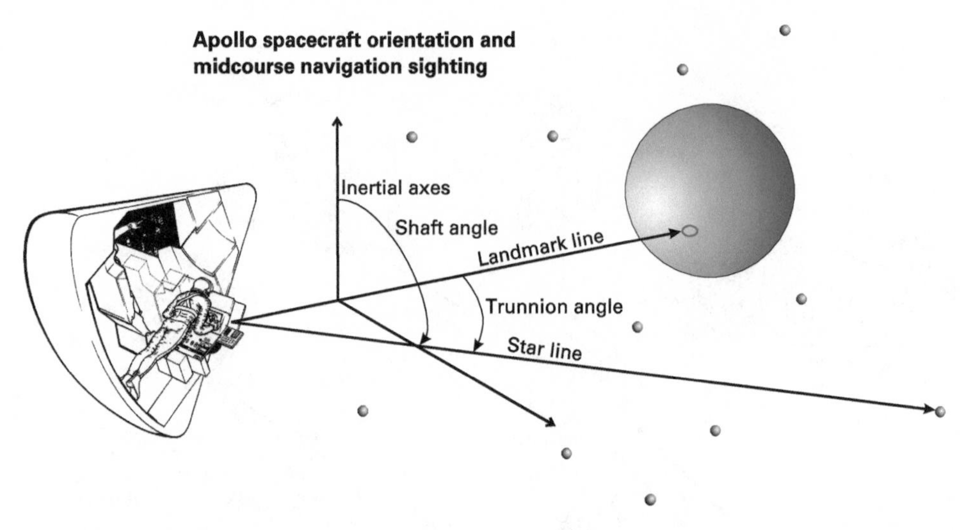

Figure 5.3
Use of landmark line and star line to take navigational fixes or alignment sightings with Apollo guidance optical system. (Redrawn by the author from Draper et al., "Spacecraft Navigation Guidance and Control," 2-74.)

the star fix. Hence the two had to be mounted together on a "navigation base" that tied the optics to the inertial platform with a rigid frame made out of the stiff metal beryllium (figure 5.5). But North American simply could not find space for all the equipment near where the crew's eyes would be when strapped to their couches. Instead North American engineers created a "lower equipment bay" for the inertial unit, well below the couches. During flight, an astronaut would unstrap himself and float down below the instrument panel to align the platform. This arrangement meant that the astronauts would not be doing star sightings in real-time as they "flew" the vehicle (or during launch or reentry), but would rather have to rearrange themselves to take a fix in a special operation. When Michael Collins likened his command module to a cathedral, he called the navigation station the altar.[65]

Questions surrounding the mounting of the optics stressed the relationship between the IL and North American. Somewhere in the lower equipment bay the optics had to penetrate the wall of the vehicle, letting a light path through so the astronaut could see the stars. As it turned out, the optics in the lower bay penetrated on the "hot side" of the vehicle—the side that would face down during reentry and absorb most of the heat. The IL designed a door that would open on the outside of the vehicle. The optical assembly would extend out on a swiveling head so the optics could rotate around for a wide field of view. This made a "deployable" package, configured by an astronaut with a crank, like that on a bathroom window.

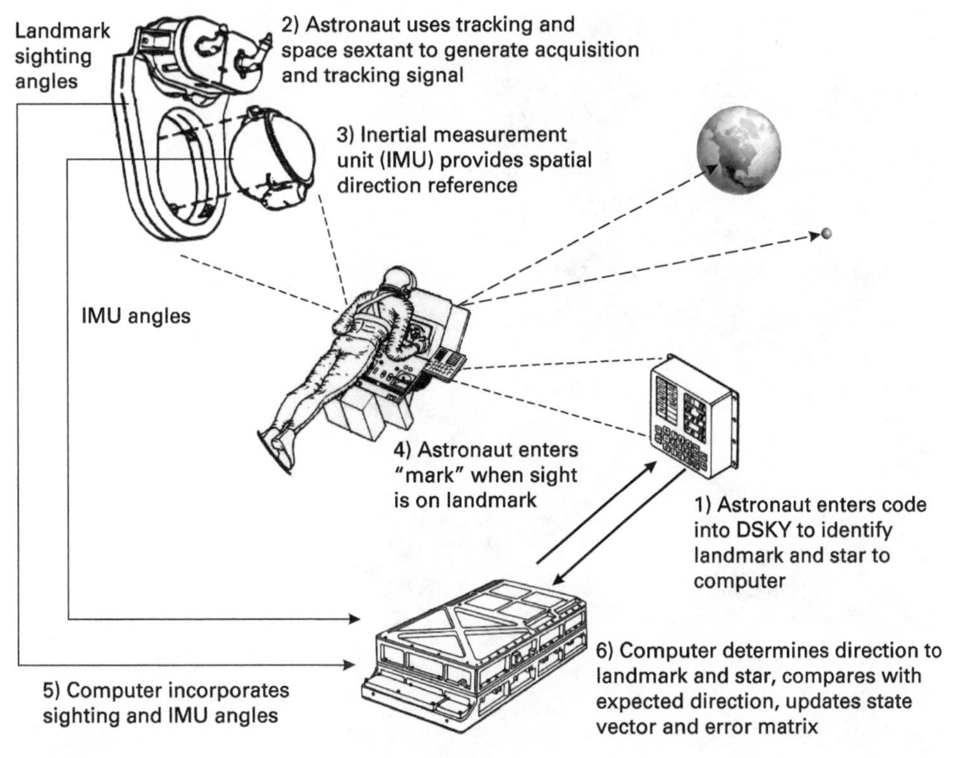

Figure 5.4
Illustration of astronaut use of optical system and computer to track a landmark for update of navigation state vector. (Redrawn by author from Draper et al., "Spacecraft Navigation Guidance and Control," 2-74.)

North Americans did not like this scheme because it was complex, added weight, and required special seals so the pressurized cabin atmosphere would not leak around the mechanism. The astronauts saw the possibility that the optics could not be retracted in the case of a failure of the mechanism, which would compromise the heat shield on reentry. They renamed the "deployable optics" the "deplorable optics." David Gilbert, the NASA program officer who had just taken over responsibility for guidance and control, remembered stepping into "a hell of a squabble" between MIT and North American over the issue. Finally, he solved it by decree—"we decided we could live with a little less field of view and ended up with a fixed optics installation." That is: no door, no deployable assembly, simply a fixed window for the telescope to look out the side of the spacecraft. If the sighting needed a different view, the astronauts would move the spacecraft itself to bring it around. Gilbert remembered a tough decision: "I guess the first bad move I made against MIT."[66]

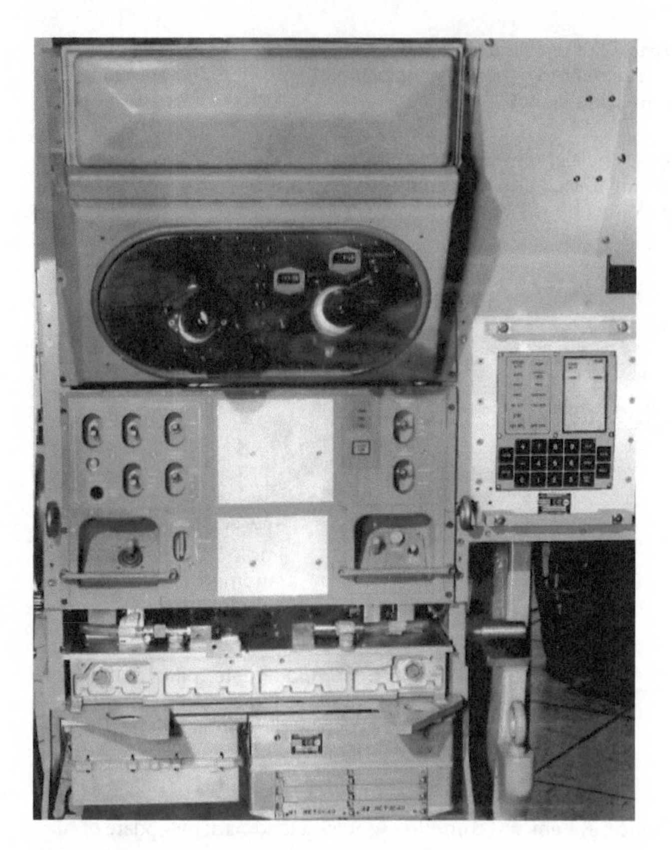

Figure 5.5
The Block II Apollo guidance computer crew station in the command module. Note the eyepieces for the space sextant and scanning telescope (top), the hand controls for aligning the sights (middle), the computer itself (bottom), and the DSKY (right). The Block II computer was sealed as a single unit. (Draper Laboratories/MIT Museum.)

Why was the issue so contentious? What was at stake for the MIT engineers? David Hoag explained that early on, they began looking for ways to quantify their system's performance—"how could we measure that we had done a good job in our design?" One way was to calculate the optimal amount of energy required to get to the moon. Based on Kepler's laws for a given trajectory, you'd need to use a certain amount of energy for a certain number of velocity changes, said Hoag. "You couldn't go to the moon with any less propulsion than that." The closer the actual guidance and navigation system could get to that minimum required level of energy—and thus the minimum possible fuel use—the better job the MIT engineers would be doing from an engineering point of view. Hence the deployable optics: if the optics could swivel around to a wide field of view, then the spacecraft's thrusters—and its precious maneuvering fuel, their measure of accuracy—would not have to burn to calibrate the optics.[67]

IL engineers' goal wasn't only to get to the moon but also to do it the best way possible, with the least amount of energy and the highest possible accuracy. Only reluctantly would they give up that goal for other considerations. In the end, they compromised on the optics (or at least accepted NASA's decree), one of many such decisions where engineers had to forgo their optimal designs for the greater goals of the project. It was a classic engineering trade-off—complexity and safety (of the deployable optics) versus a little more fuel use. Here NASA played its role as a system engineer—making decisions to arbitrate disputes between contractors, between subsystems, and between conflicting criteria of performance.

Hoag's view of these conflicts is revealing: "One of the most frustrating things was that we couldn't design alone . . . all these things came in and it was a communications chaos really but all turned out to be pretty skillfully handled. In retrospect, the experience was pretty rough."[68] In theory, the guidance problem was mathematical, even beautiful. Even when implemented in an inertial system, it retained a sparse elegance that the IL engineers had come to love. Yet when combined with other organizations, a human operator, and a real system, the abstract purity of the guidance scheme had to adapt to the gritty realities of a large project.

Gimbal Reliability

The IL group faced a similar showdown with Grumman, this time over the issue of gimbal reliability. An inertial guidance system can have any number of gimbals, or gyroscopes, depending on what the system needs to do. The gimbals keep the accelerometers fixed in inertial space so their frame of reference is not affected by changes in the orientation of the spacecraft. Three or four gimbals would be required to track a full navigational solution.[69]

When Hoag looked at the problem, he found that all of the thrusting that the command module would be doing, and hence all of the accelerations the inertial system would have to measure, were limited to a single plane. Because of this constraint, the inertial system could be designed with three, rather than four gimbals. The drawback was "gimbal lock," a subtle condition where the solution hit a mathematical singularity and the platform could not produce useful measurements for a given alignment. Certain orientations and maneuvers would just not be allowed with a three-gimbal system (fighter planes, e.g., required four gimbals to accommodate their extreme maneuvers).

Hoag realized the system simply did not have to fly about in all directions—it only had to fly to the moon and back. Put another way, the spacecraft needed to fly *around* the moon but not *over* it. So Hoag designed in three gimbals, which, incidentally, had been the number in Polaris. "We were motivated by simplicity, low-weight, reliability," Hoag explained, "and adding the fourth gimbal fights each one of these aspects." There would be just a small cone of attitudes, very far off normal, that the spacecraft could not assume.[70] More trade-offs.

Strict adherence to a set of procedures could avoid gimbal lock, and the computer could monitor itself and warn when gimbal lock approached. But the astronauts and Grumman were accustomed to systems from aircraft that had four gimbals, including Gemini, and nervous about the prospect of "forbidden attitudes" and the danger of gimbal lock. The issue hinged on the knotty problem of reliability—what was the likelihood that a gyro would fail, and hence require the redundancy of four instead of three? The trouble was, reliability is notoriously difficult to predict, and conflicts arose over how to interpret the sparse data.

The IL provided its own estimates of reliability, based on experience with Polaris, showing that three gimbals would be reliable enough. But Grumman too developed estimates, which NASA found "highly pessimistic." Grumman extrapolated reliability numbers from earlier missile programs and used them to argue for a redundant, four-gimbal platform. "It was a very dramatic forecast of how bad the reliability would be," Hoag recalled, "and we were incensed."

In January 1964, NASA manager Joe Shea brought the engineers from the various parties together, "to find the truth and punish the guilty." He gathered the men for a tense standoff. As they entered the room, Shea announced that the meeting was being tape recorded. "I intend to force a black-and-white conclusion to this meeting. . . . Someone, Grumman or MIT, will have to leave this meeting admitting he was wrong." Grumman presented their case, noting its data came from Minneapolis-Honeywell and the Titan missile program.

To counter, Hoag presented the IL case, meticulously exposing errors in the Grumman data, showing how the reliability predictions improved as he corrected the errors. "I'll never forget the day," Grumman's Gavin recalled. "What happened, of course, was that Davey [Hoag] had a bunch of facts that we did not have. He wound up showing what they could do was perhaps not quite as good as what they had advertised, but it was a darn sight better than what we were saying they could probably do."

Grumman fell hard. In Kelly's words, "MIT had blown Grumman's analysis out of the water." Grumman engineers recalled the incident with bitter remorse, and felt their reputations permanently damaged. "For years afterward we endured strained relations with MIT."[71]

Ironically, the three-gimbal platform came to be one of IL's regrets, an institutional victory but a loss for the project. Returning from the lunar surface, when Neil Armstrong was docking the LM to the command module, "I flew it right into gimbal lock." Gimbal lock loomed as a constant fear on Apollo 13. "We felt we were being unnecessarily restricted in the maneuvers we could perform," Michael Collins wrote.[72] Hoag wished he had done it differently, because the conflict and recriminations it caused were not worth the reliability improvements that his design provided.[73]

Ruled Out as Passengers, Not as Pilots

By the end of 1962, the Apollo guidance system was well on its way. The IL felt confident enough in their preliminary designs to begin briefing the users and asking for their input. In December, astronauts Alan Shepard, John Glenn, Scott Carpenter, and Deke Slayton (the first three of whom had already flown in space) came to Cambridge for two days of meetings with the IL team to discuss skills, training, and automation for the Apollo system. "We expect a visit from several astronauts," Milt Trageser told his engineers. Asking them to define the "astronaut-system relationship" for the visitors, Trageser admonished, "Care should be taken to avoid accidents in presentation which give the impression of complicated operational requirements. I think we have a simple system; let's present it well."[74] The team made a broad introduction to guidance and navigation, the characteristics of the computers, and asked the astronauts to operate early simulators and mockups, discuss training issues, and explore in-flight maintenance.

The technical engineering trip became a public event, and the astronauts, who "had to fight their way through crowds of giggling girls and goggling newsmen," held a short press conference. Shepard pointed out that much of the day's discussion was taken with "the percentage of guidance that will be automatic as compared with piloted" in the vehicle. Glenn added, "Our experience has shown that human reaction, especially in decision making, is superior to automated equipment." When asked if they wouldn't be too old to fly to the moon by the time Apollo actually landed, Shepard responded "we've been ruled out as passengers, but not as pilots" (indeed Shepard would be the only one of the Mercury Seven to fly to the moon).[75] Despite the success of Mercury, the question of pilot control was still on the table. The astronauts' first meeting with the computer reminded them that the answers were not determined ahead of time, but the product of numerous engineering decisions and evolving operational plans.

6 Reliability or Repair? The Apollo Computer

If I wanted to write a philosophical novel about Apollo and say where did this technological capability come from? And what was it? You'd have to go back to these two things: you developed a group of men and an approach, a systems approach if you will, that lets you undertake high-speed flying, you developed a systems approach that lets you undertake risky missions and be able to call them ethical.

That system plus the digital computer that's the Apollo mission.

—George Rathert, NASA Control Engineer

While "mini" computers were beginning to come on the scene in the 1960s, the word was still relative, and a small computer was still the size of a phone booth. Specialized machines, like the seven-function unit that ran rendezvous calculations for Gemini, or the ballistic solvers in the Polaris missile, could be made reasonably compact, but the IL engineers envisioned a "general-purpose" computer, one that could be reprogrammed at will to do any possible task with the data at hand.

Such a machine afforded great possibilities. New, specialized functions could be added at any time, merely by changing the programs, allowing late-stage design fixes, mission changes, or improved calculations. The computer could take over routine tasks from the astronauts, improve the accuracy of their flying, and help them manage the spacecraft's systems. Imagine punching in a series of numbers, then sitting back while the computer automatically oriented the spacecraft and fired the engine, in just the proper direction for just the right amount of time.

But would it work? Digital computers in the early 1960s seemed impossibly complex and broke down on a regular basis. Impressive, highly mathematical analyses, not to mention practical experience, showed that any such machine would fail frequently. How could NASA ensure that an Apollo computer would run for an entire two-week moon flight? Mercury and Gemini were banking on redundancy, backup systems, and manual takeover. What alternatives would the computer offer? A new technology, the integrated circuit, seemed to promise great increases in capability in very small packages. But the little chip was a black box, seemingly immune to human understanding and repair—and nobody had ever tried it in such a life-critical application.

From 1962 to 1965, NASA and IL engineers struggled through these problems as they developed the hardware for the Apollo Guidance Computer (AGC). At odds were differing engineering cultures, differing technical philosophies, and differing visions of the astronauts' roles. The design decisions, while considering subtle, technical details, also stemmed from judgments about human performance. Astronauts had been hailed as the ultimate backup systems. Could they take over if the computer failed? Could they help make the computers more reliable? How could the IL convince NASA and the astronauts to trust their lives to the machine?

It has become fashionable to denigrate the computers of the past with phrases like "we flew to the moon with less computing power than I have on my wristwatch," or "can you believe the entire Apollo program fit into a mere 36 k of memory?" Simply focusing on memory size, or the computer's speed, however, misses the important engineering accomplishments of the Apollo computer. For who among us would risk our lives on our desktop computers, with all their speed, accuracy, and memory, and rely on their working flawlessly for two straight weeks? The space shuttle flies with five redundant computers. Any fully digital airliner has a minimum of three. Apollo had only one. It never failed in flight.

A Computer for the Moon

As the IL's Apollo project entered its initial design phase, the engineers began thinking about their computer. They started, logically enough, with the Mars probe computer and a good dose of Polaris. After a few prototypes, a computer design emerged that was designated "Mod 3C." Typical computers of the time occupied buildings and burned enough electrical power to light a small town. These big machines, in universities, military installations, and corporations, tended to be "numbers in, numbers out" crunchers. But the Mod 3C, like MIT's Whirlwind, was a departure. In today's terms it resembled a "microcontroller," the microprocessor embedded in everything from cell phones to automobiles, more than it did a scientific computer. It occupied a cubic foot rather than a tiny chip, but it shared with microcontrollers a modest calculation capacity melded to numerous, sophisticated channels for quickly getting data into and out of the machine. Like a microcontroller, the Apollo computer included an interruptible processor so events could command the computer's attention in real-time. It had a "night-watchman" circuit to keep the computer from crashing and locking up (a feature still included in many embedded computers today as a "watchdog"). In today's terms, the Apollo computer was "embedded."

The original Apollo machine used 16-bit words for its data and instructions, adequate for the local control functions, but the navigation calculations required "double precision," more bits with more numerical precision, a feature implemented in software. The computer had two types of memory, "erasable" and "fixed," what today are called RAM and ROM. Erasable memory was used as a kind of scratchpad for data and calculations while the programs were running (or for temporary programs), while the fixed memory contained the programs themselves. All was "core" memory; that is, bits were stored by magnetic fields on tiny cylindrical cores. The computer comprised 1,700 transistors, each in their own metal can, and nearly 20,000 metal and ferrite cores for memory.

Over the course of 1962, IL engineers refined and then actually built Mod 3C as a large prototype to enable further experiments. Eventually it had 12,288 words of permanent storage, 1,024 words of erasable memory, and a 16-bit architecture with 8 instructions. It included a variety of counters, pulse outputs, interrupts, and input-output registers for interfacing to the rest of the vehicle. For a control computer such as this, the interfaces were key—the machine would have to not only perform calculations, but also take data from the inertial platform, read the optics' angles, torque the gyros, fire the thrusters and the rocket engines, and, of course, interact with the astronauts.[1] Often, these interfaces embodied the relationships with other organizations, companies that built sensors or engines. Most of the interfaces were analog, so the computer needed ways to accurately convert the discrete digital data to the continuous analog world and vice versa.

Transistors to Chips

As often happens with a large project lasting several years, especially an electronic one, the basic technology for the Apollo computer was hardly static. Transistors, invented in 1948, had become commercially available and accepted in the 1950s, especially for military equipment. Integrated circuits (ICs) could put several transistors on a single semiconductor chip, doubling or tripling the package density. IL electronics guru Eldon Hall had incorporated an early IC into the Polaris Mark II computer and thought integrated circuits might have some application to the new project—especially since reduced size and weight would be such high priorities. The Minuteman ICBM was using integrated circuits with strict requirements for reliability, giving Fairchild credibility with military and aerospace systems. Fairchild Semiconductor put its "MicroLogic" family of ICs on the market in 1961, followed soon thereafter by Texas Instruments and Signetics.[2]

The Mod 3C prototype used a circuit design inherited from the Mars computer called "core-transistor logic." Familiar, proven, conservative, and made out of discrete transistors, it seemed just the sort of circuitry on which to bet a multibillion-dollar spaceflight. Yet in 1962 Hall asked David Hanley, a young engineer recently come to the IL from the air force, to purchase a large supply of Fairchild's MicroLogic ICs and to look into designing a computer with them. Hall bet him he couldn't build a version of the machine with ICs that would be faster than the discrete version. Hanley began to build an exact copy of the Mod 3C with the MicroLogic devices, but found that a basic redesign would be easier and more efficient. So he set about designing and building a MicroLogic-based computer that would run the same instructions as the Mod 3C, and named it AGC for "Apollo guidance computer." The name would stick. Logic designers Ray Alonso, Hugh Blair-Smith, and Albert Hopkins saw the redesign as "an unusual second chance" to tweak their work.[3] Hanley won his bet with Hall. The AGC model began working in early 1963, occupying less space and running faster than the transistor version.

Late one day in 1963, engineer Herb Thaler was working on the new computer with the integrated circuits in the IL's large computer lab. "I was debugging it when the word came that Kennedy was shot," he said. Thaler remembered the shock, and how afterward people on the project became rededicated to going to the moon as Kennedy's legacy.

Still, for all the impressive engineering, a model AGC working in a laboratory would not be sufficient to convince NASA. Hall recognized that his idea for using MicroLogic would require advocacy and persuasion. Hall based his arguments on three factors: (1) the speed of the MicroLogic computer would be more than twice as fast, (2) size and weight would be cut in half, and (3) costs would be comparable to the traditional approach. The integrated-circuit version of the computer, however, would consume nearly double the electrical power of the discrete version (about fifty watts, or the equivalent of a small light bulb).[4] Still, the integrated circuits allowed a simpler, more elegant, and more modular design. One type of circuit, a two-input NOR gate, would repeat throughout the entire computer, rather than thirty-four different modules, as would be the case with the core-transistor logic.

While the benefits to integrated circuits seemed clear, the risks were uncertain. It was new technology. Only one company, Fairchild, made the required chips. Would they remain available for the entire length of the project, or were they just another passing fad in electronics? (As it happened, Fairchild did stop making the chips, and the IL ended up buying them from Philco.) Would Apollo be a sufficiently large customer to encourage the suppliers to stay in the business? To increase the demand, Hall proposed that all the ground test equipment for the computer be built with the MicroLogic circuits as well (sometimes chips that had failed the rigorous tests for flight hardware).

NASA was convinced, and in November 1962 allowed the integrated-circuit design to go forward.[5] Integrated circuits were not limited to the computer itself, but also covered the analog circuits that brought data into the digital world. Hall soon added to the computer not only the digital NOR-gate IC, but also a new analog IC for a sense amplifier—to condition the analog signals from the spacecraft's numerous sensors.

Hall's relative ease at convincing NASA to accept integrated circuits did not mean that the decision was straightforward, or that the use of integrated circuits was "obvious" in any way. For no other major Apollo subsystem used the new technology to the same extent—not the computer that IBM built to stabilize the Saturn, not the radio or radar electronics, and not the LM control electronics outside of the computer. NASA would not allow integrated circuits in these systems, and at least one contractor envied the IL's ability to use the MicroLogic designs, having been turned down by NASA.[6] The agency's approval to use them in the AGC, then, was a testament to Eldon Hall's competence in electronics, his confidence in his own analysis, and his skill at presenting the case and persuading his sponsor.

The AGC gave a boost to the fledgling IC business. The Minuteman missile program also used integrated circuits, but over twenty types of custom designed chips. Apollo, by contrast, used a single, standard type of chip. By 1963 Apollo was consuming 60 percent of the integrated circuit production in the United States.[7] In 1964 Robert Noyce of Fairchild stated that the company had had shipped 110,000 ICs for Apollo.[8] Noyce cited the Apollo computer as evidence of the reliability of the ICs he manufactured, and of their suitability for military electronics.[9]

The Block I AGC took shape during 1963. The IL created a formal organization for the project, under David Hoag as project director. Reporting to him were the Mission Development group under Battin, Digital Computation group under Alex Kosmala, and Digital Development group under Eldon Hall. There was still no group devoted to software.[10] For the following three years, the bulk of the IL's manpower was devoted to hardware development.

Apollo's guidance and navigation team brought their system to the public in September 1963. A press conference at MIT—led by David Gilbert of NASA and including Trageser and Hoag of the IL, Ralph Ragan of Raytheon, and project directors from Kollsman, Sperry, and AC SparkPlug—reported the design as "virtually complete" and "within spitting distance of meeting our schedule objectives." Hoag presented the system itself, with the computer at its core, and described the navigation and sighting techniques. Ragan described the manufacturing operations in Bedford, Sudbury, and Waltham, all in Massachusetts. When asked about what an astronaut would do during the mission, Hoag replied: "He has to align the inertial system, set in the right sub-routine to the computer, and then let it go."[11]

"As reliable as a parachute"

However innovative the architecture, design was not the major challenge of sending a computer to the moon. The difficulty would be making this computer robust, reliable, even bulletproof. Moreover, how did one measure such things when the machine could not be tested in its real environment for years to come? Early on, the IL group recognized that reliability would largely determine the success of their enterprise. If their computer failed in testing, it would never be allowed on the mission. If it failed in flight, it could cost the astronauts' lives.

NASA manager David Gilbert made the IL responsible for the AGC's reliability and gave them a set of policies. First on the list: "Make maximum use of the man to simplify equipment." Design conservatively. Rigorously test and select parts. Require the contractors to test the components, test the final assembly. Carefully report and track each failure. Gather data about them, analyze the data, and review the design accordingly.[12]

Robert Chilton remembered that Hall and his group paid constant attention to reliability questions, though NASA wasn't prepared to give them a specification: "[Hall asked:] How reliable is this? How reliable is that? How many failures per thousand hours, or how many hours between failures? They asked me how reliable has the computer got to be, this guy in charge of computers. How reliable does a computer have to be? I didn't know anything from reliability, I said, "It has to be as reliable as a parachute."[13]

But what did "as reliable as a parachute" mean? Chilton responded vaguely because reliability requirements for Apollo were notoriously imprecise across the board. Eventually, the project settled on .999 for safety and .99 for mission success. That is, a one in a hundred chance the mission would fail, and a one in a thousand chance the astronauts would not survive. Max Faget recalled that, using these numbers, "one of the study contractors came to me and pointed out that wasn't very much different from the expected mortality from three 40-year old individuals on a two week mission if you took the standard actuary tables."[14] These were not useful criteria for design decisions.

In-Flight Repair

Hall estimated that his Block I integrated circuit computer would have a reliability of .966 (one failure in 4,000 hours), but the spec he had been given required reliability nearly ten times better. To make up the difference he proposed neither redundant machines nor exotic circuit designs. Rather, he would rely on the skills of the astronauts: they would repair the computer during flight.

In Polaris, crew members inside a submarine could remove the computers from the missiles and repair them. On long-range bombers, crews could make repairs to their guidance systems during a long mission. But on Apollo such repairs had troubling implications.

What kind of equipment would the astronauts need to isolate a fault? A meter, at least, possibly an oscilloscope. But a soldering iron?

The "map and data viewer" would provide the thousands of pages and manuals necessary for troubleshooting. Hall proposed that Apollo flights also carry a special machine, a "MicroMonitor," a smaller version of the equipment used to check out the computer on the ground. This device was heavy and took up space, required its operator to "exercise considerable thought," and required the operator to have a mere *three to six months* of training. "This device is known to be effective," Hall wrote, "in direct proportion to the training and native skill of the operator."

The Block I computer would be made of easily removable modules, which the astronaut could replace by hand with simple tools. It would require twenty-nine types of modules, so the spacecraft would simply carry one of each type. If a failure occurred, the astronauts would diagnose the problem with the MicroMonitor, slide out one of the modules, and slide a fresh one in. To repair the computer, astronauts would require a special tool kit, complete with mundane items like gasket seals and grease (figure 6.1).

In addition, the computer would need self-check circuitry and software to detect a failure and the flexibility to allow the astronauts to do something about it. A suite of software programs would self-check and aid the astronauts' diagnoses. The IL developed a series of procedures for in flight repair, all based on the assumption that there were 100 percent spares available on board.[15] Of course, making the modules

Figure 6.1
Apollo Block I computer, showing removable modules that astronauts could replace in flight to improve reliability. Such repairs required carrying extra modules, as well as specialized tools and diagnostic equipment. (Draper Laboratories photo CN-4-165-C. Reprinted in Hall, *Journey to the Moon*, 16.)

replaceable required additional electrical connectors and mechanical connections, themselves likely sources of failure. And the complete set of spares added a significant amount of weight.

Astronauts had been billed as the ultimate redundant components. Asking them to improve the reliability of their equipment seemed sensible, but it proved no simple task.

Reliability Concerns

Early on, in-flight repair was a NASA policy for Apollo overall. If the agency was to undertake ambitious space missions in the future, Apollo seemed a good time to start to make the craft "self-healing." Still, it did not lead to elegant solutions, especially for the IL, a laboratory that prided itself on the autonomy of its instruments. In fact, for the IL it highlighted remaining uncertainties around integrated circuits, and questions about design. Reliability would dominate debates about the computer as the program moved away from a prototype in a lab to a real computer on the way to the moon.

The state of the art for inertial guidance systems on military aircraft at the time was something like fifteen hours mean time between failure (MTBF)—that is, the average amount of time between failures was fifteen hours of use. Aircraft just didn't generally fly much longer than that, and missiles were rarely in the air more than a few minutes. Yet to have a great chance of succeeding on a two-week mission, Apollo would require 1,500 hours MTBF—an improvement of two orders of magnitude.[16] And that was just for the flight, but did not include the many hours of testing, alignment, and checking prior to launch or intervals between scrubbed launches.

Much of the technology designed into Apollo's guidance had been developed for ballistic missiles, and hence the reliability data that existed related to those types of missions. The short-lived missiles, however, had comparatively low reliability. The air force and navy's solution to these problems was simply to build many more missiles than they needed and assume that some proportion of them would fail in the event of war. For a manned system the country could not afford (politically or financially) to mount several extra missions to ensure one would get to the moon.

For the guidance and navigation system, reliability depended on two major elements: the mechanical systems that ran the inertial platform (the gimbals and gyroscopes) and the reliability of the electronics. Gimbal reliability and the difficulties it generated were discussed in the previous chapter. It is difficult to picture the problems with electronics during its adolescent decades (when was the last time you had an actual, electronic failure, as opposed to electrical connections or software, in your computer?). Vacuum tubes, which were still fairly common, were notoriously failure-prone (and had given all of electronics a bad name), and transistors, which promised better reliability in principle, were still in their infancy and had not yet proven themselves.

IL engineers pointed out to NASA that their Polaris Mark II guidance unit, then in the final stages of development, was meeting its reliability goals. NASA replied that the Polaris system only needed to work for about three minutes at a time, as a missile was fired, whereas the Apollo system would require at least fifteen hours. This estimate assumed that the computer would be turned off when it was not being used. If the computer ran continuously (which it eventually did), it would need to work for nearly two hundred hours.[17]

Some of these problems stemmed from a technique becoming increasingly fashionable for analyzing large systems: statistical failure analysis. With a new, complex system with so many untried components and techniques, it was not difficult to "prove" statistically that it would not work. NASA's stipulation of .99 for mission success, and .999 for safety amounted to looking at the reliability of the individual components, with reliabilities like .9994, and then combining all the probabilities together in "fault trees," or models of how particular failures might compromise the mission. Statistical approaches, with their complex equations and multiple decimal places, had an air of precision that lent them credibility in technical settings, and even more so in nontechnical settings.

While a useful technique for identifying critical components, statistical failure analysis has the drawback of being contingent on the accumulation of numerous low-probability events. It also assumes those numbers are well known, and accounts only for individual component failures, not for failures from interactions between components, or "system failures." Indeed Apollo would experience numerous systems failures ranging from merely annoying to dangerous and fatal, which had not been predicted by the component-failure techniques.[18] Debates over statistics versus systems approaches to reliability pervaded the Apollo program, reaching into the lunar module, the command module, and numerous other areas. Furthermore, reliability had direct implications for the human role. If humans in space were to be the ultimate backup system, then who the astronaut was, and what his duties were, depended critically on how engineers approached the question of reliability.

The reliability question was not an academic one for the IL—the entire program was at stake, and IL's managers spent much of their time on the defensive. Members of Congress even wrote to Webb demanding explanations of reliability.[19] They questioned the performance of the IL, a group of academics in a game otherwise reserved for major defense contractors, working from a noncompeted sole-source contract. Indeed, IL management was haphazard in the early years: the ICs really did have some reliability problems, and the IL was perennially late releasing drawings to the manufacturers in ways that threatened the flight schedule (at one point the lab began issuing blank pieces of paper so the manufacturers would at least have a drawing number to begin procurement). These delays compounded the software problems explored in the following chapter.

Designing and Fabricating Reliability

In fact, Hall countered the statistics of component failure analysis with an entirely different approach. Statistical analyses, he argued, were only helpful when one already had a large base of experience with a particular technology. Moreover, statistics black-boxed the failures: not all occurred for the same reasons; chips from different companies failed for different reasons under different conditions. Without understanding the processes by which parts were produced and assembled, he claimed, "the [statistical] technique is highly questionable."[20] The same held true for building redundancy into a system—it had to be done with a deep understanding of how and why the parts would fail, and only under some circumstances would the added reliability offset the added risk from the complexity of redundancy.[21] Simply adding more computers was not the answer. You could not calculate, analyze, or graph reliability into a machine. It had to be designed in. The debate went on throughout the Apollo program, and pitted local engineering cultures against the statistical analyses of NASA headquarters.[22]

Hall and his colleagues proposed another solution entirely, one based on basic engineering—"the fabrication of parts into a system." They started by questioning the assumption that underlay the statistics: that failures would be random. Instead, they took the position that there is no such thing as a random failure; rather, failures always occur "based on cause and effect principles" (an approach credited with success in other parts of Apollo as well). All failures had a source, and for electronic devices, most of those were "the result of poor process control or the vendor's lack of complete technical knowledge of his process." Reliability was not simply a matter of statistics, but also "always an integral and basic part of design, or procurement, and of operation," best left to the "judgment and wisdom of the engineers."[23] Key to this approach was standardization—build systems out of the smallest possible numbers of different parts and focus a great deal of effort on improving every aspect of the process of producing them. The technique built on the classical tenets of American manufacturing: economies of scale, detailed control of process, and standardized components. Ultimately, the skill, reliability, and management of the workers on the production line would ensure the Apollo program's success.

Implementing this philosophy required careful control, especially of the industrial suppliers who fabricated the parts. Here the contract structure is revealing: the major guidance components, such as the inertial unit and the accelerometers, were supplied under specific subcontracts. The integrated circuits, which proved as critical as the larger components, were simply bought from vendors. At the outset, nobody realized what a critical role these circuits would play; otherwise they might have been formally contracted, like the gyroscopes or accelerometers. Instead, they were simply purchased.

The only leverage Hall had over the suppliers, then, was as a customer. He hoped that by buying large numbers of a single part he could persuade them to improve their processes. He certified the vendors by prequalification testing of parts. For any batch that the vendors delivered, the parts were screened and burned-in, and then tested for extended-life. If too many components failed, the entire lot was rejected. One test involved immersing the chips in a Freon solution, then meticulously weighing them one by one. If a chip got heavier by more than .00050 grams, then Freon had leaked into it and chip's seal was compromised, forcing a reject.[24]

Large volumes of standardized parts also lowered prices, which freed up funds for testing, evaluation, and monitoring, and allowed the suppliers to focus on a continuous production flow and constantly improve their process. At Fairchild, one manager reported, "Apollo really taught us a lot about reliability," because workers had to account for every single circuit failure. The company eventually developed separate production lines for Apollo, with workers selected for high motivation and attention to detail. NASA and the IL arranged for the astronauts to regularly visit manufacturing plants and meet the workers who were assembling guidance systems, to impart a direct sense of the importance of quality in their work.[25]

Human presence in spacecraft did indeed improve reliability, although not always in the ways its advocates intended. Before a single astronaut went into space in an Apollo spacecraft, the very anticipation of their presence improved the machinery. Imposing the criticality of human life into development and manufacturing forced engineers and production workers to emphasize robustness of design, attention to detail, and quality of workmanship.

The IL gradually grew a base of experience and collected data. By March 1966, Block I computers had operated for 66,000 hours, experiencing twelve failures among seventeen computers over the course of three years, making for an MTBF of about 3,000 hours.[26] "In the final analysis," Hall wrote in retrospect, "only successful missions proved the product."[27] After all of the Apollo flights, Hall calculated the MTBF for a computer operating in the command module environment at 50,000 hours.[28]

Shea's Systems Approach

As the AGC hardware evolved, so did NASA's management of the project. In early 1963, Brainerd Holmes, who headed Apollo at NASA headquarters, left the agency after a disagreement with Webb. That July, Webb consolidated management at the agency and hired a new man to bring coherence and control to the program. George Mueller came to NASA from air force systems engineering contractor TRW, taking over Holmes's job as NASA deputy associate administrator for manned spaceflight. Mueller had classic systems-engineering credentials: a Ph.D. in physics, time at Bell Labs, and experience in adopting systems-management techniques while working on the Minuteman missile for TRW. "One of the things that became apparent immediately," Mueller recalled about taking over the Apollo program, "was that there wasn't any management system in existence."

The statistical naysayers supported his position. Mueller received a pessimistic report that under the current program, the chances of landing on the moon were one in ten. The probability of failure provided the justification Mueller needed as he quickly moved to consolidate control of Apollo in NASA headquarters and to better coordinate among the NASA centers at Cape Canaveral, Houston, and Huntsville. He also created ASPO, which had dispersed sites but was part of the headquarters structure. "The laboratories were going to have to become a support to the program offices or else we weren't going to get there from here," Mueller believed.[29] One result of Mueller's changes was the famous "all up" testing approach, wherein the entire Apollo system was tested as a unit, rather than the more conservative, step-by-step approach favored by the NASA centers, Huntsville in particular. He also hired Bellcomm and General Electric to provide technical advice from outside the centers.

Mueller also brought in General Sam Philips, who had directed the U.S. Air Force's Minuteman ICBM project. Philips came with a number of his managers from the air force, and together they imposed their systematic management methods on a NASA organization used to the freewheeling style of research. The Space Task Group was already straining under the demands of Mercury and Gemini (which were both running late and over budget). Apollo, while still progressing quietly in the background in the early 1960s, was beginning to show strains. Philips employed techniques like configuration control and interface control, formal methods of documenting the hardware and their connections.[30] While these were appropriate for a large organization, the craft-oriented engineers sometimes interpreted them as simply layers of bureaucracy. "I've been identified as a procedures and methods man," Philips lamented, "with a management manual all written that I intend to force on the Apollo program and down the throats of the existing 'good people' base."[31]

Putting these controversial methods to work on the ground was Joe Shea, a tough, focused man, not always diplomatic, but deeply committed to Apollo. Like Mueller, he represented the new kind of engineer that had emerged during the ballistic missile projects of the 1950s: the systems expert. He had bachelor's, master's, and Ph.D. degrees from the University of Michigan in engineering mechanics, and had spent much of the 1950s as a mathematical analyst and development engineer at Bell Labs. From there he went to work for AC Sparkplug on the Titan missile program, where he brought systems methods to bear on the troubled program and pushed it toward success. He then moved to the West Coast bastion of systems expertise, TRW, whence he was recruited to NASA headquarters as "Deputy Director of Systems Engineering." Taken to quoting Milton and citing Greek mythology in engineering lectures, Shea took an unusually broad view of Apollo and its significance.

Shea's mission was to translate Kennedy's all-encompassing mission statement into precise engineering specifications. Along with George Low, Shea soon managed a 400-person organization to oversee NASA's projects. Here he wrote requirements for safety, reliability, and the probability of mission success, and studied how those measures manifested themselves in hardware specifications. Shea also managed Bellcomm as it was looking over the shoulder of the Apollo contractors, including the IL. Shea and his reliability analyses played a key role in the LOR decision, although he came to distrust a pure statistical approach. As he put it: "So we finally concluded there was no way to assure statistical reliability. You had to really use engineering confidence and do a little wishing and lucking and hoping."[32] Shea's methods focused on "interface control," formally defining in great detail every possible link among the spacecraft components, essentially creating a virtual model of the system in a mountain of paper.

Shea actually mixed the analytical methods of the systems men with an appreciation for focused, insightful, creative engineering. One understanding of systems methods calls for breaking a large system down into component parts defined as "black boxes" by specifying their input and output interfaces. Then the black boxes can be contracted out, and then recombined by a systems integrator. Shea indeed employed these methods, but he also brought an approach that adapted to the ultrareliability required for human spaceflight, what might be described as "white box" systems engineering. No box could be truly black, because every component was open to scrutiny (hence "no random failures"). An assemblage of black boxes made a brittle system, a house of cards. By contrast, in a white box system, systems engineers always had the ability to peer into every subsystem, to examine every component. Nothing inside the Apollo system could be unknown or taken for granted. As an engineer, Shea saw how the system held together as a whole, but he could also see inside each piece and make tradeoffs between widely separated subsystems.

In October 1963 Mueller sent Shea to Houston, where he replaced Charlie Frick as manager of ASPO ("I was exhausted so much of the time I just couldn't hardly take it," Frick said of his Apollo tenure).[33] Even though the official contract with North American for the CSMs (command and service modules) had barely been signed, unofficially the program was nearly a year late and plagued with problems.[34]

Shea's arrival in Houston marked a sea change in Apollo culture. Mercury and Gemini were largely being run by old Langley hands in Robert Gilruth's group who had made the move to Houston. These were aeronautical engineers, raised in research and test flying. Headquarters felt they were running Apollo too informally, like a research program, rather than a large-scale development and operations project.

Once in Houston Shea found himself in the middle of a classic NASA conflict between headquarters and the centers. Though he was Mueller's man, Shea tried to fit into the center. The Space Task Group engineers saw him as a spy; understandably so, as Shea, with his fancy specifications, insistence on documentation, and white-collar style was wresting control from them. "I found a mess when I went to Houston," he recalled: deficient management of the contractors and poor control of the spacecraft and the interfaces between their various components.[35] "The spacecraft basically needed a major redesign."[36] At North American, it was even worse. He found a "lack of configuration discipline," meaning that engineers and technicians were changing the design without complete documentation. This sloppiness, he confided to Mueller, was more than a matter of procedure: "In the limit, we are really discussing the competence of their [North American's] people. I am convinced that we are not dealing with a first rate engineering organization."[37]

Plenty of people resisted Mueller's centralization and Shea's style. "I thought it was wrong and I still think it was wrong," complained Caldwell Johnson, one of the original Space Task Group.[38] Powerful Max Faget, with nearly a thousand people in his design bureau, froze Shea out from gathering data. Chris Kraft viewed Shea as "an outsider and an enigma . . . his close tie to George Mueller made him a stepchild among Mercury veterans."[39] Their objections were understandable: the close, hands-on engineering style inherited from Langley was slipping from their fingers, never to return. "The growing mountain of Mueller's bureaucracy looked to us working troops as unnecessary, cumbersome, and expensive," Kraft wrote in his memoir. "It turned out to be all three, but it never went away."[40] These tensions, between centralized control and decentralized management, between hierarchical systems and craft-based approaches to engineering, would remain throughout the Apollo program. They mirrored the human-machine tensions being played out in the hardware design: systemization versus intuition, professionalism versus informal organization, documentation versus personal trust. They would come to a head in the aftermath of the Apollo 1 fire.[41] But that was still years away, and in 1963 the Apollo program had barely gotten started and already it was behind.

Block II Is Born

Despite the resistance, Shea made at least one major contribution that helped create hope that the drifting program would actually get to the moon. The number of design changes required to produce a spacecraft was so overwhelming and constant that the contractor could not make progress in the face of the evolving requirements. Shea decided that the existing spacecraft design would be frozen and designated "Block I." It might fly a test mission or two in earth orbit, but not all the way to the moon. It would not be suitable for lunar landing because it would lack the LM docking adapter and a tunnel to transfer the crew between the two spacecraft. After Block I was completed, the required changes would be incorporated all at once into an entirely new design that would become the hardware that would really fly to the moon. As Shea wrote to Mueller, "Block II was born."[42]

Aaron Cohen, though part of the Houston culture, recalled Shea's decision "like coming out to really realizing we really had to go build a piece of hardware to go to the moon." Cohen gave Shea great credit for getting the Apollo program serious and moving: "He kept us on the right track. He didn't let us overdo it and he didn't let us underdo it."[43] Cohen believed that the Block II decision represented the turning point in Apollo.

In far-away Cambridge, these changes in the Apollo program would have direct bearing on the engineering of the guidance system and on the astronauts' roles in flight. Because headquarters was increasingly concerned about the IL's ability to handle the program, its managers realigned the contract to offload some of the work to the contractors so the IL engineers could take on the job of redesigning Block II.[44] Now AC Sparkplug was the prime contractor, reporting directly to NASA Houston, and Kollsman, Sperry, and Raytheon would report to them. The IL would also report to Houston, and serve as designer and systems integrator, but without cognizance over the contractors or authority over production hardware.

"Weak systems thinking"

The Block II redefinition had great import for the Apollo guidance computer (AGC) and hence for the human role in the missions. First to go was the idea that the astronauts might repair the system in flight. Shea found North American's principle of in-flight maintenance "another example of weak systems thinking."[45]

All during this management shakeup, engineers at North American and the IL still struggled with the problems of in-flight repair as its implications became more and more cumbersome. For their part, the astronauts were none too happy about having to learn maintenance techniques—the hottest test pilots didn't want to be repairmen in space. One IL engineer recalled Alan Shepard responding to the prospect of in-flight repair with insightful sarcasm: "Yeah, and we should all train to be brain surgeons so we can operate on each other."[46]

To make such repair practicable, the capsule would have to carry an extra copy of every electronics module in the computer, which amounted to carrying an entire additional computer. By October 1963, that's exactly what the IL decided—to include two complete computers in the capsule and eliminate the possibility of in-flight repair. That option, however, either doubled the volume or halved the computing power that would be available, neither of which was acceptable.

Meanwhile, Mercury was flying, and in May 1963 a new data point came in. When Gordon Cooper orbited the earth during the last Mercury flight, his control system failed, forcing him to orient and fire the reentry by hand. While it was easy to hail the flight as a triumph of manual control, something had gone deeply wrong. Upon examination, engineers found that the urine collection system had failed and globules of urine had migrated into the electronics, which, when combined with moisture from Cooper's own exhalations, proved to be rapidly corrosive. North American began a study of the effects of humidity on Apollo electronics and recommended design changes to protect the circuits from moisture. The Apollo electronics must be sealed shut.[47]

This change ruled out in-flight repair. Under Shea's direction NASA removed the requirement from the North American and Grumman contracts for the spacecraft, instead requiring the companies to install redundant systems in the vehicles.[48] Bellcomm showed the increase in reliability from two computers to be negligible or nonexistent, and the IL engineers agreed. Each Apollo spacecraft would have only one computer.

IL engineers would now make the single computer aboard the spacecraft ultra-reliable—an uncertain proposition given the poor record of most electronics. But what if it did fail? What would be the backup? Without a second computer, the astronauts might have to navigate home from the moon with a failed computer. Could they do it with paper charts?

The answer came from the ground. During the early 1960s, with experience gained from the Mercury and Gemini flights, advances in electronics and signal processing, and the advent of atomic frequency standards for very precise timing, NASA learned that it could track a spacecraft with great precision from ground-based antennas. A transponder on the spacecraft listened for radar interrogation signals from three huge antennas on earth (in Spain, Australia, and California) and echoed them back. Precise measurement of time delays and Doppler shifts, improved by averaging over time, allowed NASA to calculate positions in lunar orbit to within ten meters, and velocities to 0.5 meters per second. These numbers became so accurate that greatest uncertainty in the ground-based navigation fixes became the knowledge of the coordinates of the antennas on the surface of the earth, which could only be pinpointed to a few meters.[49]

So in the words of NASA engineer Cline Frasier, "Why not free up the whole mission planning and everything, and just go ahead and do it from the ground as the primary mode?"[50] Primary navigation from the ground would not only eliminate the problem of a backup, it would also significantly reduce the workload of the astronauts, who could turn their scarce attention to other things. The primary navigation system, then, became ground-based tracking, backed up by the onboard computer. The computer, of course, would still be primary when the spacecraft, on the far side of the moon, lacked access to earth-bound signals. Implicit, of course, in this decision was a demotion of the MIT effort—for most of the flight now the onboard navigation would be a backup system. Furthermore, it meant abandoning the original requirement of "autonomous" navigation—the IL engineering value of being able to navigate missiles and aircraft with no references to the outside world. Their approach seemed reasonable enough in wartime, but by 1964 the idea that the Soviets would try to jam navigation signals for peaceful moon shots began to seem far-fetched.

The Digital Autopilot

Shea made one other change in the control system for the Block II design. When he came to Houston from headquarters, the Apollo guidance computer was just that—a computer that figured out where the spacecraft was and where it ought to go. Another system actually took care of commanding the thrusters to move the vehicle and was under development at Minneapolis-Honeywell, the same company that had made the control systems for Mercury and Gemini. As in those earlier spacecraft, if the astronaut wanted to orient the spacecraft, he would command the Honeywell servos to go to a particular point, and then analog feedback loops would command the thrusters to hold it there. As in Gemini, an indicator would show velocity to be gained to the astronaut, who would command thrust while the indicator counted down toward zero when the maneuver was complete, a scenario appealing to the pilots. But the Honeywell project was behind schedule and having technical problems, and NASA engineers began to think about simplifying the situation.[51]

The IL system plus the Honeywell system meant that the Apollo spacecraft contained two nonredundant control computers, two sets of gyros, two sets of electronics. Why? Cline Frasier was the NASA representative on the project: "I was always nervous about all of these mechanical parts running around." His first idea was to put the MIT and the Honeywell computers in parallel rather than in series, so that if the analog one failed the digital one could still fly the craft.[52] But Frasier went one step further and convinced Shea to eliminate the Honeywell autopilot and incorporate its function into the digital computer. Shea had been the program manager for the Titan missile, which used digital computers for flight control, so he was no skeptic about their capabilities.

ASPO manager Clifford Duncan made the digital autopilot decision in June 1964. "So Cliff [Duncan] sent me to the Instrumentation Lab to go up and tell the guys," Frasier recalled, "that instead of two computers there is only going to be one. Instead of just doing the guidance, they were also going to do digital autopilot."[53] Though not eliminated, the Honeywell SCS was much reduced in complexity to provide a simple analog backup to the digital autopilot.[54] For the IL engineers, this change amounted to doubling the responsibilities of their computer, and hence doubling the work to be done before the flight (fortunately, NASA ordered the memory doubled as well, but the IL still had to write and test the programs).

The digital autopilot made the IL computer central to the Apollo spacecraft—some compensation, perhaps, for no longer being the primary source for navigation. Now, the controls of the CSM and the LM would go through the AGC and its software. A software-based "state estimator" could derive the motions of the vehicle and automatically compensate for thruster failures and other anomalies. Shea cemented the decision in a memo at the end of 1964. From now on, the "digital autopilot control mode," would be primary, with the analog Honeywell system as a backup.[55]

Putting a digital autopilot into Apollo was a radical step. Today, some commercial airliners use this fly-by-wire technique, but in the 1960s, no aircraft had yet demonstrated digital fly by wire, and only a few research aircraft had used an analog form. The astronauts were none too happy. "Why don't you guys quit wasting time," David Scott recalled thinking about the decision, "go back to MIT and think."[56] Frasier believes the digital autopilot never would have been approved were the flight crews not too busy with Gemini to notice. He remembers running into Pete Conrad one day in a hallway, and Conrad "just chewed me up one side and down the other and said that we're crazy. That his friends at Grumman had told him this stuff wasn't going to work, and it was his life on the line, and kind of on and on." Frasier, a junior engineer at the time, was genuinely intimidated at being told off by an astronaut, but the decision had been made.[57] The astronauts could make their peace with it, or they could choose not to fly—which none of them did.

As Davey Hoag put it, "The way we look at it, we actually have four autopilots."[58] The CSM and LM each had two separate modes: free fall mode, which provided attitude control while the vehicle was coasting or in orbit, and thrusting mode, which controlled the large engine on the command service module for major maneuvers. Hoag might have added two more for the CSM: a boost mode, where the AGC monitored the Saturn rocket's performance, and even allowed an astronaut to manually takeover if the Saturn's guidance failed in the second or third stage (a vestige of the piloted-boost debate), and a reentry mode, which worked similarly to the Gemini computer. For any particular mission mode, the navigator would select the program using the computer's keypad.

In Gemini, the computer calculated velocity to be gained and the astronaut controlled thrust with his stick, according to the display. In Apollo, the digital autopilot oriented the spacecraft automatically and then commanded a program to fire the main engine. The digital autopilot also trimmed the large engine's gimbals on a regular basis while it fired, ensuring the thrust was aligned with the vehicle's center of gravity. During coasting flight, the computer could hold a particular attitude, or change it constantly at a defined rate. The digital autopilot also had "jet selection logic" in the software to determine which of the sixteen thrusters on the spacecraft should fire at any given moment. If any of the thrusters failed, or if an entire "quad" cluster failed, the autopilot would automatically sense the situation, compensate accordingly, and avoid doing anything dangerous. It limited the rate of motion in any axis, to avoid tumbling, and it could automatically recover from a tumble by slowing the rates in each axis. Like the X-15 adaptive autopilot, the digital autopilot contained a model of the spacecraft's motion and compared its actual motions against the ideal in the model, taking into account the effects of body bending and fuel slosh in the vehicles.[59]

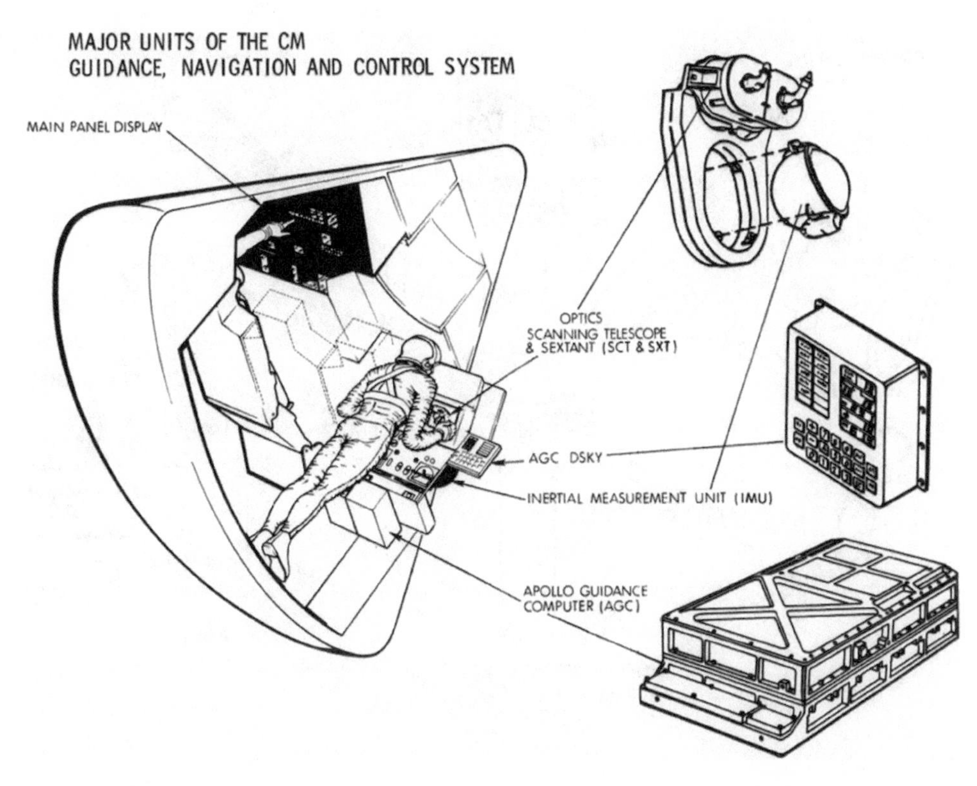

Figure 6.2
Physical units of the Block II guidance and navigation system in the command module. (Draper Laboratories/MIT Museum.)

This new role for the computer had great impact on the IL engineers. Suddenly they were responsible not only for alignments with the stars, but also for detailed interactions with the other components in the system. Suddenly they needed reams of highly accurate data—on the vehicles' dynamics, bending modes, actuator performance, engine gimbals—creating a whole new level of complexity in their relationships to North American and Grumman. About 10 to 30 percent of computer capacity would be required to run the autopilots, capacity that was far from plentiful as the software edged close to its performance margins (figures 6.2 and 6.3).

The digital autopilot also confirmed the decision to use a general-purpose computer in the first place and underscored the intimate links between systems engineering and digital computing. Engineers could move particular functions out of hardware devices and into computer programs, saving critical quantities of weight, money, and hardware complexity. In one example, Shea nixed an expensive program to add a heat shield for the side of the command module facing the sun. With his knowledge of

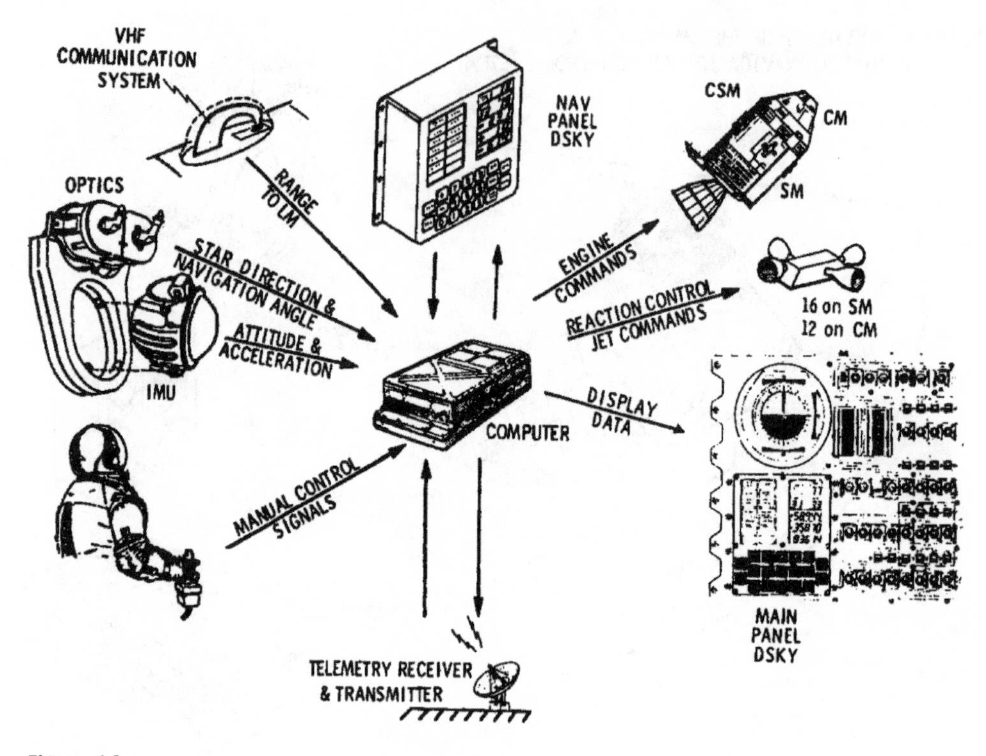

Figure 6.3
In the Block II system the computer linked a variety of systems, including the astronauts' control of the spacecraft. Note how the manual control signals go through the computer instead of through a separate controller, as in Block I. Not only the computer but also the MIT Instrumentation Lab had to integrate these diverse components, from a variety of contractors. (Hand, "MIT's Role in Project Apollo, Vol. I," 52.)

control systems and the digital autopilot, he simply suggested replacing the insulation with a software routine to keep the spacecraft rotating like a rotisserie, distributing the heat load around the craft. A few lines of computer code replaced a heavy mechanical structure.

In a 1967 interview, Hoag expressed wonder at the flexibility of digital controls: "All this, and everything else too, in this tremendous computer and we are doing it, by golly, we are doing it."[60] Yet there were costs, mostly in software, in key areas of complexity, reliability, and schedule. All three would raise problems down the line. Still, Aaron Cohen considered the inclusion of the digital autopilot a "milestone in the Apollo program." Frasier believes that without "the combination of the digital flight control and the navigation from the ground, we would have been at least a couple of years later [getting to the moon]."[61]

Peak of the Hardware Effort

September 1964 saw the delivery of the first Block I guidance system, slated to fly on the unmanned missions and the early manned ones. NASA and the IL team spent that year developing and finalizing Block II with the digital autopilot, in a series of meetings with North American and Grumman. 1965 saw the peak of hardware design activity, with more than six hundred engineers at the IL (though not all worked on hardware).[62] The Block II requirements were complete by February 1965, and soon thereafter the IL received a $15 million-dollar contract to build the Block II.[63] The new design was released in July 1965, a production prototype delivered that November, and the first flight qualification model delivered to NASA in July 1966. In the fall of 1966 Block I production was terminated (Block II production would last until the summer of 1969). Block I would fly the unmanned AS-201 (August 1966) and Apollo 4 and 6 missions, while Block II computers flew on Apollo 5, 7, 8, 9, 10, and all the lunar landings.

The new computer would be sealed from moisture in the cabin, with no provision for repair. To accommodate the digital autopilot, the Block II computer's program memory was increased from 24 k words to 36 k (16-bit) and the erasable memory from 1 k to 2 k. Other aspects of the computations speeded up as well, with new machine instructions built into the architecture (where the Block I had eleven instructions, Block II had thirty-four) which allowed for more efficient code, and the clock now ran at 1,024 kHz, allowing a double-precision multiply in about 1 ms (by contrast today's desktops will do the operation in a few nanoseconds). The number of logic gates went up, although the packages for the ICs were also changed, from small metal cans to flat packs much more like modern chips. The flat packs contained two logic gates instead of one, doubling the packaging density, which halved the circuits' size and brought the AGC's weight down from eighty-seven to seventy pounds, and reduced the power from eighty-five to fifty-five watts.[64]

In the five years from the granting of the contract to the preparations for the first Apollo flights, the guidance and navigation system matured from a basic idea, through laboratory prototypes, to highly qualified and manufactured flight hardware. Despite its small size and modest processing capacity, the AGC was cutting-edge technology, incorporating the latest in process control, reliability, circuit design, and packaging. In what would become a familiar pattern in the computer world, memory capacity had doubled, and then doubled and doubled again. At the same time, the presence of the computer meshed with systems management methods, especially as the project became more complex and behind schedule. Hardware devices in the cockpit were gradually eliminated to save time, weight, or cost, and migrated into the flexible computer or into the astronauts' brains. But as the Block II computers began rolling off the production lines, NASA and the IL began to recognize the stress they had added to a new part of the project, one barely envisioned when Apollo began: software.

7 Programs and People

Quest oculus non vide, cor non delet
What the eye does not see, the heart does not regret
"A lot happens that we are not telling you about."
—Opening lines of Apollo software source code

Programming the Moon Flights

With the exception of integrated circuits and extremely high reliability, the hardware for the Apollo guidance computer represented the state of the art when Apollo began. The same could not be said of the software and user interface. An aspect of the system barely envisioned when the program started, software turned out to be among the most difficult, and the most critical, components in the Apollo system. The software would carry all the burden of Richard Battin's complex guidance schemes. It would embody the goals and constraints of specific missions, and hence would require special finesse (sometimes completely new programs) for each one. It would control the flight at critical moments and stand between the astronauts, their machines, and the ground controllers and their tasks. Most important, it would interact with the astronauts, who might do unpredictable things at strange times in seemingly random order. As the Latin aphorism quoted above warned, the code contained secrets and hidden truths not apparent to the user.

Apollo began in a world when hardware and electronics were suspect and might fail anytime. It ended with the realization that as electronics became integrated, computers could become reliable, but that software held promise and peril. It could automate the flying, eliminate black boxes from the cramped cabin, and make the subtlest maneuvers seem simple. Yet it was also hideously complex and difficult to manage. If it went wrong at a bad time, it could abort a mission or kill its users.

When NASA first let the contract to the IL for Apollo guidance, programming was barely mentioned. In fact, the word *software* does not appear in the original statement of work. The word had only come into use in the late 1950s (*The Oxford English*

Dictionary lists the first usage of the term as 1960).[1] The original work statement referred to programming only twice, the most extensive reference being: "The onboard guidance computer must be programmed to control the other guidance subsystems and to implement the various guidance schemes."[2] Software was not included in the schedule, and it was not included in the budget. The absence of software in the initial plans reflected not only an underestimation of the effort, but also the belief that writing code was part of engineering the system, not a discrete activity. Dick Battin likes to recount that when he told his wife he was in charge of Apollo's "software," she found it unmanly and joked "please don't tell our friends."[3]

At the start, there were fewer specially trained programmers than engineers who could turn their ideas into executable code. As the IL's official history put it, "It was believed that competent engineers with a credible, solid mathematical background could learn computer programming much more easily than programmers could learn the engineering aspects of the effort."[4] And it was difficult to find people who were skilled at both. Programming was new enough that it was not broadly recognized as a technical, much less engineering skill. Jack Garman managed the software project for NASA for a few years, and he recalled an attitude that "the real problems [in spaceflight] are in propulsion or in . . . plumbing. I mean pressure vessels and stuff like that. That's where the real macho kind of space biz is."[5] Just writing instructions for the digital circuits still seemed a soft afterthought. The IL was fundamentally a guidance and control organization—mathematics, electronics, and the high-precision gyros were the dominant technologies. Obscure lines of machine code didn't have their own place of prestige in the organization.

Nevertheless, in the summer of 1961 a few engineers at the IL got started programming. Battin was in charge, with a laid-back management style typical of dedicated academics. "He didn't really run it," one of his engineers recalled, "he just nodded." Another thought of him as "a real hands off manager, just a guru in his corner."[6] In the early days NASA's attention was on Mercury and Gemini, and the IL engineers had a great deal of autonomy. Dan Lickly, who worked on reentry guidance, remembered a loose environment.[7]

Fred Martin, a calm, intense engineer, started at the IL in 1956, working on autopilots and fire-control systems before winding up in the Polaris program. He then went back to school at MIT, though he stayed at the IL and joined Battin's guidance and navigation group. Martin soon became the project manager for the CSM software, which, before the LOR decision, comprised all of the Apollo flight code. Martin remembered the "small group, seat of the pants effort that was unfettered and unhindered by NASA and devoid, pretty much, of real bureaucracy." He divided up the program, assigned tasks to different people, and kept track of their progress.

Some who worked for Martin remembered his subtle, hands-off style, "like a psychiatrist."[8] During this early period, the group focused on the fundamental engineering issues associated with getting to the moon, particularly the subtle mathematical models and numerical filters required to robustly solve the guidance equations in real-time.[9] Some referred to it as the "systems engineering phase," when budgets and schedules were loose and the group could afford to look at the big picture.[10] Guidance to the moon seemed an interesting mathematical problem, and a calculating machine should be able to handle it without much trouble.

People at the lab worked long hours focused on a clear goal with compelling technical challenges. As in the rest of Apollo, sixty- or seventy-hour work weeks became the norm. "There was a feeling of national urgency, you might say, of national responsibility," Martin recalled. "You had the astronauts coming to the lab regularly.... You had the pressure of what was going on." The IL group felt that NASA and the astronauts held them on a pedestal: "Everybody was a guru and everybody was a genius." Lickly remembered "a very privileged position," as the programmers regularly received visits not only from astronauts, but also from von Braun and the top NASA management, "all those people we got to talk to as a small sort of elite organization."[11] Some IL engineers never found such creative excitement again in their careers.

Apollo software would be the product of many people's work, integrated into a coherent whole. Martin's group had physicists and mathematicians and engineers, "probably had a literature major here and there too," he said, but no computer scientists. As few people were coming out of school with computer science degrees (MIT did not even create its department as part of electrical engineering until 1968), the IL group included people like Alex Kosmala and Margaret Hamilton. Hamilton, one of the few women engineers in all of the Apollo program, had a mathematics degree and had already worked as a programmer on the SAGE air defense system at MIT's Lincoln Labs. Lickly worked on the reentry programs for the command module. Others focused on navigation, or the inertial platform.

The LOR decision was not made until a year after the IL began working, so software design and programming began in a vacuum of detailed mission goals and specifications. Engineer Ed Copps remembered an exciting time as they contemplated the possibilities of a small computer onboard a spacecraft flying to the moon: "We had automata doing things that we had not planned for them to do.... We were the new guys coming to grips with new capabilities, which were now placing conceptual issues within the reach of actual implementation.... We designed it in a very abstract way."[12] One day Kosmala, a programmer, went to the MIT library and looked up a series of star coordinates to align the inertial platform, which he then embedded into the code. That star catalog survived into the final version and became the basis for the astronauts' navigation to the moon.

The official IL report on the software effort credited its success to "an intricately-tuned interaction among men and machines."[13] Interestingly, this statement refers not to the interactions aboard the spacecraft, but rather to the coordinating and scheduling of engineers and programmers on the ground, and their mainframes and simulation machines. Early on, of course, no actual Apollo computer existed to test the programs. The IL had a number of mainframes, including a Honeywell 1800 and the new IBM 360, which ran simulations of the Apollo computers. Indeed, throughout the program the software ran in simulation for validation on these machines, as well as on actual Apollo hardware. A digital simulator could analyze program operation in great detail, step by step, with a host of tracing and reporting tools, but at a comparatively slow speed. The simulator included routines like UNIVERSE, LUNAR, and TERRAIN to model various environments, and even one called ASTRONAUT that simulated the human operator. To complement these digital models, analog computers simulated the spacecraft's dynamics, from center of mass and rocket thrust to parameters for structural bending and fuel slosh, connected to models of the AGC that would run in real-time. Hybrid simulators mixed the two, and even included a sextant, an inertial unit, and a full user interface, allowing engineers and astronauts to exercise the system from the front panel.[14] Together, they amounted to building the Apollo spacecraft and traveling to the moon in a completely numerical, virtual environment, an electronic equivalent of the wind tunnels of an earlier era.

During the 1960s, just across the same campus, project MAC at MIT was incubating much of the technology and culture that underlay computing today, from graphical user interfaces to the nocturnal hacker culture. Students and young engineers were becoming enthralled with the creative potential and practical difficulty of interactive computing. Apollo programmers at the IL, while a separate group in distant buildings, shared some of this ethos. They tended to believe that any program could be written by a few smart people. As the project grew, they began to realize the need for more help, but were reluctant to jeopardize their small-group culture. "We've got to do this with either five people, maybe, or we've got to do it with 150," Jim Miller remembered thinking.[15] The group, he said, developed a closed, proprietary attitude toward their system: "We couldn't figure out why anyone else needed to know what was in the guidance computer besides us. What does NASA need to know for?"[16]

Of course, NASA needed to know because that was how its managers practiced systems engineering, where every subsystem had to be a white box open to scrutiny. And they were the customer. Martin later realized that MIT's hard-headed, creative exuberance appeared to NASA as exclusionary. "I think that that arrogance started to irritate people," Martin said, "who were tired of hearing that MIT knew best and knew better." NASA began to feel the need to "rein in" the IL, "to get them on this program not on their own program."[17]

Only in 1964 did NASA begin a series of meetings of a "G&N system Panel" to coordinate requirements among the contractors. Another series of meetings began to specify the interfaces—how the computers would read data and command actions of the spacecrafts' various sensors and actuators.[18] The computer, it turned out, especially

Block II, would tie together numerous different systems in the vehicle—from clocks and needles to hand controllers and rocket engines.

Software as Systems Integrator

Apollo's software derived from the basic design of the Mars mission. Designer Hugh-Blair Smith created a language called "Basic," a low-level assembly language of about forty instructions (distinct from the high-level BASIC programming language developed at Dartmouth at about the same time). On top of Basic was "Interpreter," brainchild of Hal Laning, a language that was really a collection of routines to do the higher-level mathematical functions associated with guidance and control, in the high-precision data format. Interpreter also had vector functions like matrix multiply and cross products that were useful for calculations of vector-based quantities in navigation and control.[19] A master program, "Executive," also a Laning creation, decided which programs ran when, including responding to high-priority interrupts.

Not by coincidence, Executive and Interpreter were both named after human roles. The software code reflected the social system of the lab. Certain programs were written by certain people or groups and they all had to work together and talk to each other, whether inside the machine or inside the laboratory. As Apollo flew, each particular operation depended on computer code written by a particular engineer, eagerly watching from Cambridge, Massachusetts, as a piece of his brain flew through space with the astronauts. Hamilton learned to think of an entire mission as one system, "of which part is realized as software, part is peopleware, part is hardware." Everything was linked, and an "error" in software could not be easily isolated to a particular line of code, as distinct from a specification, an interface, or a crew procedure.[20]

Executive and Interpreter comprised what we would today call the operating system, which also included code like program sequencers and interrupt-response routines. They reflected how the Apollo computer allocated its time, in a manner still novel in the 1960s but quite common today. The great innovation in computing in the 1960s was "timesharing," the idea that many users (or programs) could access a computer simultaneously, in real-time, as though each had the machine to him or herself. Timesharing worked by allocating a slice of time to each user or program, and then switching between them many times per second. When designing Apollo's software architecture, Laning eschewed this fixed-time-slice mode of operating. Instead, he made the Executive program asynchronous: it used a system of priorities, focusing its attention on whatever task was the highest priority at a given time. When that task was finished, Executive moved on to the next one, and so on down the priority list. This scheme had the virtue of always making sure the important things got done and of not taking resources at a critical time for operations that were less important. Updating the displays on the astronauts' keyboard, for example, was less critical than making sure the LM's thrusters kept it stable and upright. Autopilots, counters, timers, sampling the inertial sensors, telemetry, and computer housekeeping events went on regularly in the background even as a major computation took place.

The asynchronous executive had one major drawback: the program flow was unpredictable. No one could say exactly how it would respond in a given circumstance, because not everything was scheduled in advance; subroutines ran in real-time in response to events or as they rose up in the priority list. One might make the analogy between the asynchronous executive in the Apollo computer and the Apollo "program" overall: numerous different tasks, working cooperatively but competing for attention and resources.

As one example, consider the critical program P40 that controls the guidance and navigation during countdown, ignition, and thrusting of the main engine on the CSM, used to change the orbit of the spacecraft. A previous program calculates the desired time of ignition and the parameters of the existing and desired orbits. P40 then calculates the direction of the thrust and velocity to be gained, and calls a host of other programs to calculate gimbal angles, display thrust values to the astronaut, maneuver to the appropriate attitude, update matrices, and count down. All the while the computer is running autopilot calculations, exercising the reaction control jets to keep the spacecraft pointed, performing system maintenance, and even responding to other programs of higher priority. The net effect is an apparent cacophony of programs within the computer, all vying for priority and sharing memory, displays, and processor time.[21] When it worked, it proved an intricate, well-coordinated symphony of code.

The worst thing that could happen to a computer on board a flight was for a bug to occur during a critical operation and cause the program to "hang" or freeze up (a common enough experience for computer users today). Rather late in the program, NASA made a decision that the software should allow the computer to be restarted, literally in the middle of a maneuver, without corrupting the process. This feature, known as "automatic restart protection" helped protect against transients on the power supply (such as from a lightning strike) or software problems that stuck the code into an infinite loop.[22] One programmer described it as "giving the computer an enema." A clever idea, to be sure, but it forced the programmers to rework every program, and every subroutine, to keep track of its current state in a permanent way, so if a restart occurred, it could pick up again without interruption. As Copps put it, "It was actually the right thing to do...but it really made things a bit more complicated. I would say a lot more complicated."[23]

The asynchronous executive also faced deep resistance. Synchronous scheduling was accepted practice, especially among those in the aircraft industry whose computer experience tended to be with timesharing mainframes on the ground. "It was very important to them," Martin recalled, "to know exactly what was happening every instant of time." He characterized the debate over synchronous versus asynchronous executives as "almost a religious war."[24] Again, the analogy of project management is instructive: should everything be directed from a central location on a precise schedule? Or coordinated in a looser, almost organic form? This seemingly arcane technical debate had implications that would surface dramatically in the final seconds of Apollo 11.

Building Programs

For about four years, the programming effort trundled along at a comparatively leisurely rate while the hardware design proceeded apace. No more than a hundred people worked on software until mid-1965, when the hardware effort peaked with more than six hundred people. Once the hardware design was completed, however, the manpower began to decline, and by the end of 1966 there were more people assigned to the software effort (about two hundred and fifty), which peaked at more than four hundred in mid-1968.[25] These numbers reflect the last-minute programming crises that arose before the flights (figure 7.1).

Apollo programmers faced an irony: the early missions were unique test flights requiring custom programs whereas the later, more challenging flights with landings tended to have more uniform requirements. For each mission, the IL developed a "guidance systems operations plan," or GSOP, that had all the specifications, parameters, and custom routines (the CSM and LM each had their own GSOP for a particular mission). Essentially a series of equations and flowcharts, the GSOP theoretically contained everything required in advance to produce the program code for each mission. In practice, the GSOPs tended to document code after it was written, rather than specifying it beforehand.[26]

For each mission, the infrastructure for running and managing the programs came together with the GSOP and generated code that implemented the equations for navigation, targeting, and trajectory calculations, based on Battin and Laning's work. Early missions, unmanned test flights, tended to be sequence-oriented, commanding the spacecraft to do a series of specified operations, or allowing control from the ground. Final versions were named after the sun (Apollo being the sun god): ECLIPSE, SUNRISE, and CORONA. Some were just collections of routines for testing, others, like SOLARIUM, actually flew and executed real maneuvers. SUNDANCE evolved into the program for the LM. These programs were divided into two categories, for the two spacecraft. A naming contest for programs beginning with "C" for the command module and "L" for the LM rejected "coughdrop" and "lemon." Instead, COLOSSUS flew on the command module, and LUMINARY flew on the LM, each in various

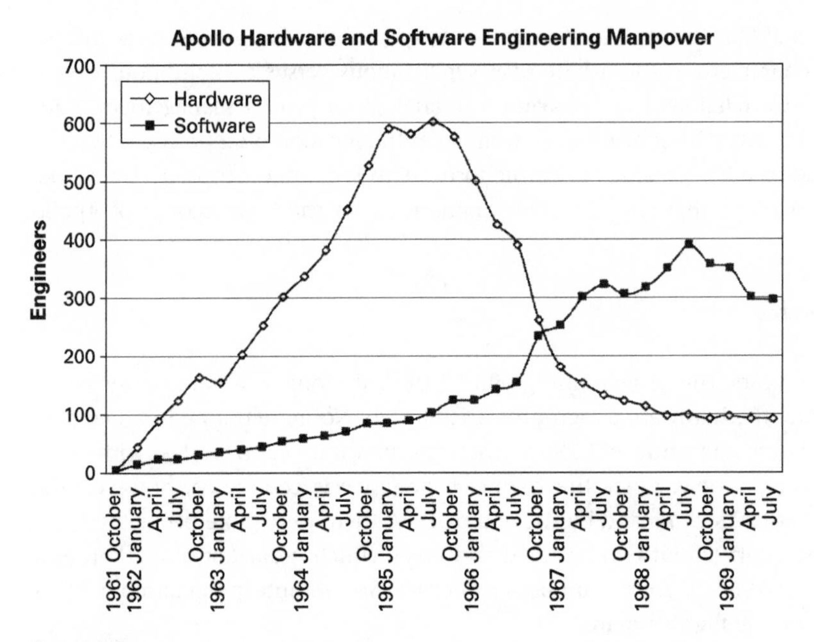

Figure 7.1
MIT Instrumentation Lab Apollo manpower plots, showing the time difference between the peak of the hardware and software efforts. Note the rapid rise in software manpower during 1966 when five core ropes (i.e., flight programs) were being developed simultaneously, and when Bill Tindall from NASA intervened. These figures do not include subcontractors. (Redrawn by the author from Johnson and Giller, "MIT's Role in Project Apollo, Vol. V," 19–21.)

versions (LUMINARY IA, for example, ran the Apollo 11 landing). These had full display and keyboard functions and allowed the astronauts to select and run individual programs (figure 7.2).[27]

Programmers labored under strict constraints. All of the programming was done with punch cards—there were no "online" terminals for rapid interaction with the computer. In typical hacker fashion, stacks of program cards were assembled at night. Managers used this rhythm to advantage and began to hold nightly "configuration control boards" that would incorporate any changes that had been made during the day into a single overnight computer run. Cumbersome as it might seem today, the nightly builds proved an effective means of keeping control of the software changes.[28]

Along with schedules, limited onboard memory imposed the tightest constraints. NASA and the IL had to come up with strict priorities about what was important enough to include in a flight. Jack Funk at NASA felt that from the beginning of the program, the IL "made a fantastic error in judgment" in their estimate of the memory

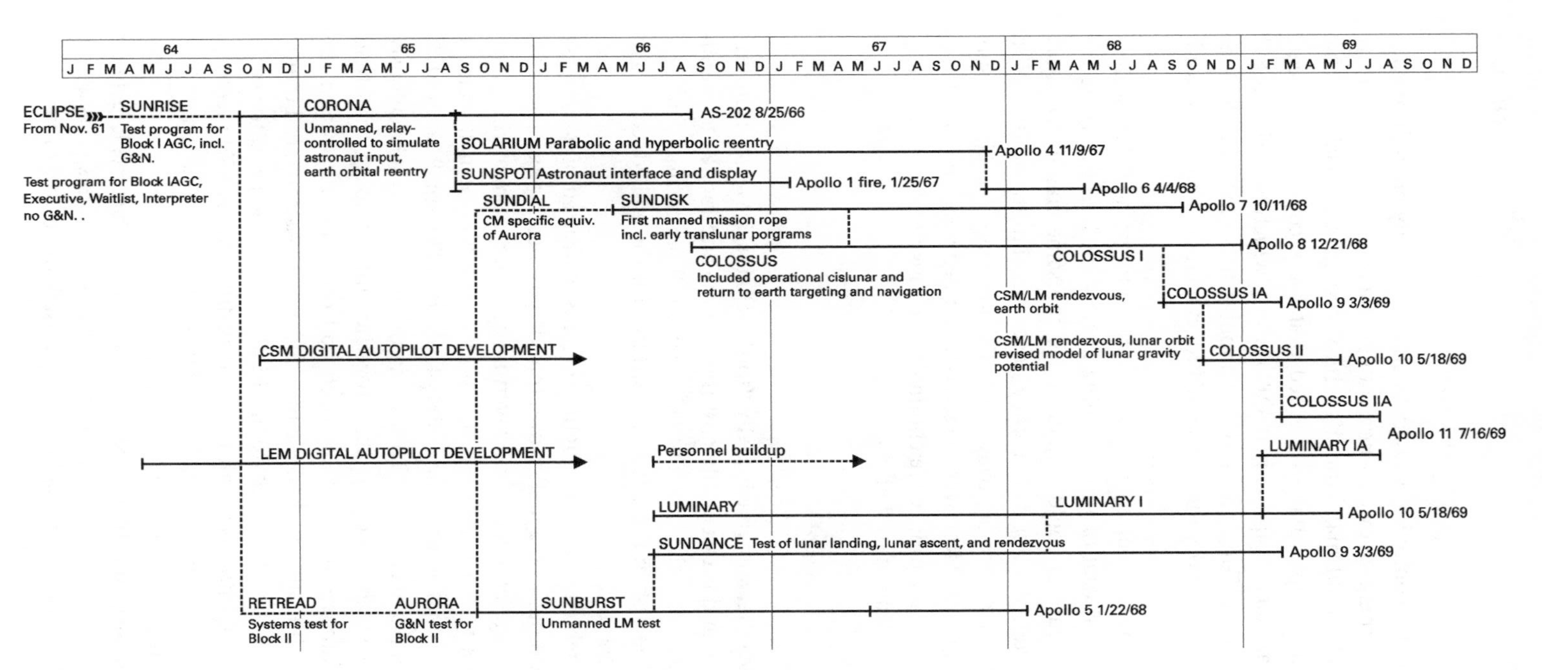

Figure 7.2

Timeline of software versions and Apollo flights. Note the busy years of 1966 and 1967, when five and six ropes were in development, respectively. (Redrawn by the author from Johnson and Giller, "MIT's Role in Project Apollo," 22.)

capacity required for the Apollo computer, in part by including only enough in their estimate for the calculations themselves and not enough capacity for the executive and other infrastructure that would make the life-critical system ultrareliable. "That was some battle, to convince people that we needed that big a computer to begin with," he said.[29] The original Mars computer had 4,000 words, which was what the IL proposed for Apollo. As built, the original Apollo computer had 8,000 words fixed memory, then it was doubled to 16,000, then to 36,864 fixed memory and 2,048 words erasable for the Block II.

In today's world of prodigiously cheap memory, it is easy to forget how dear storage space was in those days. A great deal of programming effort went into shrinking the memory required. Different routines could actually share the same erasable memory for storing parameters, for example, but only if they never ran simultaneously. With the asynchronous executive, that was not always perfectly predictable, leading to subtle problems with overlapping data. Moreover the software tended to accumulate functions as time went on and hardware units were eliminated or consolidated, so memory and processor time became ever more stuffed.

Manufacturing Code

Today we are used to the idea that software is really "soft"—that is, it weighs nothing, is easily and perfectly copied, and is essentially cost-free in time and money to manufacture. Download a new piece of code into your PC or your cell phone and you're off and running. This was not the case in the 1960s, at least not for the embedded system of Apollo. Software became hardware.

The permanent memory, which stored the flight programs, consisted of a complex series of wires running in and out of magnetic cores that determined if a particular bit in a memory location was a one or a zero. At a given memory location, a wire going through a core represented a "1," a wire going around a core a "0" (actually, clever wiring meant that each core could actually store sixty-four bits). The small amount of erasable memory used a similar technique. Thousands of these cores, meticulously threaded with thin, hair-like wires, were packed together into "ropes" that held Apollo's programs. While cumbersome, this approach had one great advantage: the program was indestructible, literally hard-wired into the ropes. Astronauts became grateful for this feature when lightning struck Apollo 12 just after launch and the computer perfectly rebooted itself. Apollo spacecraft had no disk drives, no FLASH memory, not even any magnetic or paper tapes. The software for Apollo was an actual thing. You could hold it in your hand and it weighed a few pounds (figure 7.3).

This "firmware" meant the programs had to be manufactured in a factory, largely by hand. Raytheon, which built the flight computers for Apollo, was also responsible for manufacturing the ropes. This process entailed precise, painstaking sewing of very fine

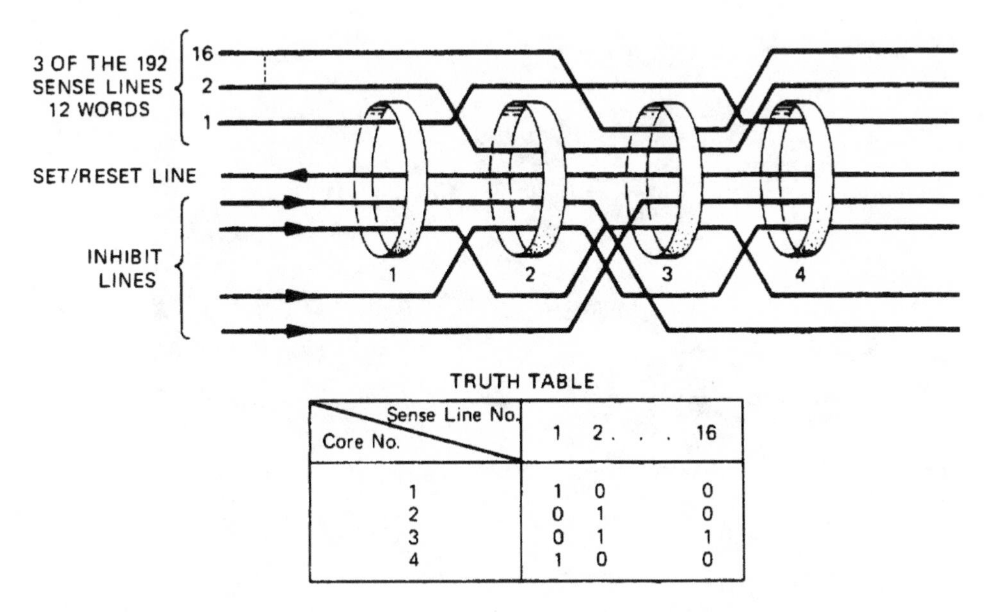

Sense Line No. / Core No.	1	2 . . .	16
1	1	0	0
2	0	1	0
3	0	1	1
4	1	0	0

Figure 7.3
Simplified schematic of principle behind core rope storage. The inhibit lines select a particular core "address," and the sense line returns a "1" or a "0" depending on whether it threads through or around the core. Each core thus stores 12 bits. (Hall, "MIT's Role in Project Apollo, Vol. III," 90.)

wires through thousands of tiny cores. The particular pattern of the wires, and which direction they went through each core, determined the pattern of 1s and 0s stored in the memory.

Raytheon did the manufacturing in its plant in Waltham, Massachusetts. The town had a history of precision machining (the Waltham Watch Company was nearby), and drew on an industrial community familiar with weaving and textile manufacturing: "we have to build, essentially, a weaving machine," Raytheon manager Ralph Ragan told the press.[30] Raytheon assigned the work to older, female workers. Engineers nicknamed them "little old ladies," and actually referred to them as "LOLs." Core rope weaving was a specialized skill, and Raytheon paid the women to sit around and do nothing if the software ran late, so they would not be called to other projects that would degrade their currency.[31] NASA was well aware that the success of the flights depended on the fine, accurate motions of these women's fingers, so they sent the astronauts through the plants to make the workers aware of the human impact of their work (as they did with many of Apollo's factories). One engineer fondly recalled that the women at the plant "adopted" the astronauts, and took great care and pride in the quality of their manufacture (figures 7.4 and 7.5).[32]

Figure 7.4
Sewing software: rope manufacture with a numerically controlled machine. The paper tape in the reel at right controls the machine to place an eyehole over the appropriate core, and the worker threads her needle through with a wire attached. (Raytheon photo CN-4-22.)

Rope manufacture did become partially automated. Another New England firm, the United Shoe Machinery Company, created a machine to speed up the process. The IL would convert their programs to a series of paper tape instructions to read into the machine. A rigid rack held an assembly of cores, and the machine used the numerical data from the punch cards to position the core assembly so the worker could quickly run her needle, with a wire attached, through each core. Then the machine would reposition the cores and the worker would run the wire through the other way. A series of folding, soldering, and potting operations then turned the core rope into a "module" that could be plugged into the Apollo computer.

Ideally, this process would happen once, producing single core rope that would program any possible mission. Each mission had its own unique features: vehicle mass, propellant volume, orbital parameters, and the like, that would go into the erasable memory. It never worked out that way—each mission was sufficiently different, and each program was sufficiently modified from the previous one that no two flights ever flew with exactly the same code (other than AS-501 and AS-502, which were unmanned).[33] By the later lunar landings, however, the programs did become relatively stable, and the changes increasingly minor.

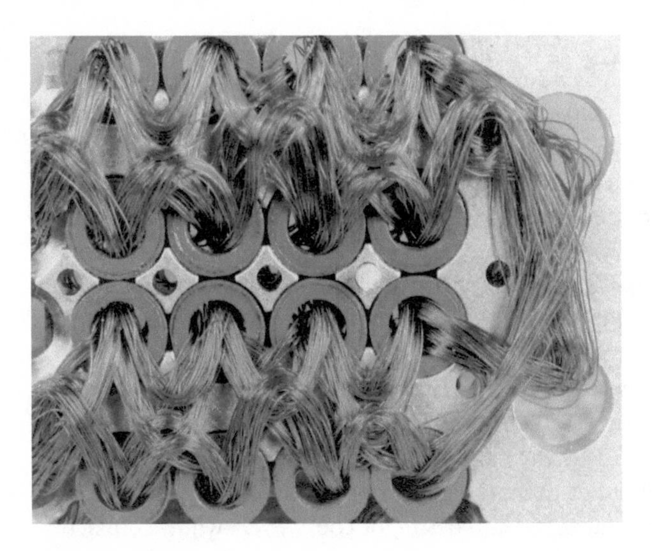

Figure 7.5
Apollo's software: close up of a core rope. (Raytheon photo CN-1156-C. Reprinted in Hall, *Journey to the Moon,* 15.)

Cumbersome and time-consuming, the rope manufacturing process took many weeks to go from functioning code to workable ropes. The pace had a critical implication for the programmers: the software had to be "frozen" at three to four months in advance of a flight.[34] IL engineer Ray Alonso remembered his initial shock when learning of the lead time: "What? I can't walk up to the launch pad and change whatever the program I like?" After all, what was the point of a programmable system? Sometimes the flights were sufficiently close together, in fact, that lessons learned from one flight could not be included in software fixes for the following one. A process for changing the ropes after manufacture did exist, but it required removing the potting that encapsulated the wires and surgically changing one bit at a time, in Alonso's words, "an incredibly small and arduous process to go through."[35]

NASA and IL management recognized that the schedule could help them impose a kind of discipline on the otherwise free-wheeling programmers. "Rather than a disadvantage," wrote David Hoag, the nature of the core ropes meant that "risky last minute changes of the program just before flight were physically prevented."[36] Manufacturing the software forced the engineers to make commitments, to stand behind their programs and get them into a form that would be suitable for testing months in advance.[37] It helped bring the young, unruly art of programming into the purview of program management and systems engineering.

But Apollo's programmers had to face more than the factory floor. They also had to face the users.

Astronauts and Automation

As eventual users of the Apollo guidance system, the astronauts had values, opinions, and goals that complicated the programmers' work. Most of the IL staff remember the early astronauts as resistant to electronics, computers, and automation. They were used to a world where electronics routinely failed. "The fact that we were going to do things in a digital system, an inertial system," Cline Frasier recalled, "they absolutely believed to their core that stuff wasn't going to work."[38] Sensitive about their level of activity in the spacecraft, astronauts found it difficult to accept the idea that the digital computer would be even partially in control.

As Alonso remembered it, the astronauts saw the computer like a calculator, a disposable appendage to the flight. "Of course we'll shut the computer off as soon as we're up there" he was told (indeed the early plans did involve shutting the system down between maneuvers).[39] Other times the astronauts wanted to approve, step by step, every move the computer made. Such a desire, of course, was wildly impractical with a computer running thousands of instructions per second and would lead to an impossibly high workload. Similarly, the users wanted direct control of the systems—right down to the valves themselves, asking, for example, for hand throttling of the LM's descent engine.[40] Nonetheless, the astronauts were not being reactionary, just hard-headed about which systems they were willing to stake their lives on. Lickly remembered his first lecture to the "new nine" on automatic control of reentry: "I was just overwhelmed by the smart questions. Neil Armstrong was all over it. . . . They knew the right questions to ask. They weren't Luddites."[41] Eventually, the word came down from NASA that the astronauts could suggest changes and provide input to the systems design, but that they could not dictate the basic controls philosophy.

Battin remembers a story he told to astronaut and Apollo 12 Commander Pete Conrad: "One day an airliner was on the field, and wasn't able to take off. An announcement came over the intercom, 'Sorry passengers, we can't take off, there is an indication of a problem with the airplane. We've got to replace a piece of equipment and it's going to be very difficult and take many hours.' Just half an hour later, another voice announced that the flight was ready to go. One of the passengers asked, 'I guess it was easier than you thought to replace that piece of equipment.' The airline representative replied, 'No, we replaced the pilot.'" Battin remembered Conrad being impressed by the story and repeating it to others.[42] The message: if the astronauts weren't happy with the automation, there were plenty of other people willing to go.

IL engineers gave regular courses about their systems to astronauts and other NASA personnel. In general, they found the astronauts disinterested in the computers, or even in the guidance per se. Research engineers at the IL had a habit of teaching the fundamental principles, but the crews wanted operational techniques and checklists. Indeed, much of the IL's training proved more academic and theoretical than the astronauts could absorb. Michael Collins found it sufficiently off-putting to record his response to one course in his diary: "This course was designed for someone who is going to either (a) build a better computer, or (b) repair and replace components of the existing computer. It was not a course for the pilot, who needs to know how to operate the computer and how to detect malfunctions."[43] Collins both rejected in-flight repair and made the point that the IL lacked experience in communicating with operators.

Still, astronauts made numerous suggestions to the IL engineers, most of them fairly concrete. For example, they wanted velocity units displayed in feet per second, even though the AGC used metric units internally. The software thus did a whole series of conversions between metric and English units for the user interface. Procedures and even trajectories were modified to allow the astronauts to monitor, and possibly interrupt, the automatic sequences.[44] A PRO button, for "proceed," allowed the astronauts to approve major actions (like firing the engines) before the computer commanded the hardware.

Difficult problems still arose, like the "attitude indicator" in the spacecraft, the primary display of which way the vehicle was pointing. Pilots were used to the artificial horizon in aircraft, a gyro-driven display that maintained orientation to the vertical, with a dark region on the bottom representing the earth and a light-blue region above representing the sky. In the spacecraft, this unit needed to be a ball, because of the wide range of possible attitudes. It became formally known as the "flight director attitude indicator" (FDAI), but colloquially as the "eight ball."

But how should the eight ball be oriented? In aircraft, a gyroscope drove the artificial horizon directly, which seemed elegant to the IL engineers, who proposed driving the Apollo eight ball off one of the gimbals in the inertial platform; like the platform itself, the ball would then be fixed in inertial space (i.e., in relation to the stars). This solution did not appeal to the astronauts, however, because it meant that as the spacecraft orbited the earth, the ball would seem to rotate as well and would not track the horizon. As an alternative, the ball could track the horizon, the "local vertical" (on earth or the moon), but that also created problems. A ball tracking local vertical would be essentially useless, for example, as the spacecraft pointed skyward on the launch pad and during launch before it pitched over toward orbit. A similar set of problems, in reverse, affected decisions on the eight ball for the LM. The astronauts got their way because they would have to use it, but it offended the IL engineers' sense of elegance and accuracy. "I do think that somehow they are being too much dominated by their experience in flying airplanes," Hoag said at the time, "to go over this [flat] plane which is the earth surface, rather than going out and away and trying to go between this earth and the moon."[45]

Another critical question: should the computer allow the astronauts to do something dangerous with the spacecraft? Initially, IL programmers put in a series of caveats and constraints. Jim Miller remembered Alan Shepard objecting to this approach on his 1962 visit. "Take out all those inhibitions . . . if we want to kill ourselves, let us. It may involve saving ourselves," Shepard said. "Of course he was right," Miller remembered, "and it made the software a lot easier."[46] Yet Hamilton fought hard to get user-error-checking in the code. "We were very worried that what if the astronaut, during mid-course, would select the pre-launch [program] for example? Never would happen, they said." (But exactly that did happen on Apollo 8.)[47]

NASA insisted these were the most expert, highly trained users in the world. Of course the astronauts were very well trained, and experts in their systems. But they also got very tired, and building performance limits in the software could mitigate the effects of fatigue. Astronauts swore to Lickly they were going to fly the reentry manually, without using his automated program (this before Gemini's success with automated reentry). "As far as I know, none of them ever touched a manual stick," on reentry, Lickly remembered, for "they were so beat" after a two-week flight."[48]

At the 1963 press conference, Hoag put the issue more colorfully, describing the computer as "a young maiden who is asked something improper, and she'd also respond in that fashion. The computer, too, will not do things that it shouldn't." If an incorrect key or code were entered, the computer would signal "operator error" so the astronaut could try again.[49] In the end, the software came up against the limits of memory usage, and most of the self-checks had to be eliminated anyway. It did, however, check basic keystrokes and allow astronauts to recover from keyboard errors, analogous to a modern "escape" key or "undo" function.

The "Go to Moon" Interface

In 1965 a training session with the astronauts at the IL left one engineer confused. He was responsible for training the astronauts on the command module simulator for guidance and navigation, and noted how "proper operation of the G&N [guidance and navigation] essentially makes the astronaut a passenger." Then he asked for advice. "The astronauts seem to have an 'I want to fly it' attitude. This would occur only if G&N failed. Are we to train them to operate the G&N system, or train them to use everybody else's system when/if G&N fails, or both?"[50] How could you teach someone to use a machine that required nothing of him when working, and everything of him when broken? The appropriate user interface for the Apollo computer was far from obvious: chauffeurs or airmen?

The IL had built its expertise with inertial guidance systems, largely for missiles. Nobody flew an ICBM. With Apollo, the computer would have a user to input commands, request data, and ask the computer to do things that might be difficult to predict in advance. "We had a hard time getting used to the concept of somebody sitting there," Copps recalled, "banging on the keyboard, getting answers back, banging something else in and asking questions."[51] The Apollo computer would have to interact.

Initially, Kosmala pictured the spacecraft with one button: "The astronaut goes in, turns the computer on and says 'Go to moon' and then sits back and watches while we did everything." Another version has the computer running two programs—"P00" to go to the moon, and "P01" to return home. These humorous descriptions capture one extreme of the engineers' view of the computer, one that didn't last long once the astronauts got involved in the design.[52]

John Miller remembered a constant battle. "The astronauts on one side wanted to fly the vehicle, I mean, they were test pilots.... On the Instrumentation [Lab] side, were, you know, automatic control... we're going to run this thing and the computer will run the thing. That battle went on kind of constantly."[53] An IL cartoon humorously captures the chauffeurs versus airmen extremes: one panel shows full automation, the astronauts smoking cigars and falling asleep, staring at the abort button. The next panel shows no automation, the astronauts overwhelmed by dials, indicators, charts, printouts, and inputs (figure 7.6a, b). The Apollo computer would have to operate somewhere between these extremes. But where?

Early in the 1960s, Charles Stark Draper himself and some colleagues had begun looking at the appropriate human role in aircraft and space systems, "as an off-line, parallel, complementary observer and actuator." To be an effective monitor, the human had to have some means to intervene in the system in order to take corrective action when problems arose. Draper and his colleagues used the word overseer; but the term supervisor became increasingly popular. Both evoke the relationship of a master to a slave, or a manager to a worker.[54]

"A transition in the art of piloting"

In 1962, the task of building the human interface for the Apollo guidance system fell to Jim Nevins, an IL engineer with a background in control systems. For Mercury, NASA had turned the job over to psychologist Robert Voas and to "human factors" experts at McDonnell, but Nevins began looking for control systems engineers with experience in ergonomics. He found one in Tom Sheridan, an MIT assistant professor with expertise in mechanical engineering and psychology who would come to define ideas in supervisory control and telerobotics. The group eventually grew to about thirty people. They began with traditional aircraft controls, giving the spacecraft hand controllers and throttles, then adding the gyroscopes and an artificial horizon (the eight ball), and inputs to control pitch, yaw, and roll. Each solution raised more questions: how would such machines be operated in a weightless environment? How by people wearing bulky pressure suits? Could an astronaut in a helmet put his eye to the sextant?

Figure 7.6a, b
MIT Instrumentation Laboratory cartoon showing the extremes of automation. Too much automation (above) leaves the astronauts bored, awaiting an abort, while too little (right) overwhelms them with work. (Draper Laboratories/MIT Museum.)

Nevins's group drew large pictures with columns labeled "astronaut" and "computer" and "ground," writing flow diagrams that detailed the transactions of data between these entities. "It basically described the decision that the astronaut had to make, the question that he would put to the computer, the answer that the computer was expected to give" and so on, Nevins said.[55] The designers were less interested in the mathematics than in the interactions between the astronauts and the machines. They detailed these tasks in verbal form, coming up with numerous "transaction schemes" for each type of operation (figure 7.7).[56]

They began looking at the information flows. What did the astronauts need to know at different points in the flight? What would display that data? Where would it go? How big should the displays be? How many numbers needed to be displayed at any

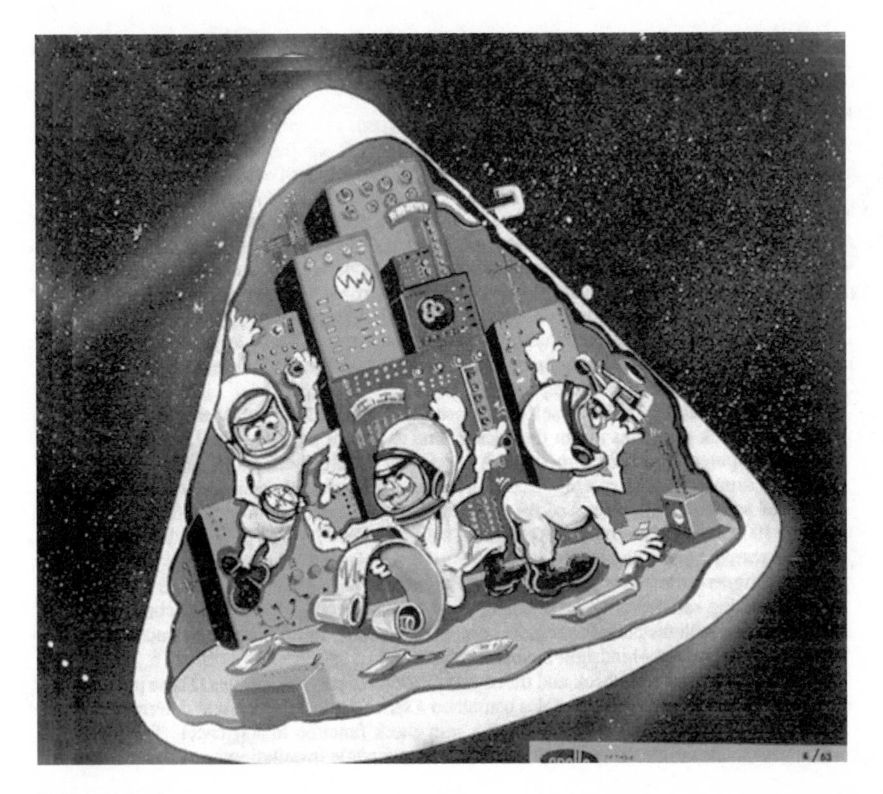

Figure 7.6a, b
(continued)

one time? Should there be a picture tube to display data? Eventually they settled on three lines of numeric display, because a picture tube would be too heavy and consume too much power, and because engineers were used to seeing vectors that had three components, three dimensions.

Nevins articulated his philosophy for astronaut-computer interaction based on their fit into the Apollo system overall. Aware of the broader implications of this approach, he called it "a transition in the art of piloting a vehicle." By "transition," he meant that astronauts would rely on interactions with the computer, and with controllers on the ground, to an unprecedented degree that would change flight forever. "The flight management system [in Apollo] instead of being an onboard operation is actually a highly integrated system of airborne and ground-based equipment." Overall, the astronauts and the computer formed but one part of "a finely structured multilevel monitoring and decision process." Ground controllers, Nevins predicted, will feel as though they are inside the vehicle, monitoring for slow degradations and trends while

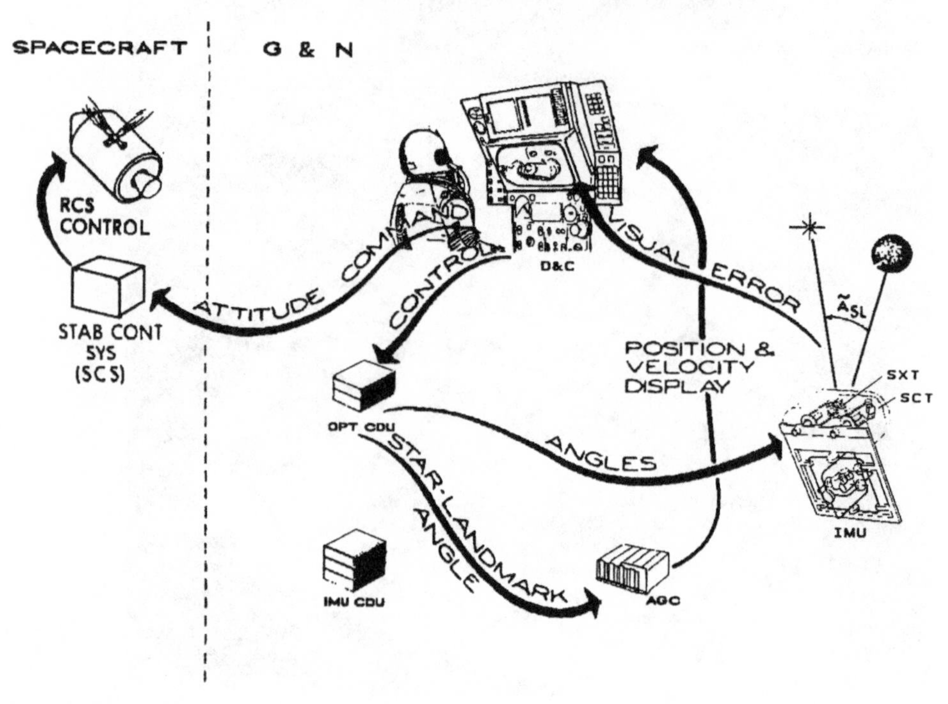

Figure 7.7

Interface flow diagram of a mid-course navigation illustrating information exchange between pilot and guidance and navigation system. (Draper Laboratories graphic 25381. Reprinted in Hall, *Journey to the Moon*, 62.)

the pilots remain alert for events that require quicker reaction. Flying to the moon would be a long way from piloting in the classical mode (although Apollo paralleled new developments in aviation where pilots shared control with controllers on the ground).[57]

For Nevins, the crew played a critical part by managing systems and occasionally aligning the gyros. Their tasks consisted of (1) monitoring and decision making in guidance and control; (2) sequencing and initializing automated systems for guidance; navigation, and propulsion; and (3) optical pattern-matching tasks associated with tracking and identifying the stars for aligning the inertial system.

Even so, "as technology improves ... many of these tasks will be automated," Nevins believed. By contrast, the computer would (1) monitor sensor data; (2) determine thrust times and vectors, trajectory parameters, and lines of sight; (3) maintain attitude control; and (4) guide the vehicles during thrusting maneuvers. Punching a few buttons would bring the spacecraft to a particular attitude; punching a few more would fire the main engine to propel it in or out of orbit. The computer's job looked more like traditional flying.

Because of the complexity of the task, Nevins continued, it would rely heavily on human control to adjudicate among flexible, redundant systems. Most critically, the crew would monitor the primary and backup systems and watch for indications of failure on one or the other. Nevins illustrated his philosophy with a detailed exploration of crew tasks during the final phases of landing, which we will examine in the following chapters. His approach amplified Chilton's original vision of astronauts as redundant backups, and placed a heavy burden on the crews in training and workload, especially when something went wrong. Nevins noted that the Apollo spacecraft had 448 switches and indicators, as opposed to 150 on Gemini and 102 on Mercury. Much of the crew's efforts would resemble operating a telephone switchboard as they flipped switches and actuated valves to configure the spacecraft for various phases of flight.

Ironically, Nevins seemed almost apologetic that the astronauts had to be so busy, explaining that only "technological gaps" required their involvement at all. Ultimately improvements in computers would automate the monitoring tasks and allow the human role to become "more purely administrative, or supervisory." Future machines would "relieve man of the necessity to play such an extensive role in either piloting or supervising his vehicle" and would allow humans to spend more time doing scientific experiments or exploration.[58] Of course, if the crews were no longer involved in the flying, NASA might send different kinds of people, perhaps scientists instead of test pilots.

Display and Keyboard

"How do you take a pilot, put him in a space ship, and have him talk to a computer?" astronaut David Scott succinctly put the question facing the IL. Nevins and his team embodied their philosophy in the interface to the Apollo computer. They developed a "display and keyboard" unit, abbreviated DSKY (pronounced "dis-key"). Somewhat akin to an early calculator display, the DSKY had a numeric keyboard with plus and minus keys, and seven additional function keys like ENTER, CLEAR, and KEY RELEASE (figure 7.8). The display included three signed numbers for numerical data like navigation coordinates, and three shorter numbers to identify functions. Each line of display was merely five numerals, with plus or minus but no decimal point (similar to how a slide rule displayed its results). Astronauts needed to know independently, for example, that time would be displayed in three digits of seconds with two digits of hundredths, while gyro angles would be displayed in two digits of degrees plus three digits of thousandths. A series of warning and indicator lights (including the feared GIMBAL LOCK)

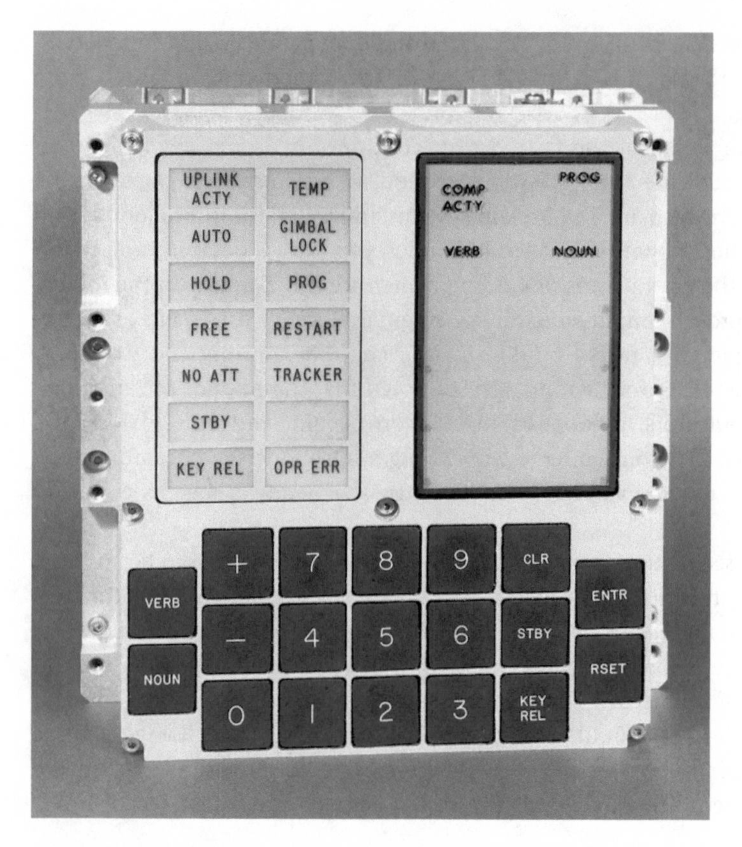

Figure 7.8
DSKY—display and keyboard unit—for the Apollo guidance computer. Note the space for digital displays on the right, with program, verb, and noun indicators, and three line numerical display below. Also note status lights on left, including warnings for gimbal lock, program alarm, and operator error. (Raytheon photo CN-4-268.)

signaled anomalies or faults. The displays used the "seven segment" format for alphanumeric data. While the style did not originate with Apollo, its use in the AGC foreshadowed the ubiquitous LED displays of the 1970s and helped make the boxy digits icons of the digital era. In the computer, a routine called PINBALL GAME BUTTONS AND LIGHTS ran the DSKY interface (see cover image).

Alonso began thinking about the command protocol: "It occurred to me that the sort of dialog between astronaut and AGC could fit into a rudimentary sentence structure, such as 'Display IMU Angles,' or 'Display Time,' or 'Fire Rocket' or 'Align IMU.'" Working with Albert Hopkins and Herb Thaler, Alonso developed a temporary design they could use for testing and demonstration while waiting for an official solution. Their stopgap became permanent.[59]

Astronauts entered data and commanded the DSKY with this "verb-noun" syntax. To enter a command, the astronaut would press the VERB button followed by a two-digit code for a command. Then he would push NOUN and the code for a particular type of data, and then ENTER. Noun 37, for example, referred to roll, pitch, and yaw data and Noun 89 referred to the coordinates of a particular landmark. The verb and noun codes appear on the display as they are entered. If the command requires further data to be entered, the verb-noun display will flash, as the program awaits further data entry on the keypad. It displays data in one of the five-digit numerical display lines. Each data number is then followed by depressing ENTER. Various illegal combinations of keys are rejected by the DSKY as invalid data. If an incorrect key is pressed, the CLEAR button allows the data to be reentered. The interface was a bit cumbersome and not simple. A task like aligning the inertial platform would require anywhere from thirty to one hundred-thirty keystrokes; an entire flight to the moon would take more than ten thousand keystrokes.

Certain verbs simply displayed data. Verb 01, for example, would display a selected value on the display, and Verb 11 would "monitor" that value, that is, update the display once per second. Other verbs ran specific guidance programs, such as Verb 41 for "Coarse align IMU," or Verb 46 for "Activate digital autopilot." To some degree, the astronauts could customize the displays for their own preferences, instructing the DSKY to display particular variables on particular lines. Any verb-noun combination, or any mix of DSKY keystrokes could also be entered from the ground, via the radio telemetry link (although the astronauts had a switch that could lock out remote control). The ground controllers would frequently update the "state vector," or the onboard computer's knowledge of its position and velocity, through the telemetry channel.[60] One of the ground controllers even had a DSKY display on his console and could push the buttons directly to command the onboard computer. Thus any DSKY command could be entered by the astronauts or from the ground.

While the astronauts could command the computer through the DSKY, the computer could also command them. Occasionally the display would flash a command, asking them to perform a certain item in a checklist.

As an example, Verb 37, "Change major mode," selected a variety of operating programs, depending on particular phases of flight. By pressing VERB and then "37" and then ENTER, the display would flash, after which the astronaut would enter a program number for a particular phase of flight. Program "00" (pronounced "poo") was the idling state for the computer. Additionally, programs starting with 0 related to preflight checkouts (e.g., P01 = "pre-launch initialization"), programs beginning with 1 performed boost monitoring (e.g., P11 = "earth orbit insertion monitor"), and the 2-series included "rendezvous navigation." In the command module, programs beginning with 6 specified reentry, whereas in the LM the 60-series programs controlled lunar landing. As we shall see in the next chapter, beginning with P63, "Braking phase," the LM computer cycled automatically through a series of programs until the astronauts would enter P68, for "landing confirmation."

IL staff remember having a difficult time getting their management and NASA to accept the verb-noun structure. "It's not serious enough," they recall hearing, along with the objections, "It's not military enough. It's not scientific enough."[61] Among the astronauts Nevins found two reactions: "One reaction is that we don't fly spacecraft through DSKYs, we don't fly planes through DSKYs." As for the other reaction, he said, "they appreciated immediately what they were looking at and that they had all this capability and power. That everything and anything that was dynamic in the spacecraft was under the control of the computer. And so those people appreciated it."

Nevins remembers having more difficulty with the earlier, Mercury astronauts. "And then we got our spies," people like David Scott who worked with Battin on guidance and received a master's degree in 1962; Charlie Duke, who received a master's degree in 1964 for studying human performance during Apollo navigation; or Ed Mitchell, Ph.D. 1964, who studied interplanetary guidance. These men learned how to think about guidance and control in the MIT way. "They all got brainwashed," as Nevins put it.[62]

Astronauts generally had good things to say about the DSKY on their missions, the result of long hours of training. Hoag recalled that "an early reticence by the crew members was in time replaced by enthusiasm and confidence in their ability to use the computer to manage many aspects of their mission."[63] Scott added sardonically, "It was so simple and straightforward that even pilots could learn how to use it."[64] Certain operations generated their own Apollo jargon, as when an astronaut aligned the inertial platform. The computer would display "residuals," or an indication of the quality of the alignment, in degrees. A display of "00000" indicated no difference, a perfect alignment, which the astronauts greeted with an exclamation of "five balls," a clear statement of success in the masculine culture.

Nevins's group put together a series of simulations, from simple jigs to evaluate sextant pointing to mockups to verify users' reach to the controls, to a full AGC interface in the Johnsville centrifuge. They also installed their systems' components in NASA's CSM and LM simulators.[65]

One day NASA headquarters called to inform Nevins that a formal complaint had been levied by the astronauts against the IL because the DSKY interface was too complex for a safe trip to the moon. Nevins responded by putting together a demonstration unit, called the space navigator, that would enable a user to "navigate" the earth as though it were a spaceship on its way to the moon, using "a physical marriage of the man, a complete G&N system, a real stellar environment, and a pseudo space craft motion generator." Adding some old radar-tracking hardware to swivel the simulator, the IL mounted it on the roof of their building as a demonstration and training device.

Nevins put together a three-hour training program for NASA managers, including Robert Mueller, Robert Seamans, and Chris Kraft, to teach the basics of the Apollo guidance, navigation, and control and train the astronauts. "They came to the conclusion," Nevins recalled, "that it's very complex, it's expensive to train—time and everything else—but it's doable."[66] A photo of Doc Draper in the machine with the Boston skyline and stars in the background became a popular image of the lab's participation in Apollo.

At MIT Scott had studied the statistical interpretation of celestial fixes for interplanetary navigation. When he became an astronaut he was assigned to monitor the MIT contract for the astronaut office. "I spent many nights up on a roof in Cambridge looking at the stars and working with a sextant, telescope and computer," Scott recalled. Some of his colleagues, he said, never got used to the keypunching, and requested programs that would combine the keystrokes into larger functions. Yet for Nevins, the turning point came when he observed Jim Lovell, Frank Borman, and Jim McDivitt operating the guidance system in a command module simulator at Kennedy Space Center shortly before their selection for the Apollo 8 mission. Lovell had mastered it, and was making it sing. "The system wasn't flying him," Nevins observed. "He was flying the system. And wow, he's flying."[67]

Change in Culture

For the first few years of the Apollo program, IL engineers defined tasks, built prototypes, conducted experiments, and did much of the creative, exciting part of flying to the moon. In the middle of 1966, however, everything changed. At the end of the Gemini program, Lickly recalled, "NASA descended on us."

"We awoke several years later realizing we had a big programming problem . . . and now we needed a new organization," Martin recalled. "Instead of having 30 people, we needed 200 people . . . we needed all this documentation. And we needed all these review meetings. And we needed all this NASA supervision." The whole team became "professionalized," Martin said. "We were going to have something called project managers." At NASA, Martin saw enormous management charts laid out on the walls, and "some contractor who would come up everyday with his blue tape or red tape or yellow tape and mark where the schedules were. . . . It was magnificent to see it." Martin seriously tried to adopt NASA's management methods but the increased oversight brought a new era to the lab, and "much less fun for a lot of people." NASA's managerial sophistication impressed a younger Ray Alonso differently: it was the first time he'd seen an electrically powered eraser.[68]

Part of the reason the programming proved so difficult was that writing serious, detailed requirements for the software amounted to defining the mission. Hoag felt that not all of NASA's criticism was fair, as the programming was often hamstrung by lack of data, specifications, or procedures from elsewhere in the program. "It has turned out that MIT has had to do a large measure of the mission planning—how the test flights and lunar flights will be performed, in order to complete these tasks," Hoag said, and therefore the programming task "turned out to be far greater than originally estimated."[69]

NASA began considering turning creation of the actual mission programs over to one of the contractors, which would be a huge blow to the IL. Battin and Ralph Ragan successfully fought the move, and in mid-1966 NASA asked the IL to program not only the basic system, but the missions as well, a task that brought them intimately into the challenges of flight planning, including its severe time pressures. Feeling a late-Apollo budget squeeze, NASA also asked the IL to reduce its staffing and to transfer most of the less-technical work (training, ground support, etc.) to contractors.[70]

At this time the IL's software efforts received the attention of Bill Tindall, the most famous unknown player at NASA in the Apollo story. As part of the Mission Planning and Analysis Division (MPAD) in Houston, Tindall had been instrumental in the Mercury and Gemini programs, helping to turn the theory of orbital rendezvous into reality.[71] A true space enthusiast with a talent for clear, straightforward writing, Tindall began looking at MIT's software effort in early 1966 and he didn't like what he saw. His "Tindallgrams" became well known within NASA for their technical insight and blunt language. Some of the earliest concerned the Apollo software effort.

Tindall started making regular trips to the IL to see what was up. At first, nobody took him seriously. Martin remembered him as "an object of real derision.... We were rolling along doing all of this fantastic work building software, doing this, doing that, when NASA sort of somehow woke up and decided that these guys were totally out of control." NASA's concerns: the documentation was no good. There weren't any real schedules.[72] The IL group was having great fun, but was insufficiently attentive to testing. "Testing is the name of the game when you're playing with mission safety, critical kinds of systems," Jack Garman remembered, but the IL folks treated it like busy work and didn't understand why they had to write up numerous different test reports for NASA.[73]

Tindall began a devastating series of missives in May 1966. These Tindallgrams began with the alarm: "There are a number of us who feel that the computer programs will soon become the most pacing item for the Apollo flights." A strong statement. Previously, nobody had paid much attention to this novel thing called "software," the pet project of a bunch of academics up in Cambridge. Now software would make or break the schedule for the firm, end-of-the-decade deadline.

Tindall studied the organization, with Battin in charge and four subunits reporting to him. Yet he still complained: "I still do not have a clear understanding of how their work is broken out between the four units." Tindall had brought some NASA engineers with him who were working with IBM on the Real-Time Computer Complex (RTCC), which did the heavy number crunching at Houston, and he hoped that MIT would learn from them and adopt their management techniques. Indeed the IL began assigning a person responsible for the program for a particular flight, for managing it through the entire process from coding through manufacture and test. In typical IL style, they called this person the "rope mother" (though it was usually a man).

Tindall was particularly worried about bloated program size, some of it brought on by an obsession with precision by the academically oriented IL engineers: "I am still very concerned about unnecessary sophistication in the program and the effects of this 'frosting on the cake' on schedule and [memory] storage. It is our intention to go through the entire program, eliminating as much of this sort of thing as possible. I am talking about complete routines, such as 'Computer Self-checks,' as well as little features, such as including the third and fourth harmonics of the earth's oblateness and drag in programs for the lunar mission."[74]

IL engineers all remember Friday the thirteenth of May 1966, "black Friday," when Tindall called everyone together to cut their favorite programs out of the software so it would fit into the limited memory. In the dry language of the official report, "these meetings became emotional because of disagreement about what was, in fact, nonessential."[75]

Two weeks later, Tindall was again writing, this time a memo titled "Apollo spacecraft computer programs—or, a bucket of worms." He began, "Well, I just got back from MIT with my weekly quota of new ulcers." Tindall was pleased that the rope mother for AS-204, the first manned mission, seemed to have adopted some of the IBM techniques (like a program development plan). Still, the mission was exceeding its allocated memory by as much as 500 bytes (at that time a large margin, or 2 percent). More distressing, the rope mother was trying to put together the total program without having tested all of the units individually. "Certainly a very unsatisfactory situation," Tindall remarked.

He concluded that the program for AS-204 "will be of less than desirable quality. It will not have undergone sufficient verification tests and will very likely still contain program bugs" when it flies into space. Tindall intended to put Houston people at MIT to watch over the IL group day by day, "with no alternative but to march along with our fingers crossed."

Programs for the later missions, especially the Block II and manned missions, were in even worse shape. "After recovering from our complete shock," about the schedule for AS-501 and AS-502, Tindall and NASA again concluded the programs would have to be ruthlessly culled and accelerated. Said Tindall: "The program paring must be done, I feel, solely for schedule reasons, which is really kind of weird when you think about how long the programs have been under development. It will mean that we fly to the moon with a system which does not minimize fuel expenditure nor provides the close guidance tolerances which are ultimately within its capability."[76]

The eternal compromises of system engineering now compelled trading weight and fuel for schedule and memory capacity. Who would have thought that a few bytes of memory could compromise the accuracy of a moon landing? Who could have foreseen that imperfections in the programs would consume the scarce resource of fuel—programs nobody even knew they needed when Apollo began?

Over that summer, and for months afterward, the bad news continued to flow. There were not enough programmers, so the IL began hiring contractors. The number of people working on Apollo software began to rise steeply in the middle of 1966 and continued until shortly before the nearly Apollo 11 launch. Costs for the second half of 1966 were double those of the first half of the year, nearly a million dollars per month.[77] The IL did not have enough computers to support simulation on anything but the next mission, putting the efforts behind for subsequent flights. Rope mothers were assigned for each mission, and they each had to assemble a series of subroutines from the programming groups into working mission program. "In most cases," Lickly recalled, "the engineers did a so-so programming job and it had to be redone and put into a form that was all consistent for a particular flight."[78]

Despite the software turmoil, in August 1966 the unmanned, suborbital AS-202 flight tested a Saturn IB with a Block I command module aboard, the first flight of the Apollo guidance and navigation system. This was a high lob intended to give the CM heat shield a reentry test. It carried a Block I spacecraft and computer, with a "mission programmer," a set of relays to mimic the actions of an astronaut. The craft flew on all-inertial navigation for an hour and a half. The guidance system performed successfully, monitoring the launch, aligning the command module for firing of the service module's engine, and guiding the command module on its reentry trajectory.[79]

After this success, in September 1966 George Low wrote to Robert Gilruth of his concern that MIT was "very far behind in the preparation of software . . . the general conclusion is that we must put the MIT programming and scheduling on a more business like basis."[80] The program for the first manned mission, AS-204, caused the most serious problem. The crew had recommended a number of software changes to improve the flexibility of the data in their displays, but they would have to be left out because of schedule and testing requirements. Programs for at least the first four missions were late, too large, not reliably tested, and full of bugs. IBM and NASA programmers had developed ambitious plans for the ground computers to interact with the spacecraft, but these would have to go "straight in the trash can," in Low's words, because the flight vehicles would not be prepared to provide them adequate data. The IL was pushed to release a program that Kosmala remembered "was just so lousy, so full of bugs."

On a fateful January day in 1967, that program, and the three astronauts who were scheduled to fly it, burned up on the launch pad.

Recovering from the Fire

The tragedy of the Apollo 1 fire stemmed from a series of failures in the program, from quality control to configuration management, and from NASA's relationships with its contractors. It brought congressional hearings and a traumatic reevaluation across the Apollo program. The high-profile crisis it provoked concerning the larger Apollo spacecraft meant the software problems could be worked out below the glare of publicity. In the fallout, Joe Shea was forced out, and Chris Kraft rose to renewed prominence; his operations-oriented approach would dominate the remainder of the program. The balance of power shifted noticeably toward NASA Mission Control, toward Houston, toward the astronauts, and away from ballistic missile-style systems engineering.

None of this was lost on Bill Tindall. Barely two months after the fire, he was already acknowledging, "It is possible to take advantage of the stretch out of the Apollo flight schedule in the manner in which we develop the spacecraft computer programs at MIT." For Martin, "it was a pretty tense time," and he acknowledged that "we were not building that program in a way that was disciplined, and organized, and had traceability in it."[81] Battin, too, realized that he was too technical, too interested in the guidance itself to be the right kind of manager. "He [Tindall] really did apply good discipline to our shop, and we did learn how to get people. It was . . . a management technique which I, frankly, was not equipped to handle. . . . I wasn't really up to trying to direct the whole orchestra when I wanted to concentrate on a piece of it."[82]

Copps remembered the turning point. "One day Tindall just gave us hell. . . . He really beat us up. 'How can you possibly do this? Here you sit at the very center of the success or failure of this extremely important program. You're behind. Get it through your head you are fucking this up.'" By this time, however, the MIT group respected Tindall enough that they could hear his message.[83]

By March 1967, Tindall was calm for the first time, his remarks conveying a dramatic difference in tone. "It is my feeling that no major problem exists any longer in this area. MIT has an organization and facilities geared up to handle the workload in an orderly, professional, unharried manner."[84] It was just two months after the fire, but already things were looking up. The number of separate versions of the software was reduced. In principle, only two were needed—one for the CSM and one for the LM. But several others had been planned for earth-orbital flights, and several others when the intended programs were not ready. The flight schedule had slipped sufficiently that software was no longer the pacing item, nor was crew training, nor testing. Furthermore, the value of quality had pervaded the organization. Techniques like "code inspection," a fancy name for a human being closely reading a printout of code, had proven effective in finding bugs. Still, Apollo software never flew completely bug-free; known problems remained in the code, which hopefully remained free of unknown errors.

In addition, there was a new realism about whether the programs could be prepared far in advance, or even whether they should be. "Instead of releasing the flight program for rope manufacture at the earliest possible date," Tindall directed, "we should release it at the latest possible date." Then any changes that might come along in testing would have the longest possible time to be included.[85] A "Software Verification Plan" clearly laid out the steps for verifying, simulating, testing, and qualifying any new programs and changes, its flowcharted procedures for organizations indistinguishable from computer programs.[86] While struggles continued into the fall of 1967, with delays and quality problems in software, and NASA continued to issue schedule emergencies, Tindall and the IL engineers found the software on increasingly solid footing.[87]

When Tindall retired, his colleagues wrote a poem for him. One of its five stanzas read:

Yes! Gemini was a hard act to follow!
But you did it again when we got to Apollo.
You became the world's foremost authority
On an agonizing process called data priority.
You said the onboard software was a GD bag of worms
And that was one of your milder terms.
You gave those programs a real thorough wash,
While MIT was out playing squash![88]

The Early Missions

To qualify Apollo for a lunar landing, the early missions had to test the systems. Basic hardware, performance profiles, reliability, abort modes, and a host of other parameters needed to be tried, measured, and analyzed. Guidance and navigation, of course, were critical to these tests. Could the astronaut control the spacecraft? Could the computer? Could the ground controllers? Could the crew track the stars and landmarks on the earth or moon? Early Apollo missions collected data on gyro drift, accelerometer performance, and ability of the inertal platform to keep stable track of position and velocity. They also began to refashion the pilots' relationship to their craft in ways that would mature during the landings themselves.

After the fire, the second unmanned flight, AS-501 (Apollo 4) was launched in November 1967, intending to fly into high orbit and reenter the earth to simulate a lunar return. An AGC controlled operations on board for five hours. It determined position and velocity, controlled numerous attitude maneuvers, and fired two major burns of the big service propulsion system (SPS) engine on the service module. Ground tracking sent two state vector updates to the computer via telemetry, necessary because without people on board the vehicle could not take its own navigational fixes. The SPS engine burned for four and a half minutes, sending the vehicle into an orbit that would simulate a lunar return trajectory and reenter at 36,500 feet per second.

Unfortunately, a ground controller in Australia sent a turn-on command to the engine after the computer had already generated the command on board. Receiving the instruction, the AGC switched modes and began only taking commands from the ground. This necessitated ground control to issue the shutoff command to the engine, which it did, but 13.5 seconds late, resulting in a 200 feet per second overspeed on reentry, which actually created a more stringent test for the heat shield. Onboard computer control ended at 23,000 feet when the parachutes opened. Despite the extra reentry velocity, the computer controlled the landing to less than two miles from the aim point.

Apollo 5 flew in January 1968, an unmanned test of the LM with no command module. This was the first flight for the Grumman vehicle, the first time its engines would be powered in flight. It was also the first Block II computer flight, and hence the first use of the digital autopilot, using the SUNBURST program. About four hours after launch, the goal was to have the AGC aboard the LM fire its engine at 10 percent thrust for about thirty seconds, and then to go up to full power for twelve seconds.

When the LM began to fire its descent engine, the computer thought the ignition was late, and prematurely shut it down.[89] "Utter chaos took place in the Mission Control Center," recalled Jim Miller, who was rope mother for the flight. "Everybody was climbing all over everybody to find out what happened. Totally preventing anybody from finding out what happened." Houston sent a signal to turn off the computer altogether and assumed remote control, bypassing the computer. Miller thought that the mission controllers didn't understand the details of the software, or the subtleties and flexibility of running a software-controlled spacecraft. "I knew right away what was going on," Miller said. "Nobody had asked MIT anything. . . . They just knew better and took over." Miller suggested a way to correct the situation, but the flight directors decided otherwise, because it had never been tested in simulation.

Mission Control commanded a backup guidance system to take over, then issued the burn and the staging commands, and the mission succeeded. The LM turned around, fired its engine, and separated the descent stage. Then the LM's computer took over again, but it had not been informed that the descent stage was no longer there, so its stabilization system was calibrated for a much higher mass. The thrusters hissed and puffed, nearly going unstable.

The cause of the problem, though called a "software error," was actually an error of communications between organizations. "I had been told by the guy that wrote the descent-burn software that it had to have a very narrow window in the startup," Miller recalled, and that there was a serious problem if the thrust didn't build up immediately.[90] But a series of pressurizations and valve closures had to occur before the engine even began to light up. The computer mistook this delay for a slow thrust buildup and shut the engine down just as it was firing up. Internally, NASA acknowledged that "the premature shutdown was the result of incomplete systems integration, and not the result of improper function of individual systems."[91] In this case, "incomplete integration" stood for a miscommunication between two individuals.

Publicly NASA used the incident to further a different agenda. Rather than acknowledging the lack of communication that led to the problem, they praised the advantages of having humans on board. "If men had been aboard to fly the vehicle," George Low told the press, "the flight test might have been a different story. . . . I don't like to fly unmanned missions . . . with hardware designed to carry men." Later Sam Philips blamed the trouble on "overly conservative computer programming."[92]

IL programmers now began to face the public and official misunderstanding of their work. Because software did so much to integrate the system and depended on so many technical and organizational interfaces, social and managerial problems could be obscured by blaming the code. "We who programmed the LM's computer hung our heads in disappointment," Don Eyles recalled, "and endured a public reaction that did not distinguish between a 'computer error' and a mistake in the data."[93] Miller did feel, however, that the incident "really shook up the Mission Control people who realized that their abilities to handle things was much lower than they thought." Computers were introducing new realities into spacecraft flight and mission control, generating tensions between the people who programmed and those who used the programs.

Apollo 6, the final unmanned mission, also aimed to simulate a lunar return. It had trouble on the way up: two engines in the Saturn booster's second stage cut off prematurely and could not be restarted, causing a more elliptical orbit than planned. Then the third stage could not be started to simulate a lunar trajectory insertion. The reentry came in too slow, at 32,800 not 36,000 feet per second. A known bug in the reentry software, which was not expected to be important at the intended higher speed, steered the capsule fifty miles short of its intended landing spot. Overall, the guidance and navigation worked well: the alignment remained under inertial control throughout the mission and the onboard state vector remained accurate to within two miles of ground tracking.[94]

Apollo 7 was the first manned Apollo flight and the first Block II spacecraft (no Block I flew with people in it). Most of the primary objectives related to the astronauts working with the computer to control the spacecraft. Of the flight's nine goals, the first one read, "Demonstrate GNCS [guidance-navigation-control system] performance." Others included demonstrating the inertial system's coarse and fine alignment, determining orbital parameters by earth landmark tracking, and exercising the attitude hold modes, both automatic and manual.[95]

Launched on October 11, 1968, Apollo 7 spent almost eleven days flying 164 orbits while running the SUNDISK program on its computer. The AGC fired the service propulsion system (SPS) rocket six times, using digital autopilot for steering, and the computer monitored the Saturn boost into orbit. The crew practiced a command module rescue of the lunar module, pretending the spent Saturn stage was the LM. They tracked the stage with the sextant, from which the computer estimated range. "This may have been the first in-flight use of Kalman estimation formulations," Hoag reported with pride. The astronauts turned the guidance and navigation system on and off several times and easily realigned the IMU with star sightings. During the flight, the computer experienced three "restarts," in response to "three abnormal procedures." These were likely due to erroneous keyboard entries, so the IL refined the logic that allowed the operators to cancel commands if they hit the wrong buttons. Some problems arose with visibility through the scanning telescope due to particles surrounding the spacecraft, and the more critical difficulty of finding good horizons for tracking so close to earth (where the atmosphere blurred the crisp horizon line). Reentry started manually, then switched over to automatic control, and the computer brought the capsule to within a mile of its aim point on the ocean (all subsequent entries were under automatic control).[96] Nevins remembered this flight as a major step in NASA's and the astronauts' faith in the computer and automation. "Performance exceeded our expectations and hopes," Hoag wrote to his team in a "how did we do" memo that would become a regular event.

While a technical success, Apollo 7 exposed tensions between the astronauts' ideal of mission control and those in Houston. Through much of the mission commander Wally Schirra proved uncooperative and at odds with flight controllers on the ground. "Wally was legitimately in command of the spacecraft," wrote his fellow crewmember Walter Cunningham, "but he attempted to expand that authority over the entire mission."[97] Whatever their control over the machine itself while in orbit, astronauts still had to contend within the larger organizational system, one with its own distribution of power. None of the Apollo 7 crew ever flew again.

Apollo 8: (Nearly) Autonomous Navigation

These early flights played dress rehearsals for Apollo 8, an inspiring triumph for the project overall and the IL team in particular. The mission originated in a last-minute plan. In the summer of 1968, George Low began to despair of reaching the moon during the decade. The lunar modules were chronically late. Waiting to test them would likely eliminate any chance of flying in 1968, leaving only one year to meet the goal. The first test flight kept slipping, right up to March 1969. Low came up with the bold idea of flying to the moon without a lunar module. Such a mission would collect valuable data and provide a public relations and morale boost. After conferring with his top managers and engineers, including Stark Draper, and a successful Apollo 7 flight, in early November Low decided that Apollo 8 would fly to the moon; indeed it would orbit the moon, right around Christmas time.[98]

Mission objectives included tracking lunar landmarks to specify locations for future landing sites, calibrating the ground tracking from lunar distances, and evaluating the optical sighting systems on board. They gave pride of place to Battin's navigation scheme and the IL's computer. For much of the IL team, Apollo 8 would climax the project as the craft flew (nearly) autonomously to the moon, with great accuracy. In preparation for the mission, the *Boston Globe* assured its readers that "voyagers to the moon will not be mere passengers. Astronauts will not sit passively by while machines and controllers back on earth do all the work of celestial navigation."[99] Apollo 8 was to test the humans as well as the machines.

Apollo 8 launched from Kennedy Space Center on December 21, 1968, and orbited the moon ten times before splashing down after a six-day flight. When the Saturn rocket's third-stage engine burned to send Apollo 8 out of earth orbit toward the moon, Battin experienced the "longest and most thrilling moments of my life." On the way to the moon, guidance was so accurate that (according to the Battin decision scheme) four of the seven midcourse corrections were not even made, and the third required only a small, three-foot-per-second impulse from the reaction controls. Arriving at the moon, it was a tense and dramatic scene when the computer fired the engine, behind the moon, to enter lunar orbit. The plan called for an orbit of 60 by 170 miles. Actual lunar orbit turned out to be 60.5 by 169.1 miles, so accurate that at the IL, "the unavoidable emotional tension was broken with unrestrained cheers." IL engineers saw this move in particular as a validation of their entire approach to guidance, navigation, and automatic control.

The famous Christmas Eve moment of the crew reading from Genesis introduced the great drama of lunar flight to the public. More quietly, the inertial system and the computer functioned for the entire mission, and the crew exercised all of the guidance features using the COLOSSUS software for the first time in flight. Earlier programs for test flights contained earth-orbit code only, whereas now COLOSSUS handled two planetary bodies, two gravitational fields, and two coordinate systems. The digital autopilot controlled the slow rotation of the spacecraft for passive thermal control. Command module pilot Jim Lovell made numerous alignment checks and navigation sightings. Most of the optical alignments required updates of only a few hundredths of a degree. The computer maintained the state-vector autonomously for much of the flight. While it also uploaded state vectors from the ground, it kept the numbers in separate memory locations from the onboard solution, although the ground's numbers overwrote the onboard numbers before each maneuver.

Lovell realigned the inertial platform thirty times during the flight. Usually, he used the automatic pointing feature for realignment (P52); the computer pointed at the expected position of stars and landmarks and the user had only to mark the difference.

For position fixing, Lovell made over two hundred earth and lunar horizon sightings with the sextant while coasting to and from the moon. While tracking landmarks, the computer automatically kept the command module gently pitching over at 0.3 degrees per second. Thus as he orbited across the surface of the moon, the landmarks stayed in view while Lovell sighted them.[100] Far from "flying the spacecraft," on Apollo 8 the command module pilot spent most of his time tracking the stars, verifying his position in relation to vectors tracked from the ground.

Soon after emerging from behind the moon and heading back toward earth, Apollo 8 experienced one of the program's notable human errors. Lovell was sighting a star numbered 01 in the catalog. By mistake, he called up the program for pre-launch preparation. The missed keys did not cause the computer to do anything radical (it always asked for a PRO key press before firing an engine), but the confused program (trying to test for pre-launch while orbiting the moon) wrote to memory in ways that could have overwritten the reference matrix for navigation.[101] IL engineers spent a great deal of effort diagnosing the situation. Finally, they walked Houston and the crew through a check of the memory, which showed there was no deep problem. Still, Lovell's error had destroyed the inertial platform's alignment so he used the "coarse align" computer function to align the platform with the stars.

The computer fired the critical burn to leave lunar orbit and return, from the back side of the moon, using tracking data uploaded from the ground. It proved so accurate that only a single five-feet-per-second correction was later required to hit the earth's reentry corridor. Lovell made more than a hundred sightings on the return trip; ground tracking and onboard navigation proved "essentially identical." The automatically controlled touchdown was within one-third mile of the target.[102]

For engineer Aaron Cohen, the trans-earth injection of Apollo 8 "was the moment when I felt the program really had been its climax, because it really encompassed everything which I thought was complicated."[103] Astronaut Frank Borman called the navigation "absolutely miraculous." IL engineers considered Apollo 8 the great triumph of Apollo navigation because it worked so nearly autonomously (a number of engineers, in fact, left the IL right after this flight to form a company to write navigation software). The *Boston Herald* printed a picture of the IL engineers celebrating at an Italian restaurant in Boston's North End.[104] *Aviation Week and Space Technology*, the main industry publication, ran a headline bound to warm the hearts of the IL engineers: "Apollo 8 proves value of onboard control," and peppered its reporting with descriptions of the "accuracy" and "precision" of the lunar flight.[105]

The Rise of Software

Apollo 8's triumphant navigation raises a question: if the state vector on board was so accurate, essentially identical with the data from the ground, why did Houston insist on overwriting it with the data tracked and computed from the ground? Hoag, and no doubt many of the IL engineers, felt that the entire mission could have been accomplished using the onboard navigation (indeed, simulations they later ran with only Lovell's onboard sightings showed that the mission would have easily hit its reentry corridor with onboard data alone). But just as the astronauts were enmeshed in a system that included power relations within and between large organizations, so were the computers and the IL. Once onboard navigation was demoted to the status of backup (as the astronauts had been years before), overwriting of its data symbolically enacted the new relationship: Houston had control. Battin was tremendously proud that, in his view, Lovell had navigated himself to the moon and back. "Sadly," he lamented, "they never did that again."[106]

Apollo software, from an element of the system little understood, indeed barely envisioned, when the program began, became central to the mission, mediating much of the astronauts' critical interactions with their machine. It incorporated information from a variety of sensors around the spacecraft, tying the complex system, with its diverse equipment made by varying contractors, into a coherent whole. It unified the work of a team of programmers, all working on subsets of the problem, into a unitary "mission." It tied the astronauts into the craft, recording and analyzing their navigation and control inputs into stable-state vectors. No wonder this critical, invisible component of the project proved so complex to create, unruly to manage, difficult to test, and hard to trust. In its unusual combination of virtual plan and handmade core ropes, Apollo's software embodied its missions, and embedded assumptions, fears, and social relationships. On no phase of the mission would software, and human action, prove as central, or as ambiguous, as on the lunar landing itself.

8 Designing a Landing

We are banking our whole program on a fellow not making a mistake on his first landing.
—Pete Conrad

This book began with a detailed description of the Apollo 11 landing, focusing on the human-machine interaction. We now revisit the landings as the culmination of the debates over pilots' roles, computer engineering, software, and human abilities. This focus necessarily excludes a great deal of the Apollo flights, but the landings represented critical moments of each mission. Neil Armstrong described them as "the hardest for the system and hardest for the crews." On a scale of one to ten, Armstrong rated walking around on the moon a one, whereas "the lunar descent on a ten scale was probably a thirteen" (he noted the speed, range, and altitude covered by the LM during its descent was similar to that of an X-15 flight).[1] Such trepidation, he would find, was well justified.

An Apollo flight encompassed hundreds of complex operations, but none were as demanding, time-critical, and plagued with uncertainties as the landing, executed in extreme conditions of darkness and cold, far from home. The landing trajectory had to accommodate a wide range of factors, from constraints on the systems' performance to the abilities of the crew and the idiosyncrasies of communications. Even the very calendar of the mission, the "launch window" of a few days per month, derived from the requirement for appropriate lighting conditions on the moon so the pilots could see while they were touching down.

The lunar landings played a microcosm of the entire Apollo program in dramatic ten-minute phases. Here the tensions between human and machine, between manual and automated, between pilots as controllers and pilots as systems managers manifested themselves in a string of life- and mission-critical operations, some smooth and some surprising. Engineers with computers and slide rules analyzed and scheduled every minute, and virtually every foot, of the final descent to the lunar surface. Constant testing and negotiation defined the human role and how it would trade off with the computer. Astronauts repeatedly practiced the intense moments prior to landing and constantly updated their procedures. Designing the landing was a task of systems engineering, trading off quantities like fuel for the pilot's visibility. Like the software itself, the design of the landings embodied the dreams and uncertainties instilled in each mission.

Controlling a Spaceship

Any story of a lunar landing must of course begin with the LM, among the strangest and most interesting craft to fly in the twentieth century (figure 8.1). "Fly" in this case is a loose term—the LM never had to work within the earth's atmosphere, to fly in air. The command module carried a sleek, aerodynamic shape around the moon because it had to push through the air on the way up and burn through it on the way down. Not so with the LM, whose odd and seemingly random protrusions kept it

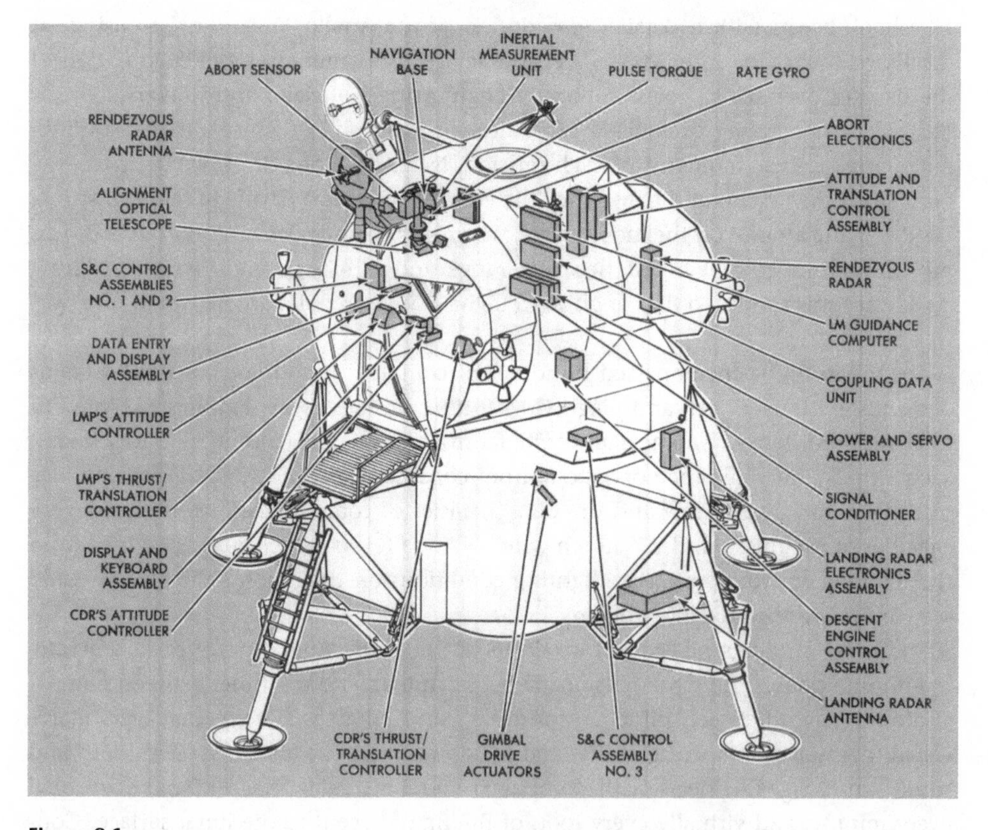

Figure 8.1
LM major equipment locations. (Grumman Aerospace Corporation, "Apollo News Reference," Bethpage, N.Y., n.d. approx. 1970, GN-11.)

from resembling the fast beauty of aircraft or missiles. Still today it is the only human-occupied vehicle built to work entirely outside the earthly environment. The LM's exterior was pure function, each bulge and wrinkle reflecting a specific fuel tank, radar receiver, or human task. A practiced eye can read it like a text and see in the LM an expression of post-Cubist engineering art.

Like its earthly counterparts, the LM was a multistage rocket, but unlike them the flight down came first and the flight up came later. The two-stage LM consisted of an octagonal descent stage, which carried a large rocket engine, power supplies, and the landing gear, coupled to an ascent stage, which carried a smaller rocket engine, the pressurized space for the human crew, life support equipment, flight controls, and the Apollo computer. The entire system would land on the moon. When the astronauts had completed their lunar duties, the descent stage would become the launch pad for the ascent stage, which would separate, rise, and rendezvous with the CSM. In an emergency abort on the way down, the astronauts could separate the descent stage and return in the top half, or abort with the entire craft. Four clusters of reaction jets, located on the ascent stage, controlled the attitude of the vehicle—a particularly challenging control task for the computer, because from landing to ascent the mass of the LM would change by nearly a factor of ten.

Engineering the LM has come to be known as one of the extraordinary engineering projects of its era. The project overall has been well chronicled by its chief engineer, Tom Kelly, and by his boss Joe Gavin.[2] Grumman built the craft in its Bethpage, Long Island, facility, and, like the IL, served as systems integrator for a variety of subcontractors.

Grumman was an old Navy aircraft shop—its engineers were used to working with pilots. Howard Sherman, for example, designed much of the human interface for the LM, and had strong feelings about pilot's roles: "The designers are more used to airplanes and the pilots are a pain in the ass in airplanes from their point of view." Sherman felt the deference shown by the Grumman engineers to astronaut input did not always produce the best solutions.[3] Gavin recalls a difficult time getting the astronauts to accept the digital controls in the LM.[4]

Where the MIT team measured its success according to accuracy achieved, Grumman held light weight as its key engineering value. Still, Grumman's navy heritage, of building robust airframes to survive hard landings on aircraft carriers, showed up in the very structure of the LM. Its heavy landing gear dominates, designed to withstand a drop to the moon's then-unknown surface. Nearly every other feature, however, was new for Grumman (as they would have been for any engineers). Initially envisioned with a round cabin where the crew was to sit as in a helicopter, the LM eventually had no seats at all; the crew stood as they flew, restrained by tensioned cables. This posture allowed the windows to be reduced to small triangles, finely tuned to the astronaut's field of view, saving tens of pounds of weight in glass. Technicians shaved

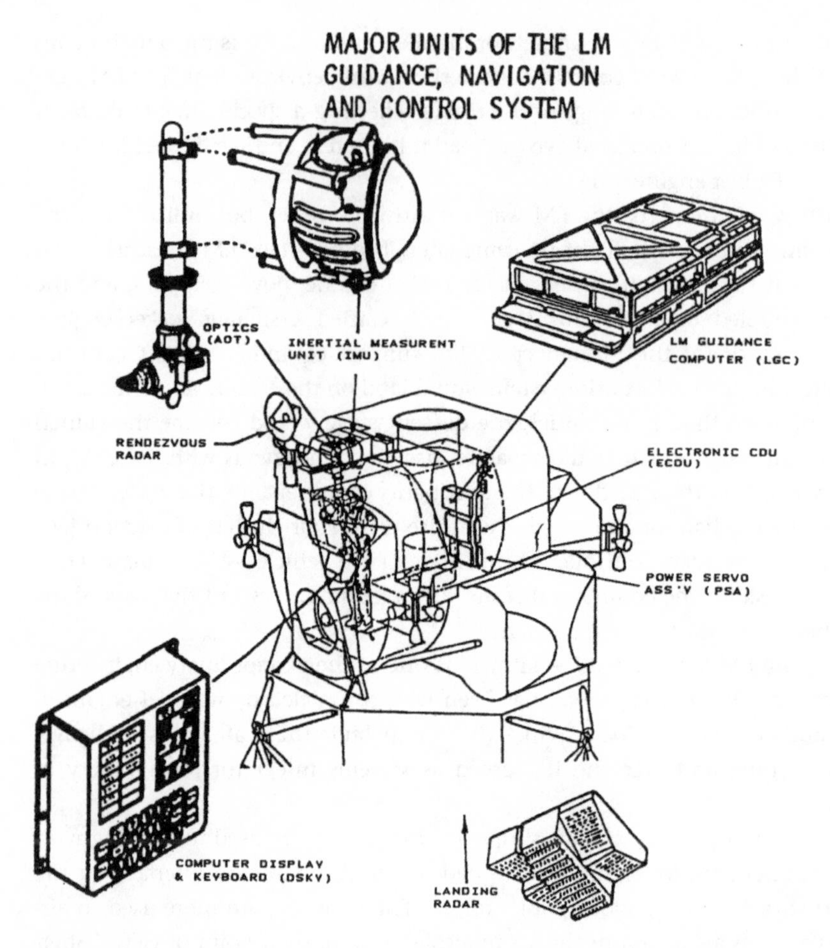

Figure 8.2a
LM guidance components MIT view (a) and Grumman view (b). (Hand, "MIT's Role in Project Apollo, Vol. III," 52; Grumman Aerospace Corporation, "Apollo News Reference," Bethpage, N.Y., n.d. approx. 1970, GN-17.)

critical ounces by whittling the LM structure to paper-thin tolerances, even chemically etching the fuel lines, for instance, to the smallest possible thickness.[5] The LM's walls were so thin that the pressurized upper stage resembled an inflated balloon as much as a rigid structure.

Since the LM was intended to carry a human crew, its relationship to the occupants was critical. Grumman Engineering Manager Tom Kelly remembered NASA's Chris Kraft convincing Grumman that "the crew's time and energy was the most precious commodity on the mission."[6] Could the astronauts stand firmly enough to operate

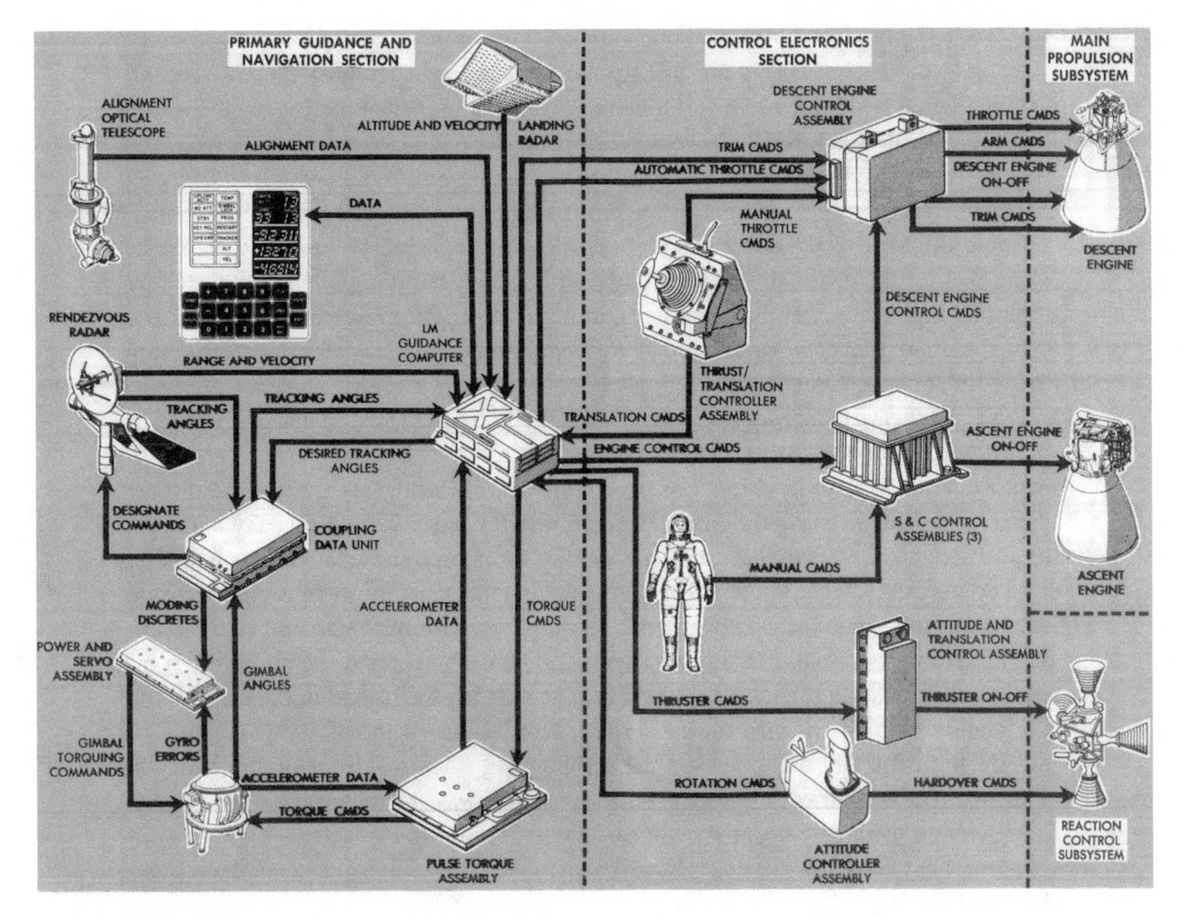

Figure 8.2b
(continued)

the craft as it rocked back and forth? Could they egress the hatch while wearing bulky space suits without getting stuck or snagged? A crewman could not, for example, carry an incapacitated companion up the ladder into the LM—any accident must leave an astronaut able to climb a ladder himself, otherwise he could not return. And, of course, how would the pilots control the vehicle? What sorts of sticks, levers, buttons, and dials would they need? How much should the computer control? At the highest level of design, the LM had to match the capabilities of the crew and the shape of the descent trajectory (figure 8.2a, b).

The first LM with people on board flew on Apollo 9, in March 1969. This test of the vehicle in earth orbit exercised its numerous computerized and backup flight modes. Inside the LM, astronauts Rusty Schweikart and Jim McDivitt separated from the command module and flew about sixty miles away as a test, separated the descent stage, and then returned for the first Apollo rendezvous. They fired the descent and ascent engines two times each, and tested the "lifeboat mode" of the LM (which would become critically important on Apollo 13), using its engine to power back from the moon. Particularly complex was testing the digital autopilot's behavior. It had to account for three major modes: (1) a docked configuration, the CSM and LM together; (2) the full LM with the descent stage; and (3) and the small LM ascent stage on its own. Each of these had different dynamics, mass, and handling qualities affected by their odd geometries, bending moments, and fuel slosh. After exercising them all, the crew reported that "the autopilot is the optimum control device for performing the entire lunar mission."[7]

While Schweikart and McDivitt were testing the LM, David Scott remained behind in the command module. After they returned, in further tests Scott got to "fly" the command module by hand, with the main engine on, following needles on the display panel, experiencing true hand-piloting (Gemini style) of a spaceship for a few minutes. In addition, Scott was familiar enough with the computer to be creative with it, asking it to do new things (causing engineer Jim Nevins some concern, because the software always had bugs in it and had been certified only for performing standard procedures). Scott took the coordinates for Jupiter from a star chart; he entered them into the computer, which pointed the telescope directly at the planet. At the end of the mission the command module jettisoned the empty LM. Scott uploaded the orbital parameters of the LM from the ground and asked the computer to calculate the sight angle for the telescope. The computer pointed the optics and Scott was able to see and track the discarded LM in the crosshairs as it swung around its death orbit 2,500 miles away.[8]

Mission Planning

The LM seemed to work in space, but how to make it gently land on the lunar surface?

Early on, Space Task Group member Donald Cheatham laid out the basic ideas for landing. Cheatham believed that the landing phase should take advantage of the crew's judgment, especially in the final moments of selecting a specific landing spot. For him, the problem came under the old rubric of "handling qualities," although with few parallels in traditional, atmospheric flight. Cheatham set up a series of simulations in Houston, with a variety of controls and displays, and asked the astronauts to evaluate them according to the standard Cooper rating scales for pilot feel.[9]

Cheatham divided the landing into phases (figure 8.3), beginning from a point 50,000 feet above the moon. His phases remained salient for the rest of the program: the braking phase, which slowed the LM out of orbit, the approach phase, which allowed the commander to assess the landing site, and the landing phase, where the vehicle hovered to touchdown.

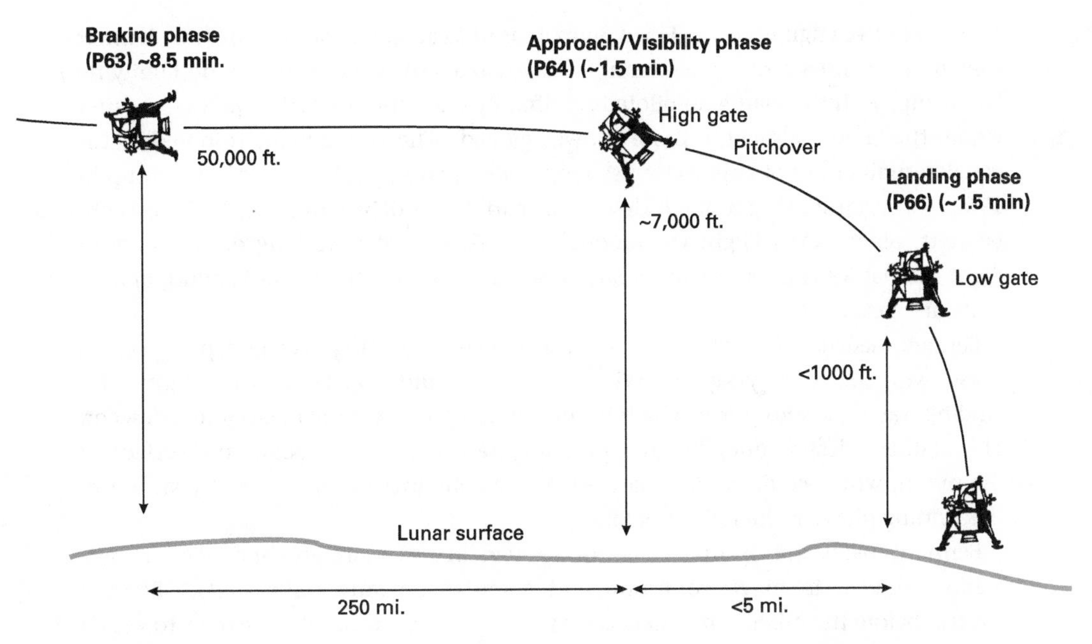

Figure 8.3
LM landing phases. Not to scale. (Redrawn by author from Johnson and Giller, "MIT's Role in Project Apollo, Vol. V," 182.)

Cheatham's ideas structured the engineering requirements for Grumman to build the LM, but as the vehicle took shape engineers raised numerous questions. How were they to transform these requirements into practical procedures, checklists, backup plans, and go/no-go decisions? This task fell to Floyd Bennett and his group in Houston's Mission Planning and Analysis Division (MPAD).

MPAD emerged in 1963 from part of the original Space Task Group and had responsibility for trajectories, orbital dynamics, navigation, and a variety of other technical functions associated with each mission. This high-morale group did the critical, difficult, but interesting calculations for the Apollo flights: they generated, defined, and analyzed trajectory data including rendezvous procedures, onboard navigation, and computer memory loads. Their offices (and hallways) stacked high with computer printouts, MPAD engineers analyzed the possible trajectories and the use of oxygen and water during Apollo 13. Bill Tindall, who applied such pressure to the IL software project, was one of the senior people in MPAD. MPAD did not directly manage the IL, but they defined the requirements and the trajectories for Apollo, which made them intimately involved in the software effort.[10]

After a series of reorganizations to keep up with the changing program, MPAD created a Lunar Landing Branch in December 1967, just a year and a half before Apollo

11, with twelve engineers, and Floyd Bennett as its head. It was no surprise that Bennett grew up to spend a career in aviation, for he shared a surname (although no family relationship) with the man who piloted Admiral Byrd around the earth's poles (and after whom the famous airfield in Brooklyn was named). The young Bennett joined NACA Langley straight out of engineering school in 1954, and then the Space Task Group in 1962. He began looking at lunar descent as part of the debate over the LOR decision. Bennett joined NASA Flight Operations in 1966, and after working on a variety of flight mechanics and rendezvous analysis positions, joined the Lunar Landing Branch as its first head.

Bennett made a logo for his group—a cartoon cowboy going over a steep cliff on his horse, with the cowboy saying "Whoa, Whoa!" (figure 8.4). Bennett's metaphor for landing was a fractious horse, which in his mind, represented the fuel-optimal descent. The mathematical solution for the trajectory targeted a point below the surface of the moon, which could result in a crash if the computer did not correctly switch to the landing phase at the right moment.

Bennett's plans and Grumman's hardware were in a constant give and take. "This iteration resulted in much confusion and many investigative false starts," Bennett wrote, "before the mission planners and system designers realized the extent to which the inputs of one affected the other."[11] The landing radar design proved particularly knotty—it depended on the reflective characteristics of the lunar terrain, the topography, and the attitude of the LM. Yet the radar signal absolutely needed to kick in to update the guidance solution, otherwise the astronauts could not attempt a safe landing. Faced with such problems, the designers were naturally conservative, but every ounce of conservatism introduced inefficiencies that cost pounds of fuel. In a system as close to the edge of performance as Apollo, excessive conservatism could push the goal out of reach.

Whatever Bennett and his group envisioned for the landing software, the IL engineers had to program. Enter Allan Klumpp, a mechanical engineering graduate of MIT and an expert in feedback control. Klumpp had been working at the Jet Propulsion Laboratory in Pasadena since 1959, when he was transferred to NASA headquarters to work in Joe Shea's systems group. Klumpp remembered seeing a presentation on Apollo guidance by Trageser and Hoag: "It was the most high-powered presentation I'd ever seen . . . they had two slide projectors going at once." Klumpp decided that the IL, and not NASA headquarters, was where he wanted to work. Before long he moved to Cambridge. Klumpp enjoyed the peaceful nature of Apollo, for he had designed weapons systems before but was disturbed by the uses to which they were put in Vietnam. War and peace were never far apart in Apollo's technical culture.

At the IL, Klumpp was first assigned to simulate the view out the window during landing. He programmed a computer to plot pictures of what the astronauts would see each second. The assembled images then formed a rough movie of what the land-

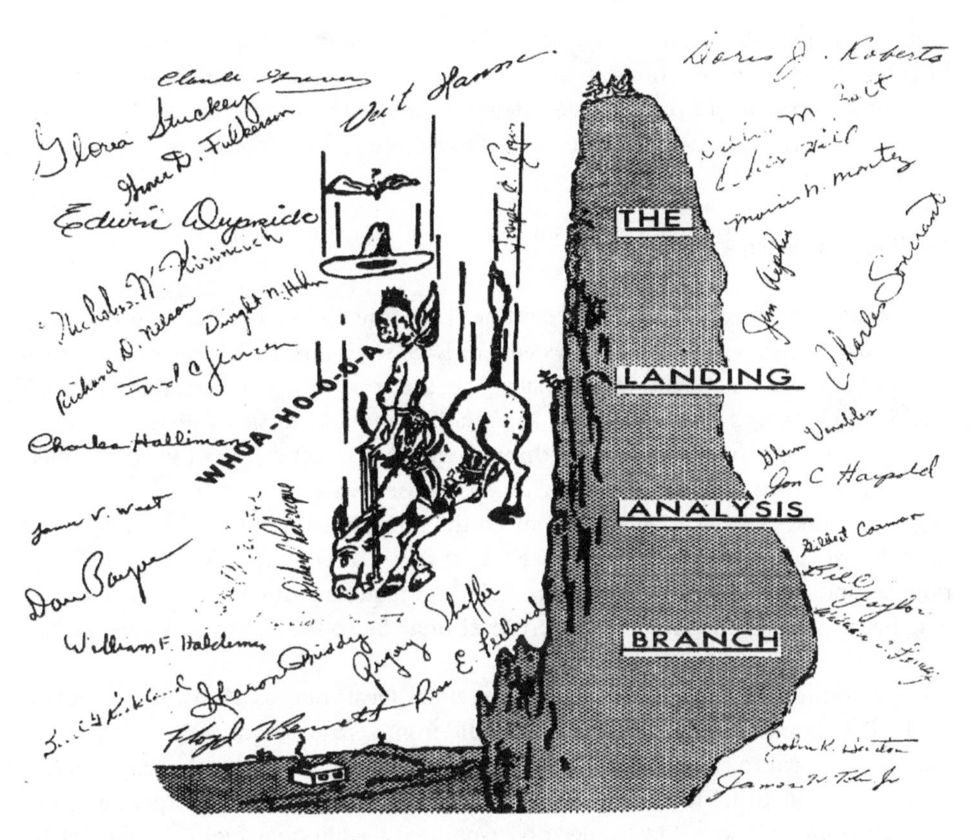

Figure 8.4
Mission Planning and Analysis Division, Lunar Landing Branch logo with employees' signatures, depicting the fuel optimum descent as a bronco out of control going over a cliff. (Mission Planning and Analysis Division [MPAD], "The End of a Great Era," June 15, 1990, JSC Archives, 336.)

ing might look like from the LM window. Before long, Klumpp was working on the guidance "equations" themselves—not just the mathematics, but also the logical flow charts for how the computer would control the vehicle during descent. A colleague, George Cherry, had built a stack of 3,000 punch cards that simulated a landing inside a mainframe computer, and soon Klumpp's version expanded to 6,000 cards as he struggled to translate the simulated landing into real-time instructions for an actual computer. Klumpp remembered that the astronauts originally wanted to fly the LM manually down from orbit. He programmed the flights on a simulator and the astronauts crashed every time. The landing would be critically dependent on the computer.[12]

Klumpp, along with his colleague, Don Eyles, implemented the landing equations in actual program code. A 1966 graduate of Boston University in mathematics, Eyles didn't even join the IL until the year before Apollo started flying. Yet the intense, bright Eyles soon found himself in the center of Apollo's most critical moments. When the astronauts in the LM interacted with the LM computer in the final phases of their descent, they were talking to a little piece of twenty-four-year-old Eyles's brain.

Landing as a System

Five major subsystems defined the landing. First, the LM descent engine (officially known as the "descent propulsion system" or DPS) stuck out the bottom of the spacecraft in the characteristic rocket bell shape. Like the enormous Saturn engines, the DPS was on gimbals, and could be angled fore and aft and side to side—useful for "trimming" the thrust, to be sure it went through the LM's exact center of gravity, which changed as it consumed fuel. The engine was unusual in another respect as well: any engine could get a vehicle *through* a spot on the lunar surface; but the LM engine had to get the vehicle *to* that point and make it stop gently, even as the LM's mass changed from second to second as fuel was burned.[13] This requirement meant the descent engine had to be "throttleable:" its thrust level must be controlled in real-time, by the computer.

For a control engineer, such flexibility seemed ideal, but, as always, there was a catch. The descent engine could only throttle from zero to about 55 percent; above that value, it had to run at 100 percent of its rated thrust, otherwise the fiery exhaust would erode the bell. This limitation added complexity to the control equations and, as a departure from the ideal trajectory, would cost additional fuel (380 lbs., to be exact).

The second major subsystem, the landing radar, bounced radio waves off the lunar surface to detect both the altitude and velocity of the LM. The radar projected four beams. Three of them detected the Doppler shift in the reflections and compared them to determine the spacecraft's lateral velocities. A fourth beam measured the travel time for the reflection to return, which indicated altitude. The antennas of the radar could be switched into two positions: one angled outward so it would point straight down when the LM was horizontal, and another pointed downward so it would aim correctly when the LM was vertical. Ryan Aeronautical of San Diego, the same company that had built Lindbergh's plane, *The Spirit of St. Louis*, built this critical device.

Problems beset the landing radar from the beginning. Only on Apollo 10 did a realistic flight test provide sufficient confidence in the landing radar, and on Apollo 11 the crew still looked down by eye to verify its readings. The device would nearly cause an abort on Apollo 14.

Communications formed the third major subsystem for landing. The LM had two types of antennas to communicate back to earth. The parabolic "high gain" antenna transmitted high-quality voice and data, but it needed to point directly at the earth to work. During landing, the computer could figure out the proper direction and point the antenna automatically, or the lunar module pilot could take over and "slew" the antenna with a knob. Two smaller, omnidirectional antennas had no such pointing requirements other than being on the side of the vehicle facing the earth, but they generated lower data rates and voice quality. Getting the communications properly lined up and free of noise proved a burden on the crew at a critical time. The LM also had a rendezvous radar for pointing at the CSM to provide range and closing rate data when the two craft were reuniting, but the device was not used during landing. Its antenna was mounted on servos so the computer could control its pointing and keeping it focused on the CSM during the maneuver. The rendezvous radar stayed on during the landing in case it was needed for an emergency abort; its computer connection would cause a startling problem on Apollo 11.

Integrating the system in real-time was the computer, the fourth critical element in the landing. The LGC (for LM guidance computer identical to the AGC in the CM) had to talk to all of the different components, each made by different companies with different specifications. It ran the digital autopilot, directed the small thrusters that controlled attitude (the RCS, for reaction control system), commanded the descent engine, controlled indicators on the pilots' panels, took data from the DSKY keyboard, and read the commander's control stick inputs. The software also measured time intervals, counted pulses, and performed a variety of other tasks. And, of course, it gathered data from the inertial platform. Together, the computer and the inertial unit comprised the primary navigation and guidance system, or PNGS, pronounced "pings." A separate, backup guidance system, the "AGS" for "abort guidance system," could come into play to get the LM back to a rendezvous if the PNGS failed.

Software had to account for the vagaries in behavior of all of these systems. The descent engine, for example, "ablated" or wore out as it burned, changing its thrust characteristics. The software thus had to incorporate the changes into its equations (a problem in this correction nearly caused an unstable throttle on the early landings).

The final components of the landing system, of course, were the humans—at once operators and cargo. A strange terminology supported their sense of control. The "commander," in the left position, actually flew the LM. The other operator, in the right position, was technically a co-pilot, but he was titled "lunar module pilot" (LMP) even though he never touched the hand controls. He served as a systems engineer and operated the DSKY. During descent, the commander verified the computer programs and landing phases, redesignated the landing sites, "flew" the final few hundred feet, and stopped the engine on lunar contact. The LMP kept his eyes on the numbers, verifying the agreement between the two navigation systems, AGS and PNGS, punched keys to issue commands to the computer, and, in the final moments, called out altitude, velocity, and fuel quantities so the commander could keep his eyes focused outside the window. Not only did the operators interface with the computer through a variety of displays and controls, they also interfaced with each other in a social relationship. Furthermore, they communicated with the ground controllers, who cleared them for the next steps and aided in managing the systems.

When Bennett and his group began thinking about how the astronauts in the LM should approach the landing site, they naturally modeled the procedure on aircraft approaches. To the pilots, the moon seemed like an unfamiliar airport, so they wanted to fly around it and get a visual read before going in for a landing (wasting precious fuel in the process).[14] Inevitably they ran into the same issue that had proved so contentious with the launches: would the landing be fully automatic or manual? More realistically, how much of the landing would be automated? How would the human and machine *trade control*?

Automatic landing offered a variety of options. Wernher von Braun's original vision for lunar landing had no human involvement, "accomplished entirely by the automatic pilot running on a guidance tape," as he put it, although it did allow the "captain" to avoid obstacles.[15] Grumman engineers had been studying a fully automatic LM mission, under a NASA contract. Selecting the landing site could be accomplished by telescope while the crew was still in the CSM, or by a real-time TV image transmitted from the surface to the LM.[16]

Bennett and his group, working together with the astronauts on "pilot-in-the-loop" simulators, developed the techniques for lunar landing. Mostly, the pilots' roles were limited to systems monitoring and abort situations, except in the final moments before touchdown. In certain aborts, especially with a failure in the PNGS, the crew could become overwhelmingly busy. To Bennett it was a given that the pilots would land under manual (semiautomatic) control, even though the computer was capable of landing automatically. "None of the crew wanted to land in an automatic system," Bennett recalled.

Bennett once suggested to Chris Kraft a fully unmanned, automated lunar landing as a test. He recalled that Kraft would not allow it, because if an initial, automatic landing failed, then Kraft felt Congress would insist on a successful demonstration before attempting it with astronauts aboard.[17] Bennett wasn't trying to undermine the idea of the lunar landing, noting "I was just being the purist engineer," trying to ensure a safe landing. With that in mind, Bennett got into the lunar mission simulator one day and instructed it to land in a fully automatic mode. He remembers the astronauts' reaction: "That's not flying," they told him with contempt.[18]

While the LM and CSM were still docked, the astronauts would enter the LM, power it up, and run through a series of checkouts. No electrical connection existed between the guidance computer in the LM and that in the CSM, so the crew had to transfer data, time synchronization, and orbital parameters by calling the numbers out from the AGC in the command module, punching them into the LGC by hand and calling "mark" to synchronize them. They would similarly initialize, calibrate, and align the PNGS. Then the two craft would separate. The LM commander would rotate the craft around so the command module pilot, now alone in the CSM, could visually inspect the integrity of the LM. If all seemed well, the LM gradually backed away and headed for a lower orbit, about ten miles above the moon.

In June 1966 NASA convened a symposium in Houston to work through the operational plans for a full lunar landing mission. MSC engineer Owen Maynard, the symposium's organizer, was one of a handful of core NASA engineers who had emigrated from the Avro Canada company after their fighter jet program was canceled. Maynard designated a series of different mission types, from the "A" missions of unmanned tests to "G," the first lunar landing; "H," further basic landings; and "J," enhanced landings with heavier payloads and longer stays. Maynard also laid out a strategy of nine "plateaus" for each lunar mission, relatively safe positions along the flight allowing careful assessments and decisions before proceeding (or aborting).[19]

Strange as it seems, for astronauts in a little metal balloon located a quarter million miles from earth, lunar orbit was a relatively safe position, and therefore one of Maynard's plateaus. From here everything could be put on hold, safety checked, and replanned. If a problem arose, ground control could send the LM around the moon for another orbit or two while they troubleshot the problem. The astronauts could return home, if necessary, with relative safety. The CSM could even swoop down and rescue the LM from its low orbit.

Beginning the landing sequence was a major commit point. It started at 50,000 feet, when the LM's descent engine fired to slow the craft, which would cause it to descend toward the lunar surface, leaving the plateau of lunar orbit. Once the engine fired the LM either had to land, hit the moon (in about ten minutes), or execute a dangerous abort. The clock was ticking.

The LM Digital Autopilot

Nothing about the LM itself would be intuitive to fly. It had sixteen RCS (reaction control system) thrusters (four clusters of four) for attitude control, but the large descent engine on the bottom pivoted on gimbals and could also control attitude and hence steer the spacecraft. Failure of a single RCS thruster, or a cluster of them, could send the vehicle spinning. The center of gravity moved as the engines consumed fuel, and sloshing of the propellants could make the spacecraft difficult to handle. These and any number of other complexities meant the LM had no inherent or natural match to a human pilot. Only the LM's computer, a set of software routines combined with a host of data, sensors, and actuators, could give the astronauts the feeling that they were "flying" the vehicle. These routines comprised the digital autopilot.

The autopilot took over the computer every tenth of a second for its calculations, which took about a fortieth of that time (twenty-five milliseconds) to complete. In addition, during powered flight every two seconds another routine would intervene to adjust the autopilot and its parameters. The digital autopilot maintained a virtual model of the vehicle inside its "state estimator" (equations similar to Battin's recursive techniques) that kept track of the various forces generated by and acting on the LM. When the thrusters fired, it estimated their effects on the LM's attitude and incorporated them into a new estimate even before their effects showed up on the accelerometers. "Jet selection logic" automatically determined which of the sixteen thrusters would fire in response to a command. If a thruster failed, or a cluster of them failed, the selection logic would automatically sense the failure and compensate with other actuators. The state estimator calculations automatically compensated for the changes in vehicle mass as the engines consumed fuel (a less massive vehicle accelerates more in response to a given thrust).

One set of digital autopilot routines related to "coasting flight," when the spacecraft was in orbit but not changing its velocity (such as during the visual inspection after undocking). A program called KALCMANU (for "calculate maneuver") could rotate the spacecraft to any specified orientation in the most efficient manner, gently leading the spacecraft's attitude servos for a smooth motion and avoiding gimbal lock.

Another set of routines controlled powered flight, when the descent or ascent engines were firing, changing the spacecraft's velocity. Powered descent to the lunar surface represented the most complex case. Ten times per second, the computer read the changes in velocity and attitude from the inertial platform as the engine slowed the spacecraft out of orbit. The guidance equations (running twice per second) then extrapolated that data to get the spacecraft's new position and velocity, and determined new thrusting commands accordingly. These routines responded to higher-level guidance routines that determined the desired position and velocity of the spacecraft at any given point and sequenced through the three landing phases.[20]

The LM User Interface

The layout of the LM reflected the design of the landing (as well as the ascent and rendezvous), and the user inputs to the digital autopilot. Each operator looked through a window on either side of an instrument panel, which had a number of common indicators and other switches and controls unique to each side. The eight ball indicator, similar to that in the CSM, resembled the artificial horizon in an aircraft and indicated the LM's attitude relative to a reference. Though an analog indicator, it was driven by the digital autopilot. During landing, the eight ball functioned like an artificial horizon, showing the LM's relationship to the local vertical of the moon. A velocity display (also driven by the computer) with x and y "crosspointers" (needles) indicated the for-

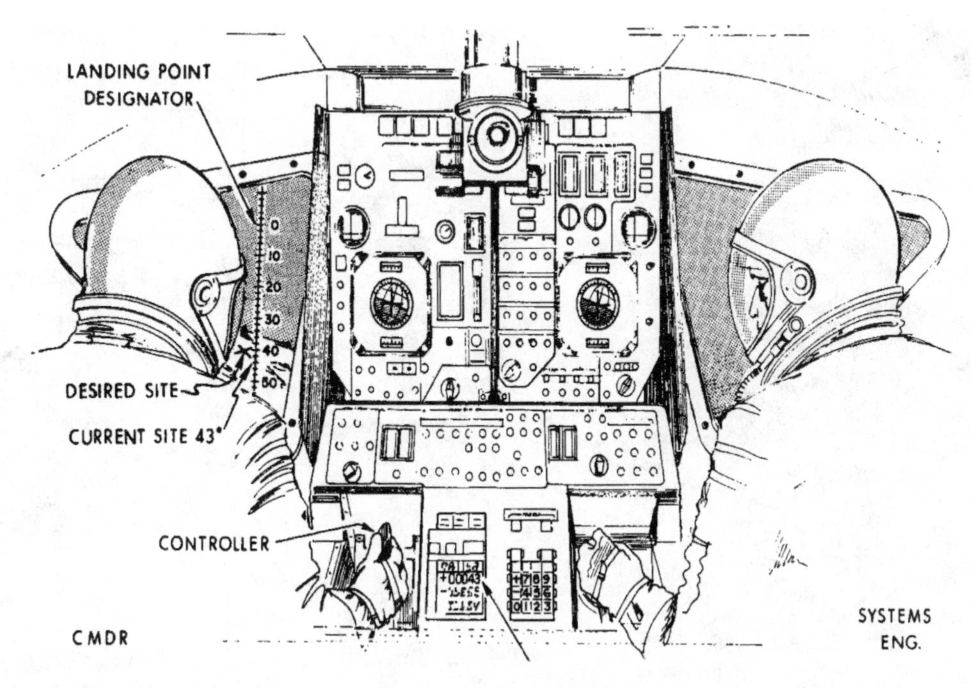

Figure 8.5
Arrangement of crew and controls inside the LM. Note that the LMP is designated "systems engineer." (Klumpp, "A Manually Retargeted Automatic Descent and Landing System for the LEM," 130.)

ward and lateral velocities. A dual tape-like indicator showed the altitude and descent rate from the digital autopilot's state vector in the computer. Each operator also had two hand controllers—one to control attitude, and one to control translations and vertical velocity (figures 8.5 and 8.6).

Under normal conditions, the commander did the hands-on flying while the LMP monitored the systems. Each time the commander moved his stick from its center position, it sent two signals to the computer, an analog voltage proportional to the amount of deflection, and a switch closure indicating that the stick had been moved.

The digital autopilot offered the astronaut several different modes to control the vehicle. These ranged from fully computerized controls (when the computer determined attitude and thruster firings), to complete manual takeover. Between them was a host of servos, mode switches, feedback loops, and software. An impulse mode, for example, enabled short, timed bursts of thrust in response to each joystick command, useful for precision maneuvers such as docking. An extreme "hard over mode" (when the operator pushed the stick all the way to its stops), bypassed the computer

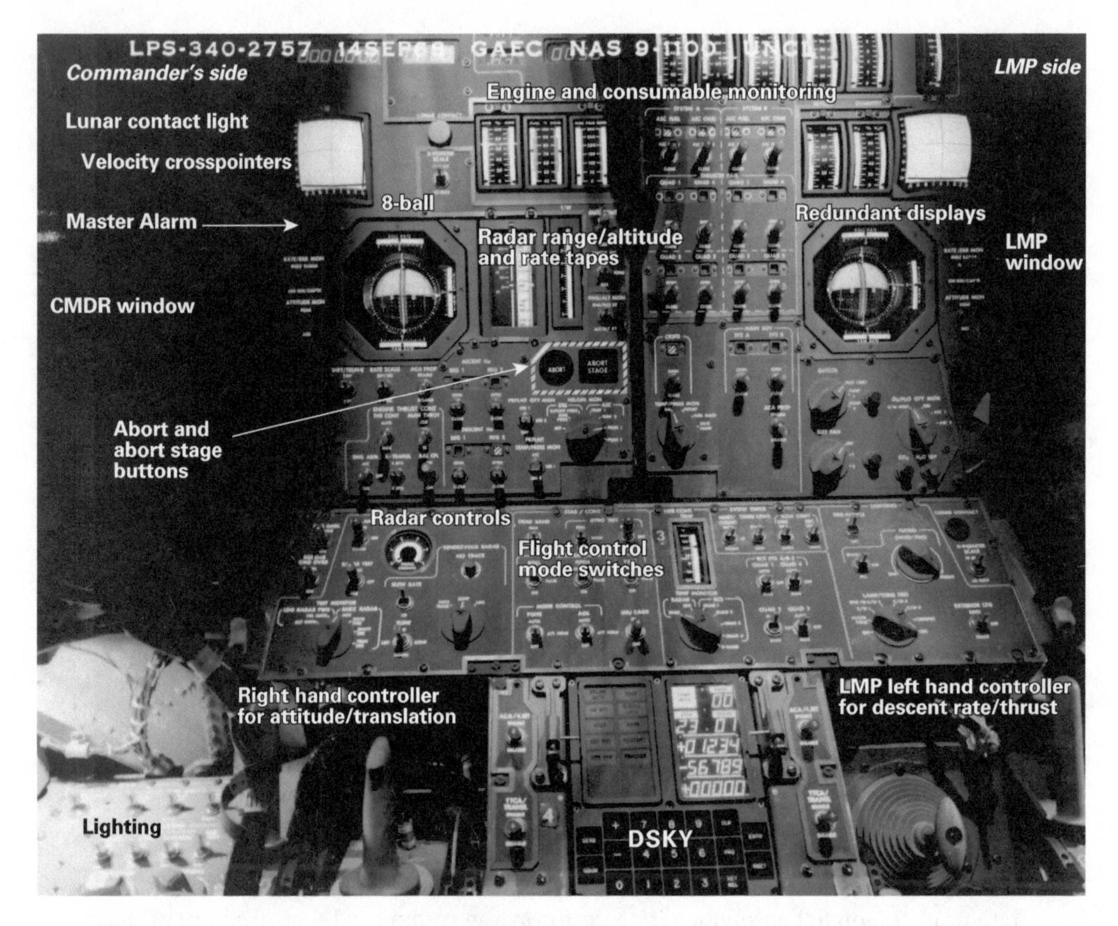

Figure 8.6
Partial view of LEM instrument panel showing indicators and flight controls. Commander stands on the left, LMP at right, with DSKY between them. Commander has additional hand controller and rate-of-descent switch in left hand. (Cradle of Aviation Museum archives, Bethpage, N.Y., Courtesy of Paul Fjeld.)

altogether and enabled direct control of the valves on the RCS thrusters. This would be an extremely inefficient use of fuel, but might work in an emergency if the computer failed and stopped accepting commands. As Jim Nevins wrote of the interface, "there is great flexibility and redundancy, but heavy burden on the crew."[21]

The most important LM digital autopilot mode was "rate command/attitude hold," entered by Verb 77 in the DSKY, or by flipping a switch from PNGS AUTO to ATT HOLD. In this semiautomated mode (similar to the "rate command" mode in the X-15), the astronaut could deflect the stick and change the vehicle's attitude. When he released the stick, the digital autopilot would automatically hold that new attitude. A similar mode (used simultaneously) for vertical control enabled the computer to maintain a precise rate of descent, allowing the astronaut to increase or decrease the rate with a switch.

The manual control modes in the LGC started out as digital reproductions of the analog loops found in the lunar landing research vehicle, or LLRV (discussion to follow), but over time they evolved to be "quintessentially digital, making freer use of the logical branches, counters, and nonlinearities which are so readily, and reliably programmed in the digital computer."[22] In this rich scheme, the computer did not replace the pilot's skill but rather coalesced the complex craft into an interface that provided both simplicity and variety.

The interface also included instructions for the crew. Figure 8.7 shows the checklist the astronauts used in the cockpit, in this case for Apollo 12. It is part of a larger timeline used for the overall mission, and one of about five pages used for the LM from undocking to landing. This piece of paper lay out on a notebook between the two astronauts just below the DSKY. The procedure started at the left one minute before the engine began to fire to slow the LM out of orbit. The first instruction was for the astronauts: "reset watch." They next read down the column and then up to the top of the next column and down to the conclusion, the landing. Areas in the dark boundaries indicated abort conditions or critical moments such as "bingo fuel." The final instructions on the bottom right indicated procedures to be taken after touchdown to "safe" the vehicle, and can be heard in Aldrin and Armstrong's words over the radio just before "the Eagle has landed."[23]

Like the programs embedded in the computer that commanded spacecraft systems to perform certain duties, these coded instructions directed the humans to perform particular behaviors and to make decisions based on data they observed. Like the ones and zeroes in the core ropes, the paper timelines in the LM cockpit helped tie the system together, binding human and machine into a single, integrated mechanism.

The Braking Phase

The landing began with the braking phase at 50,000 feet. Why did the LM orbit ten miles high before descending? The moon has no atmosphere, so a vehicle could theoretically orbit very close to the ground. In practice, however, that would not be safe: one would need a virtually perfect, circular orbit to be sure not to intersect the terrain. Such perfection was not achievable in the real world—tracking, guidance, and a variety of other factors introduced uncertainties into the orbit. The same held true for the terrain. It's one thing to orbit down low above a perfect sphere, but quite another if a hill or a mountain pops up in the way. NASA engineers decided that the uncertainty in the orbit and the guidance was about 15,000 feet, and the uncertainty in their

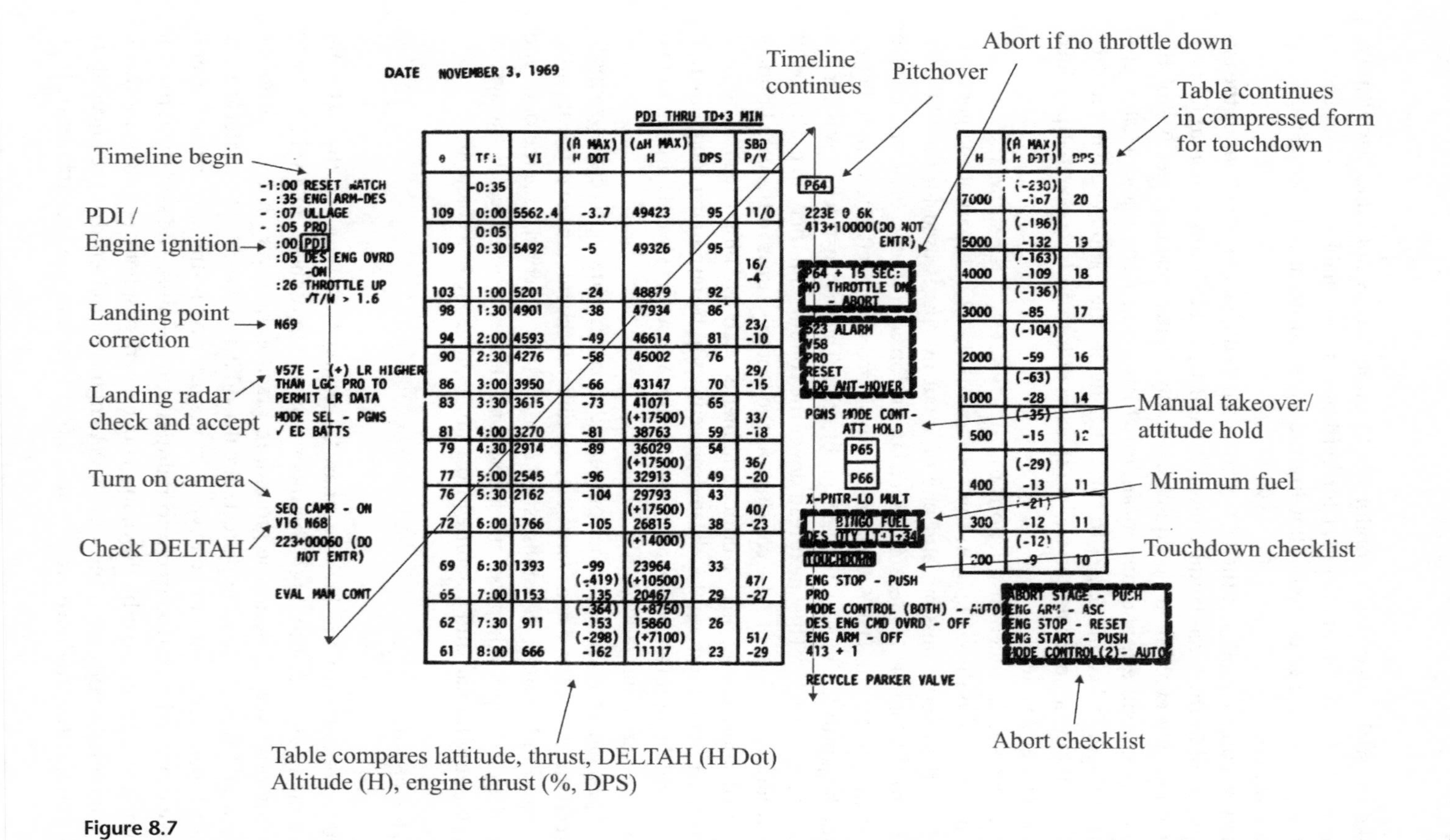

θ	TFI	VI	(Ĥ MAX) H DOT	(ΔH MAX) H	DPS	SBD P/Y
	-0:35					
109	0:00	5562.4	-3.7	49423	95	11/0
	0:05					
109	0:30	5492	-5	49326	95	16/ -4
103	1:00	5201	-24	48879	92	
98	1:30	4901	-38	47934	86	
94	2:00	4593	-49	46614	81	23/ -10
90	2:30	4276	-58	45002	76	
86	3:00	3950	-66	43147	70	29/ -15
83	3:30	3615	-73	41071 (+17500)	65	
81	4:00	3270	-81	38763	59	33/ -18
79	4:30	2914	-89	36029 (+17500)	54	
77	5:00	2545	-96	32913	49	36/ -20
76	5:30	2162	-104	29793 (+17500)	43	
72	6:00	1766	-105	26815 (+14000)	38	40/ -23
69	6:30	1393	-99 (-419)	23964 (+10500)	33	
65	7:00	1153	-135 (-364)	20467 (+8750)	29	47/ -27
62	7:30	911	-153 (-298)	15860 (+7100)	26	
61	8:00	666	-162	11117	23	51/ -29

H	(Ĥ MAX) H DOT	DPS
7000	(-230) [illegible]	20
5000	(-186) -132	19
4000	(-163) -109	18
3000	(-136) -85	17
2000	(-104) -59	16
1000	(-63) -28	14
500	(-35) -15	12
400	(-29) -13	11
300	(-21) -12	11
200	(-12) -9	10

Figure 8.7

Checklist timeline for Apollo 12 from PDI to landing, showing flow of commands down left side and then across to middle column. Tables are values of acceptable descent rates (h dot and delta h) versus altitude. (Annotations by the author from Apollo 12 Timeline Book, http://history.nasa.gov/alsj/a12/a12LM_Timeline.html [accessed January 5, 2007].)

knowledge of the moon's terrain was about 20,000 feet. Thus they chose 50,000 feet as the preferred orbit to speed along without fear of hitting anything.[24]

The LM attained this orbit in two different ways. In the early flights, after the command and lunar modules undocked, the LM fired its engine on the far side of the moon, just about opposite from the intended landing area. On the later missions the vehicles remained docked, and the CSM fired its engine to bring them both to the lower orbit, thus saving fuel and weight on the LM (the CSM could then return to a higher orbit during the lunar stay). Either procedure brought the LM to an elliptical orbit of about sixty by ten miles. Both burns occurred out of sight of the ground (and hence in communications blackout), which introduced some anxiety, for an error of a few seconds in the burn could cause the LM to intercept the moon prematurely and crash.

The landing sequence really began when the two spacecraft reemerged from behind the moon, reacquired signals from Houston, and verified that the burn had succeeded and the new orbit was within acceptable limits. At this point, nicknamed AOS for "acquisition of signal," the LM zoomed over the moon at nearly 4,000 miles per hour. The astronauts were wearing their spacesuits with the helmets and gloves removed, and they copied a series of vectors read up by voice to the ground, known as "PADs" (preadvisory data). These data provided them up-to-the-minute reference data in the event of an abort with lost communications.

If all checked out in the 50,000 foot orbit, the astronauts began the descent procedure, initiating the braking phase with an event called PDI or "powered descent initiation." Typing "V37," pressing ENTER, and then "63 ENTER" initiated P63. This program calculated the ignition time based on the navigation state vector and the landing target, and then prepared to fire the descent engine when the LM reached its ten-mile perilune (the low point of its orbit). The DSKY then displayed the time to ignition, among a few other variables, for the astronaut to review. Depressing the PRO(ceed) button on the keyboard advanced to the next step. The computer then replied with a VERB/NOUN message asking the astronaut if he would like to align the inertial platform, to which he ordinarily answered no by pressing the enter key (it should have been aligned before separation). Initial requests to approve every calculation in the computer proved impractical, but the PRO button allowed the astronauts to give final approval before major activities.[25]

Next the computer maneuvered the LM to the initial attitude for the PDI firing, though not before presenting the angles for the astronaut to review, and waiting for a PRO keypress again. Once in this position, the displays briefly went blank, about thirty-five seconds before the burn, to indicate the computer was getting ready to fire the engine. Five seconds before the burn, the astronaut again had to review the display and press PRO before the engine fired. At this point the crew could reject the burn or delay it by up to five seconds. Any longer delay would mean going around for another orbit before trying again.

The burn began with the RCS thrusters firing for a few seconds for "ullage" to push all the fuel to the bottom of the tanks. Then the main engine began firing, marking the official moment of PDI. When the burn began, the DSKY's three-line display presented inertial velocity, the rate of descent, and altitude. The computer also started a clock, and later events until landing were referenced to this moment.

At the moment of PDI the LM pointed backward, orbiting the moon in a "feet first" position; the computer fired the descent engine ahead as the spacecraft pointed back along its path. For twenty-six seconds, the engine stayed at 10 percent thrust to allow the computer to trim the engine gimbals a bit to be sure it would fire through the center of gravity and not impart any additional motion to the vehicle besides slowing it down (the astronauts could barely feel this gentle thrust).

Once trimmed, after about thirty seconds, the thrust climbed to 100 percent, for about seven and a half minutes as the LM covered nearly 250 miles and slowed its velocity from 5,500 to 600 feet per second (from 3,750 to 410 miles per hour), and from an altitude of 50,000 feet down to under 10,000. Now, at any time, the crew could abort by pressing buttons marked ABORT (to abort using the descent stage) or ABORT STAGE (to jettison the descent stage and abort with the ascent stage only). In an abort, the crew workload would rise dramatically as they selected and commanded one of numerous different trajectories to make their way back for a rendezvous.

If everything went well, however, workload was much lower, as the major job of the crew during the braking phase was to monitor the guidance system, the primary PNGS navigation and the backup AGS. They also had a third source: as the LM traversed the front face of the moon, ground-based systems had excellent tracking, monitored closely by ground controllers. Divergence in the PNGS and AGS solutions would indicate a problem. In case of a discrepancy, the ground-based solution could provide an additional "vote" to decide which was most accurate and whether to proceed.

During the PDI burn, the LM could face either up or down. On Apollo 11 the astronauts faced down, giving themselves a visual confirmation of their height above the terrain. They then turned themselves over (actually "yawed around") with the hand controller, or the computer would have turned them over at a predetermined time, about three minutes into PDI at about 40,000 feet in altitude. Either way, they were now coming in on their backs, feet pointed toward the landing area, looking up toward earth to enable the landing radar to see the ground. About four minutes after PDI, the flight director in Houston took a survey of his team and cleared the LM "go" for landing (an exercise repeated at about 3,000 feet).

Around this time the landing radar began detecting the lunar terrain, a critical moment. A light marked "altitude" went out when the radar detected range, and "velocity" went when the Doppler signals came in to measure velocity (usually a few minutes later). Until this point, the navigation solutions were inertial, based on motions of the spacecraft measured by accelerometers, calibrated by star and landmark sightings, and updated with data from earth tracking. Now the landing radar provided literal ground truth, the moment when the LM navigated as much with data from the lunar surface as from the stars. It was the first direct sensing of the two quantities that really mattered: where was the spacecraft relative to the lunar surface? How fast was it moving? Without these critical numbers, the guidance solution could be off by several miles in altitude; without them an astronaut would likely not be able to eyeball his way down.

When the landing radar locked in, the LMP then entered VERB 16 and NOUN 68, which displayed DELTAH, the difference between the radar data and the onboard state estimator. If they were close, say within 10,000 feet at 20,000 feet altitude, it provided an excellent check of the overall system's accuracy.[26] (This measure assumed a perfectly round moon. If the terrain were changing rapidly, as it would on later missions in mountainous regions, then the altimeter reading would change drastically even as the spacecraft descended gradually. Later missions actually included a rough model of the underlying terrain in the computer to account for these variations.) If the radar data looked good, that is if DELTAH was not too large, the crew "accepted" it with a VERB 57. The computer thus incorporated the radar altitude and velocity in a weighted average with the inertial estimates and issued a new navigation solution, one that should converge within a few seconds to approximately the landing radar altitude. When it occurred as planned, the appearance of the radar data relieved significant tension; the LM was now locked into its target, the moon. It also gave them a new primary means to measure their altitude, descent rate and along-track velocity, displayed on two tape meters and an x-y display, respectively.[27] Incorporating the radar data and the computer's convergence on a solution was a major milestone in the landing.

If the radar did not come in, or if it came in with too much of a difference from the PNGS, or the new solution did not converge, mission rules demanded an abort. The LMP compared the H and DELTAH numbers with a printed chart on the timeline and checked it against the chart, as well as a variety of other parameters. He was operating like a computer himself, comparing values from the instruments against the "dead man's curve" (NASA encouraged the crews to use the safer-sounding "abort boundary"). For each altitude, there was an acceptable DELTAH and descent rate that would allow a safe abort; if those limits were exceeded (i.e., if the LM were descending too fast for a given altitude), the LM could not recover in case the descent engine failed, so an abort would be mandated. Other problems might be indicated by program alarms, each with a number to indicate the problem. For instance, 1406 indicated that a guidance computation was failing, and 1410 indicated an overflow in the guidance equations.

High Gate: 9,000 Feet

At first glance, the lunar landing trajectory seems purely an engineering design, almost a high school physics problem employing the time-honored principles of Newtonian mechanics. Indeed, were the system totally automated, the LM would continue along a flat, fairly horizontal path, only orienting itself vertically as it neared the ground. But at about 9,000 feet the LM reached an imaginary point called the "high gate," a term derived from aviation denoting the beginning of an approach to an airport (the term for the next point, "low gate" has a similar origin). Here the landing changed from an ideal trajectory to one that accommodated human judgment and decision. At this point the computer would select "P64," the approach phase program.

The landing sites had been chosen with a variety of maps, made by robotic orbiters, previous Apollo missions, or ground telescopes. But the resolution of these maps was too low to depict features that might disturb a craft the size of the LM, so none could show for certain that the sites would be perfect and flat. Early plans called for a previous flight to drop a beacon or radar transponder for the LM to home in on, but the idea never made it off the drawing boards (other plans even called for clearing special landing fields for fully automated LMs carrying cargo).[28] Lacking such fixed reference points, precise landings required some sort of terminal guidance for the last few seconds. Here NASA relied on human vision and decision. Was the predetermined landing area suitable? Was the ongoing trajectory going to lead to a safe landing? The commander made these judgments. His life was on the line, so despite the variety of instruments and computers, he would verify the landing site on his own.

To do that he had to see.

But until high gate he was coming in feet first on his back, looking out into space.

Thus the LM needed to "pitch over" as it began the approach phase, rotating the astronauts to a standing position, bringing their eyes and their windows to bear on the landing site as it loomed ahead. This simple motion raised a series of questions: How close to the surface must the pitch over occur, to allow the commander enough time, and enough detail to make the decision? How would he identify the proposed landing site? How much time would he need to identify it and make a decision about where and whether to land? How much flexibility would the commander need to "redesignate" the landing site if necessary to a more suitable location (assuming he could find one)? "It soon becomes obvious," one flight plan explained, "that a strategy is needed that will trade off the system capabilities of the spacecraft and the crew capabilities against the unknowns of the lunar environment."[29] More systems engineering: trading energy (fuel) for information (vision), buying safety by reducing uncertainty with the resource of human perception.

Apollo mission planners and designers sacrificed much to save on weight and fuel, eliminating useful equipment to make the vehicle and its trajectory highly efficient.

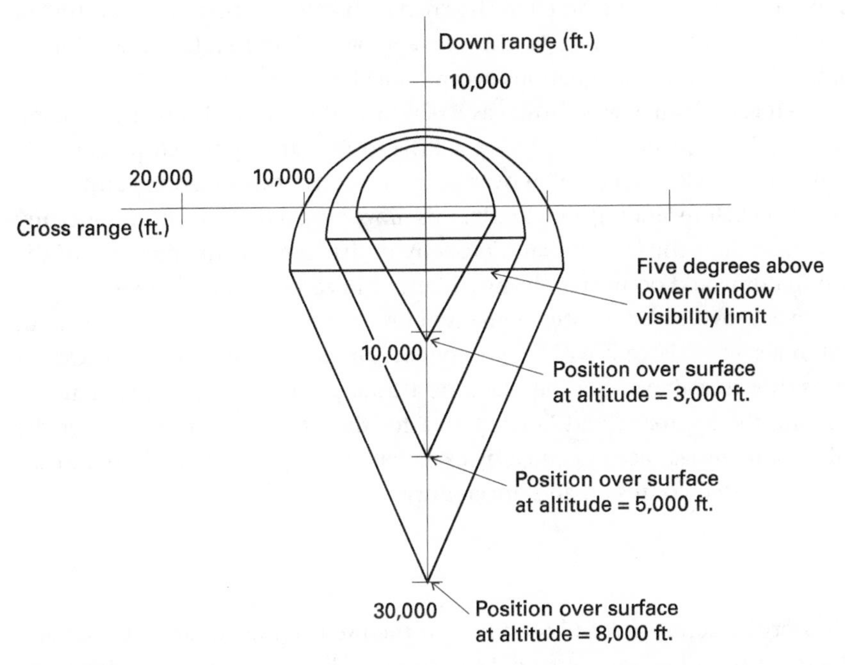

Figure 8.8
Analysis of LPD redesignation footprint from an 8,000-foot altitude. If the LM is at the point on the bottom of the graph and targeted to the center, the commander can redesignate the target and land anywhere within the top of the cone. As the LM descends, the cone becomes smaller, limiting the options for alternate landing sites. (Redrawn by the author from Cheatham, "Apollo Lunar Module Landing Strategy," fig. 36).

But saving fuel was in direct conflict with providing visibility to the commander at decision time.[30] Human vision won: "the trajectory is shaped at the cost of fuel, in order to provide the crew with visibility of the landing area."[31] They were devoting their scarcest resource, more than a hundred pounds of it, to a moment of human judgment. More fuel, more time to decide. For the first landing, the fuel budget included a hover time of about one minute; the commander began to see the landing area at 8,000 feet, and if necessary he could move the landing spot, or "redesignate" it, up to four miles away.

Figure 8.8 is typical of the many analyses done for these critical moments of assessing the landing site and "redesignating" to an alternate as needed.[32] The trade-offs of fuel were carefully plotted against the astronaut's sighting angle, as was the trade-off for altitude of the commander's assessment. At 5,000 feet, the commander could still redesignate up to 3,000 feet; going that far would require about 45 feet per second of additional velocity (fuel was measured in the "delta v budget," a precise accounting of possible changes in LM velocity). As the LM approached the surface and altitude decreased, the options for redesignations became smaller.

Thus the high gate altitude was chosen as 8,000 to 9,000 feet. At this point the computer automatically sequenced from P63 (for PDI) to P64, the approach program. At the high gate the vehicle would "pitch over"—rise from its near-horizontal attitude to a vertical position before landing. Obviously, this move would need to occur under any circumstances, so the vehicle could land properly on its legs, but the presence of the pilot meant that it would occur significantly higher and earlier than otherwise.

The pitch over provided a dramatic moment of revelation. The long flight from earth culminated in a speedy descent with the astronauts laying feet first on their backs. At pitch over, as though waking from the dead, the astronauts suddenly rose to a standing position, seeing the approach and landing area for the first time. On several of the landings this moment was accompanied by exclamations of joy and recognition. During others, it generated confusion and uncertainty.

LPD

After pitch over the commander needed to find the predetermined landing spot and decide whether it would be safe. How did the commander select a new landing spot? The approach program P64 would display a new quantity, "LPD angle," on the DSKY. This allowed the commander to identify the landing site out the window, or at least know where to look, using a clever device called the "landing point designator" (LPD), the brainchild of Donald Cheatham, implemented by Allan Klumpp. Klumpp called it a "hybrid" control system, because "the LM commander can manually steer the LM to the selected landing site, yet the trajectory he flies is produced by an automatic system" (figure 8.9).[33] The LPD consisted of a few lines of code and a very simple piece of weightless hardware—a set of markings on the commander's window that labeled the angles of his sight. Two separate grids were printed on the inner and outer windows of the LM, slightly offset; when the commander positioned his eye so the two grids lined up, his vision too was aligned. The DSKY gave him a number and the commander could look through the corresponding grid point on the window, where he should see the landing spot.

The commander could also redesignate the landing site, moving the computer's aiming point. He would find a new site, sight it through the LPD grid on the window, and read out the number to the LM pilot, who would enter it into the computer with a few DSKY commands. Klumpp found that process clumsy and error-prone, so he added a feature: the commander could also nudge the joystick and the landing site would "click" a degree or two in the corresponding direction (figure 8.10). One nudge of the

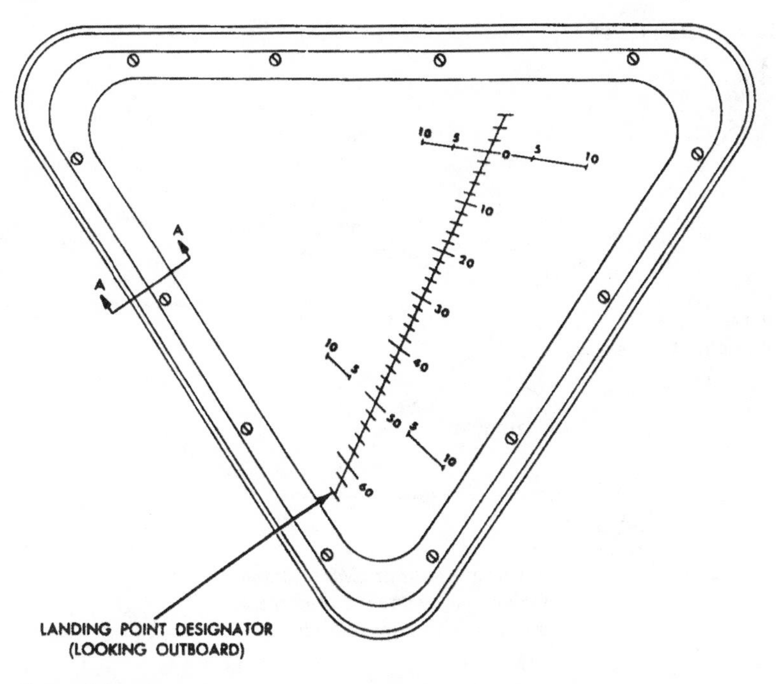

Figure 8.9
Landing Point Designator (LPD) scribed into the LM commander's window. (Grumman Aerospace Corporation, "Apollo Operations Handbook, Lunar Module, LM 10 and Subsequent, Volume I: Subsystems Data," April 1, 1971, 1-11.)

stick left or right would change the landing site by two degrees in the commander's field of view; one nudge fore or aft would move the landing site ahead or behind by a half degree (translating these angles to the actual distance change of the landing site depended on altitude: from 6,000 feet high, one click corresponded to moving the landing site 600 feet; at 500 feet altitude, one click only moved it 79 feet). The computer then recalculated the new landing position, and a new trajectory, and flew the LM accordingly to the new spot. The LM pilot could read out the number again, and the commander could look to find the new landing area, and redesignate again if necessary, in an iterative loop that should converge on the right spot. Through this repetition, Klumpp wrote, "the commander literally steers the current landing site into coincidence with the desired site," via a computer-mediated feedback loop.[34]

The pilot could continue to redesignate, evaluate landing sites, and make decisions for several minutes after pitch over. At some point, however, the LM would be so close to the surface that the LPD would be unusable. Just before the low gate of about 500 feet, the computer would not accept further redesignation and would switch into program P66 to guide the LM toward the site.[35]

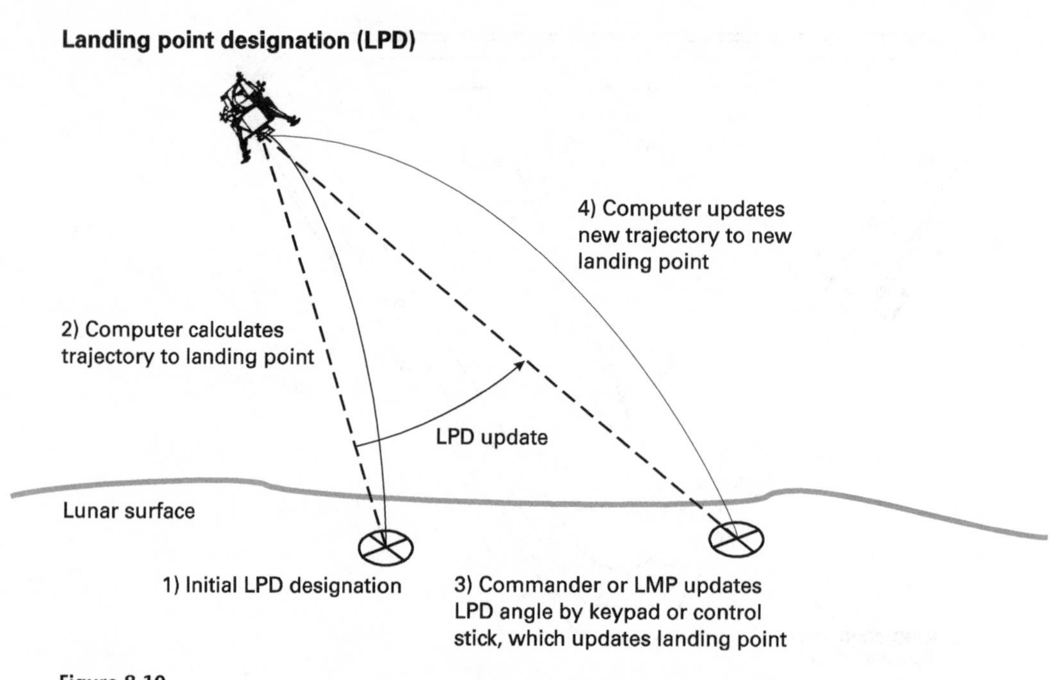

Figure 8.10

Landing point designation, showing how the commander or LMP can move the aiming point by updating the LPD angle, either by entering a new LPD angle into the DSKY or by nudging the control stick one "click." The process could be repeated an indefinite number of times. (Redrawn by the author from Cheatham, "Apollo Lunar Module Landing Strategy," fig. 40.)

The LPD formed the core of the human-machine system for landing. It allowed the commander a decision-making input and a high-level control of the vehicle's trajectory without burdening him with the actual control of the engines and thrusters, or of maintaining the velocity and attitude of the LM. It seemed an ideal allocation of tasks, using the human for what he was good at—(finding and assessing the landing site)—and the machine for its strengths (precision control and calculation). Apollo commanders did indeed use the LPD to steer them toward safer terrain. None, however, ever let the automatic feature actually land the LM on the spot designated by the LPD.

Touchdown

About a minute and a half after the high gate, when the spacecraft reached about 500 feet, it passed low gate and entered the final landing phase about a minute from touchdown. This phase was designed "to provide continued visual assessment of the landing site and to provide compatibility for pilot takeover from the automatic control."[36] The commander might wish to spend some time making decisions, so extra fuel was put in the delta v budget for the pilot to hover and inspect the landing site before actually descending to the surface. Different pilots might do things differently, so extra fuel was added in the budget for variations in pilot skills and techniques.[37]

Nobody really expected the pilots to land under control of the LPD (Bennett thought the fully automatic feature might be used on some future unmanned version of the LM). Any time after pitch over, the pilot could flick a switch from PNGS AUTO to ATTITUDE-HOLD, commanding the computer into yet another program, P66, or the computer would enter P66 automatically just before low gate. In P66, "the astronaut has a choice of four modes of control, differing in the division of duties between astronaut and guidance computer."[38] Here the computer controlled the rate of descent, but the commander could change that rate in real-time. Using a "rate of descent switch" near his left hand, one click down increased or one click up decreased the descent rate by one foot per second. Also in P66 the commander could exercise "rate control" of attitude, so he could fly the vehicle like a helicopter while the computer managed his descent rate. This mode "is manual, but is aided by automatic control loops, that is, the pilot has taken over direct control but he has stabilization loops to provide favorable responses." The earlier the commander entered P66 the further he could fly—at 1,000 feet he could fly for three minutes and head down range nearly three miles from the designated spot; taking over at 300 feet, he could travel only one-half mile in just over two minutes. P66 also had a true manual descent mode where the lefthand controller would throttle the descent engine directly and the righthand controller would control the RCS thrusters (very fuel-inefficient and difficult to fly, this mode was never used).

Now, while descending from 100 feet in final approach, the commander would "null" the lateral velocities—that is, stop the spacecraft's lateral motion. This would have been an easy enough task for the computer, but an automatic mode to slowly lower the LM in a hover was never used. Instead the commanders eyeballed the velocity needles, strained to see the lunar surface through the dust, and worked the stick to stop the motion while decreasing descent rate to about five feet per second. About 50 feet above the surface the commander slowed the descent to a touchdown rate of 3.5 feet per second. At this point the computer was precisely controlling the descent rate, and also automatically holding an attitude commanded by the pilot. Though these final control modes were often referred to as manual, they were really semiautomatic. Every LM commander landed with a significant degree of computer-aided control.

Below 100 feet, the commander could no longer abort by shedding the descent stage; he was committed to at least a momentary landing. From this point, he could land visually, based on what he saw out the window; or if his vision was obscured, land by looking at the instruments, primarily the altitude and velocity indicated by the PNGS.

Apollo missions landed in both visual and instrument modes, depending on the degree of dust they kicked up.

The final touchdown also presented delicate problems. Shutting off the engine too early would drop the LM to the surface with a crunch, possibly a splat. Shutting it off too late, however, could cause the engine to kick up too much dust and obscure the pilot's visibility. Worse, if the engine bell hit the surface and clogged, it could explode. Ideally, the commander shut the engine off at the moment of touchdown. The landing radar likely would not be reliable at these low altitudes, so an additional means was required to give a discrete indication of the shutdown time, at about five and a half feet above the surface.

Three of the LM's four legs had landing probes attached; these were long tubes with switches on the end that gave a positive indication of touchdown. When the probes touched down, a blue light in the LM cockpit illuminated, with the words "lunar contact," prompting the astronauts to manually shut down the engine. The length of the probes reflected the estimated delay of the pilot's response, the amount of time it took to close the valves in the descent engine, and the amount of time it took to actually reduce the thrust to zero. The pilot could take up to one second to respond and still allow the engine complete shutdown at touchdown.

He would then enter P68 to confirm to the computer that the vehicle had landed. He also momentarily jogged the stick to ask the computer to hold a new attitude. If the LM landed on a slight slope, for example, and the automatic control system was still trying to keep the LM level, it would be thrusting away with the reaction jets to level the vehicle. P68 turned off all the control loops and commanded the computer to display its estimate of the latitude and longitude of the landing site. The LMP would also disable the AGS by entering a command in its memory address number 413.

Training for a Landing

Like all designs on paper, the landing trajectory, related computer programs, and crew procedures seem elegant, even perfect when viewed in isolation. But long before they were put into practice, a critical question arose: how to match the human operator to the trajectory? That is, how do you train people to perform the actual task of landing the spacecraft? As the most uncertain part of the Apollo flights, the landings called for the most preparation.

Again NASA turned to simulation. Indeed the breadth and variety of simulations for Apollo exceeded anything in previous programs. During Mercury and Gemini, crews spent about a third of their training time in simulators; by Apollo it was closer to half.[39] The earlier programs had four distinct simulators each. Apollo had no less than eleven. These included procedure simulators (for both CSM and LM), translation and docking rigs, centrifuges, and partial-gravity machines to simulate lunar walking.

By far the most important, however, were the "mission simulators," one for the CSM and one for the LM, which aimed to replicate as much of the mission as possible with high fidelity. In addition to accurately reproducing switches and indicators from the two vehicles' control panels (according to computer models of the systems' dynamics), the mission simulators included thruster noises, cabin decompression, and the sounds of firing pyrotechnics. Each comprised a huge "train wreck" of computers, spacecraft hardware, film projectors, models, and analog video equipment.[40] They were supplied by the spacecraft vendors themselves, often out of parts rejected for flight hardware, and replicated conditions inside the craft and the appearance and behavior of the controls (although they were "fixed base," meaning they did not move).

Both Grumman and North American subcontracted the job of creating mission simulators to the Link Company (a division of General Instrument), which had produced the famous "Link trainers" of World War II. Link delivered one set of mission simulators to Houston and another to the Kennedy Space Center (actually two for the CSM and one for the LM) to allow the astronauts intensive training right up until launch day. The mission simulators were run by enormous digital computers, several for each, with extensive optical systems to synthesize views outside the windows during missions. Data lines connected the computers to mission control, so "integrated simulations" could be performed that included not only the vehicles themselves but also the flight directors and ground support teams.

Staying current with the rapidly developing spacecraft proved a constant challenge, and the simulators themselves faced overwhelming problems of configuration control. Trying to simulate the behaviors of the Apollo computers proved difficult because the software changed so frequently. Instead, an entire mainframe was dedicated to simulating the computer itself, which ran the actual flight software. When the IL released the code for rope manufacture, it also sent a copy to NASA for loading directly into the simulator. This third-party verification helped build confidence in the code and also revealed some errors.[41]

Simulator instructors worked long hours to get the crews ready. Simulator availability became a pacing item in crew training. Michael Collins called simulator training "the very heart and soul of the NASA system." One was not ready to fly until having proven one's skill on the simulator.[42] He might have added that the spirit of the simulator, buried inside code in the computer, would accompany the astronauts on their journeys.

The "lunar mission simulator," or LMS, ran off three mainframe computers, and included five tons of glass—lenses, mirrors, and projectors to accurately recreate the scenes of a landing. Astronauts could practice sighting landmarks, entering data into the computers, and simulate landings from about 12,000 feet to touchdown. As the pilots "flew," computer models of the LM's motions directed a small camera above a physical model of the lunar terrain, sixteen feet in diameter at 1:2,000 scale (actually, the terrain model was mounted upside down, so the camera looked up at it from below). Craftsmen made the models from Surveyor and Orbiter spacecraft images, and later updated them with data from the early Apollo missions. Their three-dimensional models recreated specific landing sites with a resolution of ten feet. A network of servos "flew" the camera over the diorama, right down to the point of landing. Accurately creating these scenes proved a particularly difficult problem—errors in one of the dioramas caused David Scott to become disoriented during his landing on Apollo 15.

A history of simulation technology in the space program has yet to be written, but it would show how the creation of virtual reality preceded, rather than responded to, the creation of real-time computer graphics. In fact, simulations during the Apollo program became so sophisticated that visual representation became their weakest link.[43] Yet the pilot's vision in the critical final moments was to be the central human function in the landings.

The Flying Simulator

The simulators provided high-fidelity training for most of the lunar landing, but astronauts found they performed less well for the final moments. Another earthly technique could actually simulate these final few feet, and it came neither from advanced computers nor from exotic visuals but from the high desert in California, and the heart of the test pilot culture at NASA's Flight Research Center (FRC) at Edwards Air Force Base. This innovation was the brainchild of Hubert Drake, Donald Bellman, and Gene Matranga—engineers with deep experience in the world of stability and control, including the X-15.

In 1960, the group began thinking about the prospect of a lunar landing. How should one design the trajectories? What controls should the pilots have? How were the pilots to be trained? An obvious solution was to have them train in helicopters—and indeed all the LM commanders had helicopter training—because helicopters, like the LM, pitch over to orient part of their thrust (from the rotor) to accelerate horizontally in a particular direction.[44] In the moon's gravity, however, one-sixth that of earth, a vehicle would need six times greater pitch angle than on earth to provide the equivalent horizontal acceleration. Put another way, the LM would fly like an extreme helicopter, its subtle pitches and yaws exaggerated by a factor of six. No helicopter could replicate that strange motion.

Drake and Matranga conceived of a free-flying vehicle, powered by a unique combination of jet and rocket engines with analog computers and an advanced control system to simulate these lunar dynamics. By coincidence, engineers at Bell Aerosystems also developed a proposal for a similar device. Drake and Matranga convinced Apollo management of the utility of their idea, and NASA FRC let a contract to work with

Bell to study it. Bell produced a design that became known as the Lunar Landing Research Vehicle, or LLRV.[45]

The LLRV was, like the LM, a non-aerodynamic craft, and just as interesting and strange. The LLRV had no wings, no sleek fuselage, no control surfaces, no giant rudder, just a clever analog computer that simulated the lunar environment. The vehicle consisted of a frame built around a jet engine on a giant gimbal, complemented by hydrogen peroxide reaction thrusters, the same type as on the X-15. The human jutted out from the front, while the computers rode far behind. The very physical structure of the LLRV depended on a literal balance of human and computer. Without both, it would not fly.

The LLRV flew in two distinct modes. With the jet engine fixed in its gimbal, it flew like an earthly, jet-powered vertical craft, much like a helicopter. The pilot, sitting forward of the center of gravity (as in the LM), throttled up the engine to take off vertically and ride up, controlling attitude with the reaction thrusters. Once he was a few hundred feet high, the pilot set up for a practice descent by pulling a lever and switching into lunar simulation mode. Now the analog computer controlled the gimbal and throttle of the jet engine, thrusting straight down exactly five-sixths of the vehicle's weight (which was changing every second due to fuel consumption), effectively creating one-sixth g. The lunar simulation mode allowed the LLRV to pitch, roll, and fly with exaggerated motions like the LM in lunar gravity, for up to eight minutes of total flight. The reaction thrusters controlled attitude, and two additional lift rockets simulated the LM's descent engine.

From the start, questions arose about whether a human could handle such a beast. Computer simulations showed that it was possible, but with a great deal of attention and skill (just the kind of exercise pilots loved). Limited fuel aboard also created time pressure on the pilot that enhanced the verisimilitude of the simulation, for which pilots liked the anatomical analogy of "pucker factor" (figure 8.11).

Because the LLRV's flying qualities originated in its control system rather than its aerodynamics, it required no sleek cowlings or streamlined shapes. Observers thought the craft was ugly, ungainly; some referred to it as "the flying bedstead" (actually the proper name of a similar English craft). But to call it ugly was simply to acknowledge the degree to which the technical aesthetics of flying machines had been shaped by the dynamics of earthly air. The odd shape reflected how the LLRV flew in a world of its own, for the computer inside created an artificial moon.

Nevertheless, the craft was quirky, delicate, and downright dangerous. As a research vehicle, it had less strenuous testing and system redundancy than the Apollo flight hardware. Like the LM, its very stability depended on properly functioning computers. Like the LM, it was a fly-by-wire craft, although analog and not digital. "It is believed that this is the only flying vehicle," wrote LLRV designer Walter Rusnak in 1964, "to which the control function has been entrusted entirely to electronics."[46] One engineer

Figure 8.11
Artist's image of the LLRV. Note the jet engine in the center on the large gimbal, aligned with the earth's gravity to cancel out five-sixths of it, and the small hydrogen peroxide thrusters, holding the vehicle at an unusual attitude. Note also the human operator in the front balanced in the back by the controls. (NASA Dryden/Bell Aerosystems.)

put it more bluntly: "if your electronics or avionics fail on that thing, that's just like your wing falling off."[47] The LLRV provided an early taste of entrusting pilots' lives to software, although the equivalent in the analog world were amplifiers, filters, and feedback loops, which could be just as scary, and seemingly random, when they acted up.

Hydrogen peroxide powered the LLRV's thrusters, ejecting steam for thrust, so while airborne the vehicle hissed, roared, and belched like a flying calliope. It first flew in October 1964. It took nearly a year of debugging, but eventually the LLRV managed to simulate a full lunar landing from about 500 feet up—the very critical moments when pilots thought they just might have to take over control.[48]

The trials attracted the attention of Robert "Cliff" Duncan, chief of guidance and control in Houston, who saw the LLRV as more than a research tool for studying the fundamentals of lunar landing. Duncan thought the LLRV might actually evaluate the LM's handling qualities, verify its controls and displays, and help train the LM commanders. As the Grumman engineers and IL programmers designed the LM's control systems, they faced the classic questions regarding handling qualities: How sensitive should the LM's controls be? How much power did the reaction thrusters need? How much of the systems should be automated?

LLRV flights soon explored the "rate control" feature (inherited from the X-15) that would prove central to flying the LM's "attitude hold" mode. Further tests verified and finetuned control authorities, dead-band settings, and controller sensitivities for the lunar mission. Engineers installed an LM-like instrument panel in the LLRV and a Doppler radar for velocity, and replaced its stick and pedal controls with a right-hand three-axis hand controller like that on the LM. They added an enclosed cockpit to simulate the LM's field of view, and a left-hand T-handle control to modulate rate of descent.

The LLRV team conducted more than two hundred flights for the Apollo program office to evaluate proposed trajectories for lunar landings. Then NASA funded an updated vehicle, the LLTV (Lunar Landing Training Vehicle), to allow the astronauts to practice their landings. NASA ordered three new LLTVs and modification of the existing two LLRVs to an LM-like configuration that more closely matched the actual lunar craft. By early 1967 the vehicles were operating in Houston at Ellington Air Force Base, not far from the Manned Spacecraft Center. Coordinating the conversion between NASA and Grumman was a committee that included Neil Armstrong and lunar landing designer Donald Cheatham.

The LLTV was a pilot's dream—a craft that was *difficult* to fly, that showed how much skill and experience would be required to land on the moon. "It gave me confidence that I knew what I was doing on the Moon," David Scott recalled in a recent interview. "I didn't have to think about things. I didn't have to consciously program myself to do things. I was automatic.... Hell of a challenge. A tough thing to fly."[49] Neil Armstrong noted at the time, "It is such a cotton picking unusual environment, so different from anything you've been in before, that you are continually amazed at how machines can fly like that."[50]

Nor did they always fly like that: Armstrong ejected from a failing vehicle in May 1968, narrowly escaping with his life. The accident investigation recommended changes to warnings, operations, and program management, but did not question the validity of the test program, despite the risk. Armstrong did not fly one again for a year, but in the month before leaving for the moon he flew eight times. "I think it does an excellent job of actually capturing the handling characteristics of the lunar module in the landing maneuver," he told the press.[53] The pilot's workload for the LLTV proved significantly higher than on the LM, because without a co-pilot he had to concentrate on both instruments and visual inputs, whereas in the real lunar landing the LMP took care of systems monitoring and calling out instrument readings. All prime and backup commanders of lunar landing missions practiced on the LLTV at Ellington, in hundreds of flights.

Realism, Risk, and Confidence

Despite the benefits, Apollo management thought the LLTVs might be too risky to continue. Painfully recalling the Apollo 1 fire, the last thing anyone wanted was for a highly qualified astronaut to lose his life in a training device. After Apollo 12, in January 1970, Robert Gilruth brought together a flight-readiness review board that included the two commanders who had landed on the moon, Neil Armstrong and Pete Conrad, and a variety of Apollo engineers and astronauts, including Chris Kraft, Max Faget, and Jim McDivitt. There had been two crashes so far: Armstrong's and another that also ended safely after a nearly fatal close call. What was the value, Gilruth asked, of the LLTV in training for a lunar landing? Should it be shut down?

The conversation not only concerned the trade-offs of risk and training with the LLTV but also provides a rare window into principal Apollo engineers' and managers' thinking about the optimum human role in lunar landing. They did not all agree.

Armstrong and Conrad were unequivocal. "Were I to go back to the moon again on another flight," Conrad asserted, "I personally would want to fly the LLTV again as close to flight time as practical." He felt the computerized lunar mission simulator was not adequate for training for the last 200 feet of the landing, nor was a large gantry-frame device built at Langley. By contrast, the LLTV gave him a good intuition for pitch attitude, which was difficult to perceive on the LM. For his Apollo 12 landing, Conrad had to make some rather radical maneuvers, pitching the LM over nearly forty degrees in a steep descent, but the confidence he developed with the LLTV allowed him to fly with no concerns. "We are banking our whole program on a fellow not making a mistake on his first landing," Conrad emphasized, and the LLTV helped a pilot with a valuable but immeasurable quality: confidence.

Armstrong, as usual, chimed in with fewer words, but supported Conrad's conclusion. He recalled the LLTV's value in helping him perceive subtle variations in lateral velocities, and in imposing the discipline of time pressure. The LLTV helped him learn how to select alternate landing areas. During training, he said, "You sort of play the game with yourself, as you fly into a touchdown area and you say no, I don't want to land there—I want to land over there." That game, related Armstrong, provided "the confidence in your own knowledge that you can fly the job in." A landing accident would be catastrophic to the entire Apollo program, and the LLTV was like an insurance policy, he said, noting, "my own conclusion is that we still can't afford not to insure against this particular catastrophe."[52]

Astronauts always supported the LLTV, and it supported them: showing lunar landing to be a difficult, risky endeavor of machine control that could be mastered by confidence, experience, and skill. Several Apollo commanders actually mentioned the LLTV training on the radio during their lunar landings. Nearly all discussed it in postflight briefings as support during the last critical seconds when they took over semiutomatic control.

In both the LLRV and LLTV vehicles, computers created the conditions that made it possible to fly at all, raising the question: if the task could be automated to that degree, why not automate it all the way? If only human perception could identify and confirm a suitable landing site that the maps could not reveal, why couldn't the humans direct the automatic system to land there? Some pilots, such as Pete Conrad, landed "blind," solely by reference to instruments, so why not entrust control to a computer? Apollo astronauts tended to conflate visual perception and manual control.

For these reasons, Flight Director Chris Kraft recommended installing an automated landing program in the LM. His idea was that once the commander had selected the landing site and flown the vehicle over it, the computer would once again take over, direct the vehicle to hover, and gently set it down on the surface under automatic control. The astronauts disliked the idea, but the LLTV review board recommended that the astronaut office study an automated landing capability.[53] After Apollo 12, the IL added the feature to P66, calling it "velocity-nulling guidance," that would automatically place the LM in a hover and descend it under automatic control. It was never used.

For the astronauts, an Apollo launch was a ride atop a fiery automaton, as they watched dials and indicators, poised for an abort while the rocket executed its sequence of staging, steering, and burns. They spent their trip to the moon largely doing systems monitoring, maintenance, and housekeeping, complemented by star sightings to back up navigation from earth. Major rocket burns were calculated in advance by computer and the ground controllers, directed and controlled by servos. Only when approaching the lunar surface would pilots do what they did best: fly and land a delicate, powerful craft. Yet even there, only the last minute or two of the ten-minute descent would be under manual control. Here "manual" meant jogging a stick that would provide new setpoints to computer-controlled feedback loops, either for attitude holds or descent rates. Yet this semi-automatic mode proved sufficient for the pilots and NASA to feel comfortable that the landing was made under human control, with human judgment and skill.

None doubted that the human eye could best assess a landing site and determine the most suitable place to put down the LM. Hence the final trajectory centered on a moment of vision, of perception, of revealing, as the LM pitched over and the commander saw his target for the first time. But what was the linkage between visual perception and manual control? Did choosing the landing site necessarily involve hand-flying?

Real landings, with skilled but fallible people flying magnificent but imperfect machines in less than ideal circumstances, would begin to answer these questions.

9 "Pregnant with alarm": Apollo 11

Armstrong, sitting in the commander's seat ... is a man who is not only a machine himself in the links of these networks ... a man somewhat more than a pilot, somewhat more indeed than a superpilot, is in fact a veritable high priest of the forces of society and scientific history concentrated in that mini-cathedral, a general of the forces of technology ... of the vast multibillion dollar technological bands which belted the very economy of the nation ... the methods of the hospital mixed with the methods of the football team.

—Norman Mailer, *Of a Fire on the Moon*

Apollo 11 was a test flight whose major goal was simply to prove the feasibility of lunar landing with the Apollo system. Most aspects of the flight to the moon had been tried before. Apollo 10 had gone right down to 50,000 feet and then returned home, only a PDI burn remaining between it and the lunar surface. Yet from that point downward everything was new on Apollo 11—accomplished many times before, but only in simulation. The Apollo 11 landing was the climax of the development program, of Apollo's methods of integrating the efforts of diverse organizations into a flight system. Not least of those components were the people, their computers, and their software.[1]

Go for Powered Descent

It is July 20, 1969, about 5 p.m. Houston time. Less than two hours before, the LM had separated from the command and service module as the two flew in similar, safe orbits around the moon, preparing for the critical descent. Then the LM initiated its DOI burn, for "descent orbit insertion" around the far side of the moon, to bring Neil Armstrong and Buzz Aldrin down from a circular sixty-mile orbit to an elliptical one, sixty by ten miles. The crew carefully monitored the burn—ready to cut it off manually if necessary, to abort if it burned even slightly longer than planned. Michael Collins, in the CSM Columbia, was also watching, tracking by radio his range to the LM. Everything was multiple, redundant, "man rated." At every turn, the astronauts were "in the loop." From the alignment of the inertial platform to the PRO button, the final check was to be executed by the people with the most to lose.

The burn went well. The first report of success heard on the ground was not from the LM but from Collins, alone in the CSM, as he emerged first from behind the moon: "Listen, babe. Everything's going just swimmingly. Beautiful." A few minutes later, the LM, nicknamed Eagle, emerged as well, also reporting the successful burn.

Now the comparatively pure orbital mechanics would begin to intersect the uncertainties and roughness of the moon. At first Armstrong and Aldrin came in facing down, feet first. This being the first time, they wanted to doublecheck the safety factors by eye. As their elliptical orbit carried them close to the surface, Armstrong and Aldrin sighted landmarks on the ground and timed how long it took them to pass across their window—in tandem with paper charts, these observations provided a way to measure altitude. The check indicated a perilune of about 53,000 feet over the moon's surface. From the CSM, Collins's range to the LM allowed the computer to calculate an independent altitude, which indicated about 50,000 feet, while the PNGS indicated 49,971 feet (later calculations showed true altitude at 51,000 feet). Close enough for confidence.

Houston began having problems receiving signals from the LM's high-gain antenna, problems that would continue throughout the coming descent. The antenna provided high-bandwidth: clear voices and many bits of data—but it had to point straight at earth to work. A servo that directed the antenna was having trouble tracking the earth, because in the face-down position it was obstructed by another part of the spacecraft's structure (also its radio transmissions were likely bouncing off the moon, creating an interfering signal). Houston asked the crew to move, to allow a clearer path for better data, in effect asking the human operators to compensate for a problem in the automatic servo: "Eagle, Houston. We recommend you yaw 10 [degrees] right. It will help us on the high gain signal strength. Over." They accomplished the yaw, although at some cost to their view out the window.

A few minutes later: "Eagle, Houston. If you read, you're GO for powered descent. Over," drawled Capsule Communicator (CAPCOM) Charlie Duke (the astronaut on the ground who communicated with the astronauts on board), in his smooth, Southern accent.

Aldrin keyed in Verb 37 Noun 63, enabling program P63 for powered descent. In a few minutes the computer would run the subroutine BURNBABY, which calculated and controlled the descent engine firing to begin guiding the LM toward the moon's surface. Verb 06 Noun 62 appeared on the DSKY, and the second display row began counting down to the burn. When it was almost time, the display flashed. The computer would not light the engine until Aldrin pressed PRO. He promptly hit the button, giving his human okay to the automatic firing (figure 9.1).

The engine came up to 10 percent thrust, barely noticeable, allowing the descent engine to trim its gimbals so its thrust would go right through the LM's center of gravity.

PDI THRU TD +3 MIN

(DES $O_2 \approx$ 91

	(OHW) θ	TFI	VI	(Ḣ MAX) -HDOT	(Δ H) H	DPS
ENG ARM - DES	(7)	-0:35				
IGNITION	286	0:00	5560.0	4.3	48800	95
DES ENG CMD OVRD - ON		0:05				
TTCA - UP @ 26 SEC	285	0:30	5500.0	4.8	48700	95
V21 N01 E	(10)					
1252 E						
2462 E	279	1:00	5200.0	14.0	48400	89
WAIT 3 SEC	275	1:30	4900.0	21.0	47900	84
TTCA - 10%						
	(16)					
12.9						
LPD ALT CHK -10.7	271	2:00	4600.0	30.0	47100	79
TM - H/H DOT 8.4	268	2:30	4300.0	40.0	46100	74
	(19)					
-9 +18	267	3:00	4000.0	53.0	44700	68
LPD POS CHK	264	3:30	3600.0	67.0	42900	63
YAW RT 174°					(17500)	
SBD P -14 Y +12	82	4:00	3300.0	80.0	40700	58
✓ ED BATTS	80	4:30	2900.0	95.0	38000	53
V16 N68 E					(17500)	
V57E - ENABLE LR	78	5:00	2600.0	109.0	34800	47
	76	5:30	2200.0	122.0	31200	42
					(17500)	
	72	6:00	1800.0	123.0	27500	37
					(14000)	
THROTTLE DOWN	68	6:30	1400.0	114.0	24300	32
				(427.0)	(10500)	
EVAL MAN CONT	63	7:00	1200.0	143.0	20200	29
				(368.0)	(8750)	
	60	7:30	1000.0	157.0	16200	25

V16 N68 E

P64

LDG ANT - HOVER

P64 + 15 SEC:
NO THROTTLE DN-ABORT

SBD P +17 Y -14

223+00020 @ 2 K

PGNS MODE CONT-
ATT HOLD

P65
P66

X-PNTR-LO MULT

TOUCHDOWN

ENG STOP - PUSH
ACA - OUT OF DETENT
MODE CONTROL (BOTH) - AUTO
DES ENG CMD OVRD - OFF
ENG ARM - OFF
413+1
~~223+00001~~ 414+2
ASC FEED 2 (2) - CLOSE

θ	TFI	VI		(Ḣ MAX) -HDOT	H	HOR	DPS
57	8:00	700.0		157.0	11500	60	22
47	8:30	500.0		139.0	7000	49	19
		TR-LPD		(186.0)			
43	8:44	99	60	112.0	5000	45	18
				(163.0)			
40	8:54	99	56	95.0	4000	42	17
				(136.0)			
36	9:06	98	52	76.0	3000	38	16
				(104.0)			
31	9:22	82	47	55.0	2000	33	15
				(63.0)			
24	9:44	61	39	32.0	1000	25	13
				(35.0)			
18	10:06	40	33	16.0	500	18	12
				(29.0)			
16	10:14	32	31	12.0	400	16	11
				(21.0)			
13	10:26	20	31	7.0	300	13	10
				(12.0)			
9	10:48	0	43	3.0	200	9	9
2	11:26			3.0	100	2	7
2	11:42			3.0	50	2	6

NO STAY

ABORT STAGE - PUSH
ENG ARM - ASC
ENG STOP - RESET
ENG START - PUSH
MODE CONTROL (2) - AUTO

Figure 9.1

Apollo 11 checklist from PDI to landing. See figure 8.7 for explication of format. (Apollo 11 Flight Data File, LM Timeline Book, Rev. "N." July 12, 1969, 9. Grumman Aircraft Corporation Archives. Courtesy of Paul Fjeld.)

About thirty seconds later, the engine throttled up to nearly full, to begin slowing the LM out of orbit. As Aldrin calmly ran through his checklists, Houston was fussing with the telemetry, asking the LM crew to switch antennas and to jimmy the tracking modes.

Aldrin's main job was to monitor the two guidance systems, AGS and PNGS, to make sure they agreed. He stared at two sets of numerical displays, comparing them in his head, and would call out if there was a problem. A simple enough job for a Ph.D. from MIT, yet Aldrin remembered his intensity: "There's a focusing of an individual's concentration and level of attention that is at the exclusion of a lot of other things. It's a kind of gun-barrel vision." He still had a great deal of trouble aligning the antennas and was trying to track the earth manually.

Despite the minor nature of the antenna problems, they added a critical piece of workload at a busy time, and made unreliable the voice and data links with Houston. After countless hours practicing in simulators, "I can't ever recall having them exercise us in the simulator with the uncertainty of intermittent comm," Aldrin recalled. "It was distracting. In simulator training, either things were working normal, or there was something going wrong. It would have a degree of polarity to it. It either was or wasn't. But the uncertainty, particularly in communications, that was exhibited here, was frustrating. You didn't know where you were—whether you were on your own, or whether you were still under the close supervision of ground control. And that sort of reality is rarely simulated in training." He later said he could have continued slewing the antenna manually for the entire landing, but it would have cost his attention to all other tasks. Aldrin considered it "a forerunner of the computer alarms which further distracted me from my task, which was one of monitoring the computers and other instruments."

Pitch Over

Now Armstrong and Aldrin were traveling feet first, face down, in order to visually verify their altitude. This was the only landing to come in face down this way, a safety feature that the astronauts had suggested during training.[2]

Then came the first indication they'd be facing something other than a perfect landing. A few minutes after the PDI burn, Armstrong recognized a landmark out the window, a crater called Maskelyne W. But it came across his window two or three seconds early. "Okay, we went by the three-minute point early. We're long." Aldrin had his head inside the cockpit, looking at the AGS and PNGS. Armstrong, again, to Houston: "Our position checks down range show us to be a little long." Flight controllers confirmed Armstrong's reading.

Why was Eagle flying long? They were going just a little too fast. A number of problems could have pushed them ahead. The lunar gravity map was incomplete and could not yet account for the strange concentrations of mass, or "mascons" that peppered the moon and would subtly pull the spacecraft and alter the trajectory. Also, when the LM separated from the command module, somehow it picked up a little extra velocity. Flight Director Gene Kranz believed it came from a little extra air pressure caught in the docking tunnel, or perhaps it resulted from the LM's maneuvers around the CSM.[3] Considering they had just come a quarter of a million miles, we can marvel they would even be aware of such a discrepancy, let alone be concerned about it. But the landing site had been precisely placed, and this extra velocity would cause them to miss it by several miles. Armstrong would miss the landmarks he had carefully memorized.

Still, everything else was going well. On the ground's intercom loop, Gene Kranz queried his team: "Go to continue powered descent?" The engineers yelled their affirmations with nervous enthusiasm: "FIDO: Go." "Guidance: Go." "EECOM: Go." "Surgeon: Go." To save time on the voice loop, Kranz just acknowledged "Rog."

Four minutes into the burn.

Now Armstrong yawed over—using his right-hand stick, he rotated the spacecraft around the axis of the descent engine, bringing the two astronauts' eyes to face the sky. They could no longer see the moon they were about to touch; now they were looking up at the earth, their backs to the lunar surface. Another problem: the rotation was moving too slowly, taking longer than planned. Armstrong realized the switch that determined the rotation rate was in the wrong position. He switched it and the LM quickly yawed around.[4]

The landing radar was on the other side of the LM from the windows, and the yaw over pointed it down toward the surface. The slow rotation meant the radar took longer than expected to lock on the ground. But it did lock on, at 37,000 feet, and thus resolved one of the major uncertainties in the program.

Aldrin keyed in Verb 16 Noun 68, to display the DELTAH, asking, in effect, "how different was the altitude reading from the radar from the one calculated inside the computer?" The computer had pulled its position literally from outer space, from its inertial positions and ground updates, whereas the landing radar was providing ground truth—the first time these ethereal calculations were checked against hard moon rocks. The state estimator in the computer ate this sort of data for lunch. As long as the DELTAH was not too large, the computer could incorporate it and adjust its solution. The display read "–2900" to Aldrin. The two solutions were less than three thousand feet apart, within acceptable limits according to the paper chart.

"A computer pregnant with alarm"

This point had a manual check as well. The computer would not accept the radar data without Aldrin's OK. He was about to tell the computer to incorporate the radar altitude into its solutions when the unexpected occurred.

Kranz was asking his guidance controller, "Is he accepting it [the radar data]?"

"Program alarm," Armstrong called, a touch of urgency in his voice.

Aldrin quickly keyed in Verb 90 Noun 50, asking the computer for the nature of the alarm.

Armstrong read aloud the display: "It's a 1202."

Aldrin approved the computer's incorporation of the altimeter data with Verb 57 ENTER (the new solution converged to within one hundred feet in thirty seconds). Then, Armstrong to Houston: "Give us a reading on the 1202 Program Alarm."

CAPCOM Charlie Duke realized that the communications problems were mostly no more than annoying, but that a computer problem could be "a showstopper."[5]

On the flight controllers' audio loop, the voice of Steve Bales called, "It's the same thing we had," referring to a recent simulation.

"If doesn't reoccur, we'll be go," Bales advised. But it did recur.

On the ground, engineers scrambled. At the IL, engineers were following the flights in real-time, sitting in an MIT classroom in Cambridge listening to the flight controllers and the LM on a "squawk box." The 1202 reading told the IL engineers it was an "executive overload." The computer was falling behind in its tasks; something was stealing processing cycles. ("Executive Overload!" Norman Mailer observed. "What a name! One thinks of seepage on the corporation president's bathroom floor."[6]) But the engineers were too far away to be of any help. "I had never seen or heard one [a 1202 alarm] in all of our pre-flight testing," Fred Martin of IL recalled.

In Houston, flight controller Steve Bales asked his back room team for some help. There, engineer Jack Garman cleared, stating "We're go on that alarm." Russ Larson, an IL employee sitting next to Garman, was too nervous and engrossed to speak. He just flipped up his thumb. Bales passed on the word to Kranz.

Within ten seconds, Duke reported to the LM: "We're go on that alarm." No problem, not to worry.

Thirty seconds later, another alarm. Another "go" from Houston.

Another thirty seconds and the engine throttled down the PDI burn, an important moment, for it meant the computer was still running. "Ah! Throttle down... better than the simulator," commented Aldrin. "Throttle down on time!" Armstrong excitedly added (Don Eyles points out that in the official transcript of communications, these are the only exclamation points).[7] One more worrisome problem that might have kept them from the moon was now behind them.

Armstrong knew he was a little long, but the computer throttled down the engine exactly on time. He realized the computer was "a little bit confused at what our down range position was. Had it known where it was, it would have throttled down later (to kill a little velocity)." The human operator understood the situation better than the computer, but the computer was driving and there was nothing he could do about it (a problem fixed for Apollo 12).

At seven minutes into PDI, on the ground, engineers were scrambling to understand the source of the program alarms. One controller recognized that the Verb 16 to monitor a variable in real-time on the DSKY was fairly processor-intensive, and that the alarm might have arisen when Aldrin commanded the computer to monitor DELTAH. "Noun 68 well may be the problem here," he told Kranz, "and we can monitor DELTAH," replacing the LM's function with human backup on the ground. Kranz asked if they should ask the astronauts not to use Noun 68, but the back room thought it unwise. Indeed, the extra computation load created by Aldrin's request for a Noun 68 was pushing the computer just over its overload margin.

And the alarms were still ringing. Armstrong did not move to abort, however, because the LM still seemed to be responding to his commands: "As long as everything was going well and looked right, the engine was operating right, I had control, and we weren't getting into any unusual attitudes or things that looked like they were out of place, I would be in favor of continuing, no matter what the computer was complaining about."[8] Armstrong explained himself as a mechanism: "In simulations we have a large number of failures and we are usually spring-loaded to the abort position. And this case in the real flight, we are spring-loaded to the land position."[9] As a pilot, if the craft was still flying, he was going to fly it.

All this took up precious time, and the controllers began to worry about fuel. "Descent 2 fuel critical," they called, but Charlie Duke just passed up "Descent 2 fuel," saying tank number two would be the first to go critical for a low level and should be monitored. One of the controllers noted about Duke that "he didn't want to say critical" up to the astronauts, who knew they were nearing trouble.

Still, Armstrong needed to make sure the LM *was* flying right. In the following seconds the LM's internal tape recorded him saying, almost to himself, "Okay.... No flags. RCS is good. DPS is good. Pressure.... Okay." He was checking the systems, trying to determine if anything obvious was wrong besides the computer alarm (telemetry recorded pitch changes in the vehicle as he felt it out).

Two minutes later, just before reaching the high gate, at 7,000 feet, the mode register on the DSKY showed 64, indicating that the computer had switched to P64, the approach phase. On another register, a new number appeared that was the LPD designator, the angle at which Armstrong should look out the window to see the landing site.

The LM began its computer controlled pitch over, starting to bring the astronauts face forward, feet down. Armstrong should have been looking through the window, scanning for the landing site, expecting the thrill of recognition. Instead he was looking inside at the instruments, trying to deal with the alarm.

Again, Kranz polled his controllers: "Go/no go for landing." And again they responded with "Go!" in an excited, nervous series.

Armstrong flipped a switch, moving the LM out of AUTO and into ATTITUDE HOLD. He jogged the joystick, testing the LM's handling qualities. "Manual attitude controller is good," he called down to Houston. According to the checklist (EVAL MAN CONT, for "evaluate manual control"), Armstrong was supposed to have done this earlier so he could focus on evaluating the landing site, but he may have delayed it due to the computer alarms. He clicked the LM back into AUTO, to continue riding the computer down.

And once again, Kranz polled his controllers on "Go/no go" for landing.

They replied: "FIDO: Go." "Guidance: Go." "EECOM: GO." "Surgeon: Go."

Kranz answered "Rog." Duke notified the *Eagle*: "Eagle, Houston. You're GO for landing. Over." Like an airplane, the LM had been cleared to land.

Then at 3,000 feet, again came the alert: "program alarm."

Aldrin: "1201."

Armstrong: "1201."

Both read out the number; both were distracted by the alarm.

One second later, Houston replied: "Roger. 1201 alarm. We're go. Same type. We're go." In the Houston back room, Jack Garman was amazed to hear his "same type" observation repeated by the CAPCOM over the transmitter, and then repeated by Aldrin. Garman's judgment had just made a half-million-mile journey to another person and back.

"We just passed low gate" on the ground loop.

Then came another 1201 Program Alarm, then another 1202. Five alarms in all. Armstrong's heart rate rose from 120 to 150 beats per minute.

In an ideal landing, in these moments after pitch over the commander would sight through the LPD, identify the landing site, and redesignate to a new site, if necessary. "This was the area we completely failed," Armstrong told the Society of Experimental Test Pilots that fall.[10] "The concern here was not with the landing area we were getting into," Armstrong recounted in the technical debrief, "but, rather, whether we could continue at all...consequently, our attention was directed toward clearing the program alarms, keeping the machine flying, and assuring ourselves that control was adequate to continue without requiring an abort. Most of the attention was directed inside the cockpit during this time period and, in my view, this would account for our inability to study the landing site and final landing location during the final descent. It wasn't until we got below 2,000 feet that we were actually able to look out and view the landing area."

Finally, the ground controllers realized it was out of their hands, up to the men in the LM. Kranz silenced his team: "I think we'd better be quiet...the only callouts from now on will be fuel."

Two and a half minutes after P64, Armstrong directed Aldrin: "Give me an LPD." Aldrin told him to look down through 47 degrees.

Finally, Armstrong looked out the window. He saw potential disaster. The landing area, already well beyond the planned point, included, he said, "a large rocky crater surrounded with the large boulder field with very large rocks covering a high percentage of the surface." (Armstrong later said the boulders were ten feet across). He was tempted to land short because of the scientific appeal of landing so near a crater. But, he noted, "continuing to monitor the LPD, it became obvious that I could not stop short enough to find a safe landing area," adding that "it's an old rule, when in doubt, land long, and I did."

The LM continued its descent. Armstrong: "That's not a bad looking area. . . . What's LPD?"

Aldrin replied: "35 degrees. 35 degrees. 750. Coming down at 23." Then: "700 feet, 21 down, 33 degrees."

Armstrong redesignated the landing site by jogging the joystick, but probably it was an inadvertent move. He likely had forgotten to switch into attitude hold mode. Four seconds later, he made that switch, which nullified the LPD, although the descent rate remained under computer control.[11]

Armstrong: "Pretty rocky area."

Aldrin: "600 feet, down at 19."

Armstrong: "I'm going to—"

Armstrong toggled a switch with his left hand, enabling P66. At about four hundred feet, he was now flying in the autopilot's rate-of-descent mode, still with attitude hold, and flew the LM past the crater. Armstrong commanded the LM's attitude with his right hand, while the computer controlled the descent at a fixed rate. In his left hand Armstrong had a rate-of-descent switch that he could click up for one-foot-per-second slower, or down for one-foot-per-second faster. For the first twenty seconds he did not change the pitch of the vehicle, but clicked the rate-of-descent button eight times to slow the descent. He pitched back to slow the LM's forward velocity, then forward toward the west to get to the other side of the crater; "I had tipped it over like a helicopter," he noted, even gently climbing at one point.

The move was quite fast, causing Aldrin to warn: "We're pegged on horizontal velocity." The LM was moving so fast across the surface (greater than twenty feet per second) that the needle on the crosspointer indicator pegged against its stop.

"300 feet, down 3 1/2, 47 forward." Armstrong was still moving across the surface at more than thirty miles per hour.

Floyd Bennett was following the action from the trajectory analysis room at Houston. "We kept watching his forward velocity and his altitude rate . . . you can't land at those speeds. And I said, 'What is he doing?' . . . He didn't have time to tell us it was a rock field out there."[12] All this maneuvering took precious additional seconds, running the fuel down to critical.

The ground called up: "Low level." Two minutes left.

Norman Mailer described the LM as "skittering like a water bug debating which pad it will light on."[13] Finally, Armstrong found a spot he liked, and "nulled the rates," that is, slowed the vehicle's horizontal velocity, a delicate task given the lack of atmosphere to damp the motions.

Armstrong later described his moves as "over controlling" and even "a little spastic in final approach," because he was "confused" about his lateral velocity. Also, he said, "my visual perception of both altitude and altitude rate was not as good as I thought it was going to be." As the engine blast began kicking up dust on the surface, it caused visual noise that made it hard to discern the ground's motion beneath (Armstrong likened it to landing an airplane through fog, "however, all this fog was moving at a great rate which was a little bit confusing").

As the LM continued its approach, Aldrin noted: "Kicking up some dust."

For Jack Garman in Houston, this moment brought the reality of the moon landing home. Until now, much of the script, even the program alarms, had been practiced in simulations. But Aldrin had never before called out a detail so vivid as lunar dust.

As the LM descended in a low hover, probes hanging below the landing pads touched down on the moon. The blue "lunar contact" light came on. Armstrong later said he never saw the light, but he shut off the engine. In the time it took the thrust to die down, the LM descended the last few feet with a gentle thud—a perfect landing, about a minute later than planned. Aldrin called out commands as Armstrong shut down the systems and told the computers to stop flying: "Mode control: both auto. Descent engine command override: off. 413 is in."

He continued: "Engine stop. ACA out of detent." He gently jogged his control stick to signal the computer to hold this new attitude, which would prevent any further RCS firing. He then chimed in with his famous summation: "Houston, Tranquility Base here. The Eagle has landed."

Charlie Duke replied, "We copy you on the ground. You got a bunch of guys about to turn blue. We're breathing again. Thanks a lot."

Duke and his colleagues had had good reason to hold their breath: touchdown was forty to fifty seconds before the fuel would have been depleted, twenty to thirty seconds before the crew would have been forced to make a land-or-abort decision. Armstrong had had the LM under control for almost two and a half minutes. His landing was 1,100 feet from the targeted automatic landing spot (and more than four miles down range from the planned landing site). An analysis by the Mission Planning and Analysis Division noted that "total flight time was extended an additional 40 seconds from the normal time for an automatic landing"[14] (figure 9.2).

After the landing, Armstrong acknowledged, "I don't think I did a very good job of flying the vehicle smoothly" in the last few seconds. In a speech to the SETP he admitted, "Well, I was just absolutely adamant about my God-given right to be wishy-washy about where I was going to land," choosing one spot and then rejecting it when he got closer, then choosing another. Floyd Bennett wanted to put this comment into a paper on the landings, but NASA disallowed it, claiming Armstrong's comments were not technical enough.[15] Armstrong's authorized biographer takes a

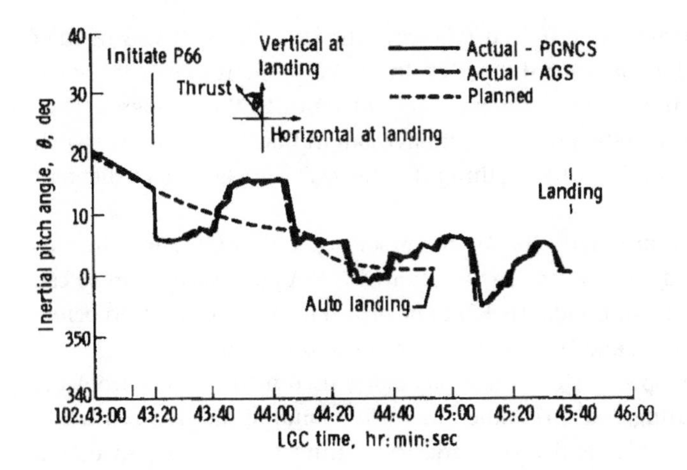

Figure 9.2
Pitch profile for landing phase of Apollo 11 and comparison with automated landing (Bennett, "Apollo Experience Report: Mission Planning for Lunar Module Descent and Ascent," 24.)

rosier view than the astronaut himself did, noting that "bringing the LM down became a matter of Neil's piloting abilities, pure and simple."[16]

Chasing the Problem

Moments after the touchdown, the phone at the IL began ringing like a 1202 Program Alarm. NASA was calling and wanted to know what went wrong, demanding an explanation and a fix before the LM lifted off the surface in a few hours. The IL engineers understood that their computer was not operating at full capacity, but they did not understand why.

They went to their simulators, in "a frantic session," trying to recreate the problem. "We worked all night and time was running short." Fred Martin recalled, "Our NASA buddies called us every 15–30 minutes anticipating, demanding a solution. We had to find it. We re-covered old ground, new ground, brainstorms, crazy ideas, anything."[17] Finally, George Silver, an IL engineer with a great deal of experience in the LM simulators, arrived at the lab. He had monitored the landing at home, heard the alarms, and rushed into work to point out that he'd seen this problem caused by the rendezvous radar during a simulation when the radar was on during a landing and in the "AUTO" position.

Fred Martin ran upstairs and pored through the telemetry data with his engineers—sure enough, the radar was on during the landing, though the IL thought it should have been off. It had to have been turned on by an astronaut, but why would he do that? The rendezvous radar was not used during landing. Checking the procedures, it turned out that the astronauts were indeed following the procedure—it was in their instructions (actually, Aldrin had put the switch in SLEW position, when it should have been in AUTO, but this would have made no difference to the problem). They had learned to do it this way in the procedures trainer. But in that simulator the switch was just a dummy, not connected to anything. In the real vehicle, it had different effects.

By now Armstrong and Aldrin had explored the moon for a few hours, and their ascent countdown was already underway. The IL phoned NASA and asked them to call the LM and ask for the rendezvous radar switch to be placed in the LGC position before liftoff. The problem was solved and the program alarms did not recur.

Why did the procedures specify the rendezvous radar, used in the ascent from the moon, to be turned on during descent? Some time before landing Aldrin asked the IL engineers if he could leave the rendezvous radar on during the landing, so that it would already be running if there was an abort and they needed to return to the CSM. IL engineer Larsen approved this step and changed the checklist. Aldrin, of course, was a rendezvous expert and he wanted to be prepared in case of an abort.

But hidden in the computer's interface to the rendezvous radar lay a problem.[18] The radar had three modes: SLEW, AUTO, and LGC. In the first two modes, the crew operated the antenna. In SLEW, they could manually direct it, and then switch to AUTO to automatically track the signal on the CSM during rendezvous. These modes operated separately from the guidance computer and displayed their data on the cockpit displays. In LGC mode, the data was provided to the software, which incorporated range and range rate as well as antenna angles into the calculations for the rendezvous guidance. Crew procedures called for the switch to be in the AUTO position during the landings, which would hold it still. Neither mode should have had any impact on the computer.

The trouble was that the rendezvous radar and the rest of the guidance system had different electrical power supplies. They both ran on alternating current (AC) of the same frequency, but had different phases (i.e., their alternating sine waves were out of sync). When the change in the switch procedure was tested in the lab, technicians connected both to the same power supply, which caused them to run *in* phase, even though they would be *out* of phase in the spacecraft. According to George Silver via Don Eyles, the problem had been recognized early on but never corrected.

On Apollo 11, the power supplies on the LM fell into a particularly unfortunate phase angle. Hence the computer and the radar were not in sync, causing the angle counters on the rendezvous radar to constantly increment or decrement in response to random electrical noise, sending nearly the maximum rate of data to the computer. The computer struggled to increment or decrement its counters for tracking the radar angles, which used up about 15 percent of its processing time.[19]

The computer had been designed with 15 percent "overhead" in processing power. That is, with all the processes running full blast, the computer would be at 85 percent capacity. But the rendezvous radar was generating so much spurious data that it ate up more than this 15 percent, causing the computer to overload.

Fortunately, the computer had a graceful way of responding to this situation. What the computer did next was not a bug in the program, but a manifestation of robustness in the software design. IL engineers were very proud of their "asynchronous executive," and when the overloads came up, this feature allowed the computer to drop low-priority tasks, meaning that basic housekeeping tasks and the DSKY display were the first to go. Indeed, Aldrin's request for Noun 68 (DELTAH) was dropped by the computer and the display returned to P63. The display froze up for short periods as well. Still, the mission-critical items—guidance equations, throttle control, attitude servos—kept running, which was why Armstrong could still feel the machine responding to his inputs.

In response to these overflows, not only was the computer generating alarms, but it was also restarting. Fortunately, these were not the cumbersome reboots required by today's desktop computers. Rather, "restart protection," the difficult new requirement imposed on the software team in 1968, was allowing the computer to restart nearly instantaneously. When the 1201 and 1202 alarms came up, the computer called a BAILOUT subroutine (ironic for a craft without parachutes), and simply restarted itself. Because of its restart protection, the computer could flush incomplete and lower-priority jobs and pick up right where it had left off. Armstrong could not even feel the hiccups.

When the computer shifted into P64 as the LM pitched over, the computer's capacity margin became even more critical, and hence the alarms continued and even increased in frequency. When Armstrong switched into P66 and took over manual control, the computing load lightened, because the computer was no longer calculating the landing point, and the alarms disappeared.

These explanations all developed in analysis after the landing. How, then, did the ground crews know to make their snap decision not to call an abort?

In the months before the Apollo 11 mission, the crew had rehearsed the landing process from the LM simulator, in contact with the flight controllers and numerous other aspects of the network. As the basic scripts were perfected, engineers began inserting a series of unlikely events to test reactions. Jack Garman, a young engineer on the ground, helped develop computer errors that would probably never happen. These included a program alarm reflecting an overload on the LM computer.

During one such simulation, just a month or two before Apollo 11's launch, flight controller Steve Bales called an abort in response to a program alarm, even though a landing could have proceeded successfully. The young controllers weren't too troubled by the incident, just one mistake among many. But NASA management was concerned, for "it scared everybody to death." Calling a mistaken abort was almost as bad as missing a real one. As Garmin recalls, "Kranz called a meeting [to] go through every program alarm, write down what could happen, what we should do about it," which the controllers subsequently did.[20]

Jack Garman made himself a handwritten "cheat sheet," which he kept at his control console in the back room under a piece of plexiglass. On the left side was a list of program errors, on the right side a series of problems and possible responses. In the section of his chart corresponding to "1201-1211 PGNCS," the alarm that appeared on Apollo 11, he wrote the following notes in the right hand column:

> PGNCS condition unknown, DKSY may be locked up, duty cycle may be up to point of missing some functions (nav. last to die) switch to AGS (follow ERR needles) may help (reduces PGNCS duty cycle signif.).

In these few words, Garman had the critical information that would allow him to diagnose the problem in real-time (figure 9.3). This was all news to Aldrin: "I was the kind of the systems guy in the LM and I was not made aware of that. And it seems as

APPLICABLE TO: IN DESCENT, AVERAGE-G ON

ALARM CODE	TYPE	PRE-MANUAL CAPABILITY	MANUAL CAPABILITY
0105 MK ROUT. BUSY 00430 CAN'T INTG. S.V. 01103 CCSHOLE-PROG. BUG 01204 NEG. WAITLIST 01206 DSKY, TWO USERS 01302 NEG. SQ. ROOT 01501 DSKY, PROG. BAD 01502 DSKY, PROG. BUG 00607 LAMB. NO SOL.N	POODOO " " " " " " " "	PGNCS GUID. LOST,. PGNCS/AGS ABRT/ABRT STG (decision how on current rules) (NO LR DATA)	PGNCS GUIDANCE NO/GO (PGNCS GO for TAPE METERS, CROSS-POINTERS, CONTROL, ABORTING) (NO LR DATA)
"O.F." = Overflow, to many. CONTINUING ⟵ OCCURRENCE OF: 01104 DELAY ROUT. O.V. 01201 EXECT. O.F (VAC) 01202 EXECT. O.F. (JOBS) 01203 EXECT. O.F. (TASKS) 01207 EXECT. O.F. (MRKS) 01210 TWO USERS 01211 MRK ROUT. INTRPT 02000 DAP O.F.	 BAILOUT " " " " " " "	DUTY CYCLE MAY DEGRADE PGNCS (AGS CONTROL MAY HELP-SEE BELOW) (WATCH FOR OTHER CUES, PGNCS CONDITION UNKNOWN, DSKY MAY BE LOCKED UP, DUTY CYCLE MAY BE UP TO POINT OF MISSING SOME FUNCTIONS (NAV. LAST TO DIE), SWITCH TO AGS (FOLLOW ERR NEEDLES) MAY HELP (REDUCES PGNCS DUTY CYCLE SIGNIF.))	SAME AS LEFT (except "other cues" which would otherwise be cause for ABORT PROBABLY AREN'T, INSTEAD IT WOULD BE PGNCS GUIDANCE NO/GO – COMPLETE MANUAL LANDING IN AGS.)

Figure 9.3

Jack Garman's "cheat sheet" for Apollo 11 landing; shaded section contains the 1201 and 1202 Program Alarms that Garman diagnosed in real time with reference to these instructions. (Courtesy of Jack Garman.)

though that was a flaw in communications. I was very much in the dark when this came up."[21]

Errors of Human or Machine?

Were there mistakes here? Was the rendezvous radar mistakenly left on? No. Aldrin had planned to do that before the flight, and it was written into the procedures. The radar's mode switch was indeed in the wrong position, but the sync problem should not have occurred regardless of the position of the radar switch (and would have occurred were it in AUTO instead of SLEW). The radar-computer interface had been tested in a laboratory where both devices used the same power supplies, rather than in the LM itself where they had different power supplies, generating a subtle, invisible piece of unreality that masked what could have been a critical problem. Still, according to Don Eyles, the lack of synchronization of the power supplies had been realized years before, but never correctly addressed.[22]

It was at least an error of communications, if not of systems engineering practice. "There were folks who knew about the RR [rendezvous radar] resolver" interface issue, ran the IL report on the matter. "There were also a great many people who knew the effect which a 15% TLOSS [loss of processor time] would have on the landing program's operation. These folks never got together on the subject." The report noted that "scrutiny of the crew checklist by the hardware personnel" could have prevented the problems.[23]

Robust processing and restart protection in the LGC saved the day, as the computer dropped only low-priority tasks while keeping the vehicle under control. Still, the 1201 and 1202 alarms had been put in only for testing, and nobody thought they would ever occur in a real situation. In light of the benign nature of the restarts, and the computer's effective responses, the "program alarm" should probably not have illuminated the master alarm, and was probably too dramatic and intrusive in proportion to the nature of the problem. In hindsight, an indicator that said something like "tasks being dropped, critical functions still OK" would have caused less distraction for the astronauts.

We can point to other aspects of the human-machine network that allowed the Apollo 11 landing to succeed despite the program alarms: Kranz's insistence before the flight that all possible program alarms be understood, part of the Apollo philosophy of "no unexplained failure." He simply was not comfortable with the possibility of anything unexpected occurring during flight. The IL team's ability to quickly diagnose the problem when it occurred depended on having simulators ready to go in the laboratory, as well as the experience built up during the long hours of testing and simulation.

Most important, in response to the program alarms the human operators continued the mission while engineers on the ground diagnosed the problem. This behavior seemed to confirm the NASA philosophy of humans as critical backup components.

But how did Apollo 11 confirm the value of keeping the human in the loop? Recall that the only real problem caused by the program alarms was demanding the attention of the astronauts; a machine would not have been distracted. Had it been set on automatic landing, the LM would have come down anyhow, with less ballyhoo, though perhaps amid a field of boulders.

Software as Human Procedures

After the Apollo 11 crew returned, President Nixon presented them with the Medal of Freedom for their work. He also gave an award to the young flight controller, Steve Bales, who had made the no-abort call on the program alarms. The citation commended Bales's "decision to proceed with the lunar landing when computers failed."[24] While acknowledging that not every critical decision that day was made in the LM itself, the wording also blamed the machine.

Despite the subtle nature of the errors, public discourse framed the episode as fallible machinery versus skilled, heroic pilots. The press took up celebrating the human factor. *Datamation* magazine reported that Apollo 11 proved that "mere mortals showed they can still put the computer to shame," and took the LGC to task for not being state-of-the-art in 1969.[25] *Electronic Design* incorrectly reported that Armstrong "seized the manual controls of the lunar module," and ran an article "The Indispensable Man," in which they interviewed engineering and scientific experts, including Isaac Asimov, on the importance of the human role in complex missions.[26] Popular press proved even more gushing, hailing a victory of human performance over the impersonal forces of science and technology.

Was the culprit a software bug? Not in the sense that there was a particular error in the programming. In fact, the IL's antibug strategies—testing, restart protection, code inspections—prevented the computer from crashing at all. IL engineers felt that they were unfairly blamed for the program alarms. In a 1973 paper Eyles attributed the problem to "excessive interface activity" on the part of the astronauts (Aldrin's calls to monitor DELTAH).[27] "The software actually saved the program," Fred Martin recalled recently, "because it, in the face of this mistake in the switch . . . was able to go on with the highest priority jobs and not tank the mission."[28] To this day it galls Dick Battin when people refer to the program alarms as computer errors.[29]

Ironically, despite the attention that the program alarms brought to the computer, one potentially fatal problem did reside in the software that was not noticed until after Apollo 12. As with the program alarms, it was not a programming error but a problem of data exchange. Reviewing flight data, Grumman engineers noticed a "castleing" effect in the engines thrust commands—dynamic variations that made the plots of the thrust commands resemble the top of a castle's turret, as much as 25 percent of total thrust. Analysis showed that an incorrect set of parameters made the servo that controlled the automatic throttle only marginally stable, which could have caused it to oscillate wildly under certain descent conditions. Only a second error in programming a related constant prevented the unstable throttle from causing this catastrophe. The problem was fixed by Apollo 14, but behind the scenes the IL and NASA engineers recognized hidden dangers could lurk in program code.[30]

From a systems viewpoint, it is not a coincidence that the program alarms (and the castleing problem) arose from interfaces between different pieces of hardware. It was an interface problem between two components of a system made by different organizations. Software in the LM's control computer, the human-built glue that held the human-built system together, highlighted relationships (successful and buggy) between groups of people.

IL engineer Hugh Blair Smith takes even a larger view. He allows that the Apollo 11 program alarms could be called a software problem, but only if one realizes that "the crew procedures are part of the software, as are the ground procedures."[31] Apollo crewmen followed carefully written "programs," in the form of their timelines, checklists, abort criteria, and mission rules. These programs governing people's behavior were as important as the programs controlling the computer, and similarly embodied assumptions and links between organizations (recall that Apollo overall was called a program). In the human-machine system of Apollo, it often was not possible to distinguish between instructions for machines and instructions for people.

Indeed, every problem on the Apollo 11 landing stemmed from miscommunication, incorrect documentation, or failure to pursue and track minute details in a complex, unforgiving system. Yet, in a pattern that began on Apollo 5, NASA and the press found it easier to blame "the computer" and to narrate the successful landing as a triumph of humanity over the machine. This simpler, less anxious story affirmed the political goals of the Apollo program and did not require acknowledging the bugs in the program's vast and impressive human-machine system.

10 Five More Hands On

We all felt that when you get to that point and you are going to land on the moon, you have to have your hands on the stick. I like computers and I believe in computers, but it ain't going to land me on the moon. I'm going to do that. If something gets screwed up then it is going to be me, it isn't going to be the computer... You are probably fooling yourself because you are still going through the computer. The stick that you move goes through the computer to fire the thrusters, which is not too different from the computer doing that itself. You feel different, though.
—David Scott, "The Apollo Guidance Computer: A User's View"

Apollo 11 accomplished a dramatic first that was difficult to replicate, both in technical suspense and public response. Still, it was only the beginning of the program, the first of six landings. Nor was it alone in raising new tensions and synergies between human and machine. Throughout the social-technical system that was Apollo, skill, experience, and risk migrated across human and machine boundaries. The social and the technical traded off, complemented each other, made up for each other's weaknesses. In the real-time pressure of a lunar landing, an extensive social network of engineers focused on two men and a computer in an air-conditioned bubble, sitting on top of a rocket engine with a telescope and a control stick.

Landing near a Robot

For all its glory and accomplishment Apollo 11 lacked one thing: accuracy. Apollo 12 set a new goal and a new task for the human pilot: demonstrate accurate, "pinpoint" landing capability, defined as within one kilometer of the aim point. As with Apollo 11, NASA chose a relatively smooth area in a flat region, called the Ocean of Storms.

Ironically, Apollo 12 would not be the first landing in this area. The goal was to land near where the unmanned Surveyor III spacecraft had come down two and a half years before in April 1967. Surveyor III was the second in its series to land on the moon after Surveyor I (Surveyor II crashed on the moon after a control thruster failed). When landing, Surveyor III bounced several times on the moon because the engines failed to shut down on time; only a command from earth finally turned off the thrust and stabilized the vehicle. Simply landing on the moon did not require human presence.

Apollo 12 was to show, however, that *precision* landing called for piloting skill.[1] NASA considered numerous changes to make Apollo 12 a more precise touchdown than Apollo 11.

What did it mean to land with accuracy? "Accuracy" for geologists might mean proximity to a known feature of scientific interest like a crater or mountain. By contrast, engineers defined accuracy by an error circle around precise coordinates. They also distinguished between accuracy, as integrity to a physical quantity, and precision, as a consistent numerical measure.

The Apollo program employed various maps based on different types of data (earth observations, Lunar orbiter data, previous Apollo missions, etc.) and their grid coordinates did not always line up. The unique requirements of the landings meant that the Apollo guidance and control system had its own coordinate system into which the Surveyor III coordinates were translated to target the Apollo 12 landing.[2] "It is not meaningful to compare stored landing coordinates with the actual site location," one mission report warned, "because of the various transformations and targeting biases which have necessarily taken place."[3] In other words, the Apollo computer's coordinate system was not necessarily compatible with those used by lunar scientists. Landing near the Surveyor, however, an arbitrary but unequivocal point (perhaps the only such point on the moon, given the monotony and homogeneity of the lurain), would serve as proof of a pinpoint landing (hence Apollo 12 is better described as a precise landing than an accurate one).

Only two changes were made to Apollo 12 software on the basis of Apollo 11, both to improve precision. A new command, Noun 69, enabled the astronauts to update the coordinates of the landing site early in the braking phase, based on improved data from ground tracking. Using Noun 69 (and about twenty-eight key presses) the crew could update the landing site position (up or down range, but not cross range) to correct any errors, thus increasing precision. By contrast, the Apollo 11 crew knew they were going long several minutes before pitch over, but could do nothing about it once the computer was in control.[4] The second software change enabled the astronauts to calibrate the LPD against a star during powered descent, fine-tuning the optics.

A few procedural changes were made as well. First, during undocking, Command Module Pilot Dick Gordon moved his ship away from the LM using the larger craft's thrusters, rather than the LM's, which had been used during Apollo 11. This maneuver avoided adding additional velocity to the lander, one possible source of 11's long landing. Second, the Apollo 12 LM would spend the entire braking phase in the face-up attitude. Unlike Armstrong and Aldrin, Commander Pete Conrad and LMP Alan Bean would not check their altitude visually against the ground—thus eliminating the necessity of a yaw-around that could introduce uncertainty into the trajectory.[5] Third, a modification to the trajectory in the approach phase allowed the crew more time for redesignation with the same amount of fuel and allowed the LPD angle to change more slowly and smoothly to make it easier for the commander to follow.[6]

An Instrument Landing on the Moon

By the time the crew of Apollo 12 arrived at their PDI point 50,000 feet above the moon, ground controllers uploaded a new state vector from the ground into the LM's computer—temporarily seizing control of the DSKY from the astronauts in the LM. The new state vector told the computer it was five miles off course, north of the intended descent track. Before the burn, a KALCMANU maneuver automatically pointed the LM back toward the track, and the guidance drove the error to zero during the descent.

As on Apollo 11, Apollo 12 began firing its engine at 50,000 feet to initiate powered descent. Where Apollo 11 had reached this point a little bit long, Apollo 12 was about a mile short.

At this point Conrad joked with the ground, "You can turn the mirror on; give me the fox-corpen, and we got the hook down." Terms from his days as a carrier pilot—the fox-corpen was the heading of the aircraft carrier's runway, the mirror referred to the "meatball" that showed his position above or beneath the glide slope, and the hook referred to the tailhook on a carrier aircraft.

Houston then called up "Noun 69 + 4200." This new command allowed the crew to retarget the vehicle 4,200 feet down range, something Apollo 11 could not do. Bean keyed it in; the ground controllers reviewed his keystrokes to check for errors, then okayed him to hit enter. A mistake entering this correction could cause a serious problem.

Ground then reported: "MSFN agrees with PNGS and AGS." The Manned Space Flight Network [MSFN] was tracking Apollo 12 from the ground, and came up with similar numbers.

Conrad replied: "Looks good. We're smoking right down there." A few minutes later he pointed out, "I have an altitude light out; and I have a velocity light out." The landing radar was sending good data, at about 40,000 feet. Conrad then instructed the computer to incorporate the radar data into the guidance equations. The radar showed a discrepancy (DELTAH) of about 1,700 feet. Within a couple of minutes, the computer had adjusted its solution so the two converged to within one hundred feet.[7]

When he incorporated the radar into the guidance solution, the RCS (reaction control system) thrusters began to fire to adjust the spacecraft to this new source of data. "I'm getting a fair amount of RCS firing," Conrad reported, "more than I think I should," and a few minutes later, "Boy, it's really giving her heck on the RCS," but the thruster firings were actually normal. (NASA later determined that there had been a software problem in the simulator. The astronauts were used to a simulation that had less thruster activity than normal, so it seemed more active when they were in the actual LM. Also, as on Apollo 11, unanticipated fuel slosh was making for a rougher ride than expected).[8]

At 7,000 feet, the high-gate, the computer automatically switched to P64. The LM immediately began to pitch over. A full panorama of the lunar surface came into view.

This sight caught Conrad unprepared, with much greater field of view and detail than he experienced in the simulator. "For the first couple of seconds, I had no recognition of where we were, although the visibility was excellent. It was almost like a black and white painting." Bean found this view, with all the craters, scary.[9]

Bean glanced down at the computer to enable the LPD. Before he could call out the number, Conrad noticed a reading of 42 degrees on the display and looked through the window at that angle. What had been undifferentiated gray, just moments before, now became a pattern.

Then came an excited shock of recognition. Conrad cried out: "There it is! There it is! Son-of-a-gun! Right down the middle of the road!"

"Hey, it's targeted right for the center of the crater!" Noun 69 had done its job; just a small redesignation would move them to the rim of the crater.

This was the critical moment missed on Apollo 11. The LM was pitched over, the commander could see out, and he recognized the features. Within about thirty seconds, Conrad began to redesignate the landing spot. He repeated seven times over the next minute (figure 10.1).

First Conrad moved the landing sight a bit to the right (north). One click. "That's so fantastic. I can't believe it," he exclaimed of the targeting.

Conrad then redesignated forward (west), with two clicks. Then one click to the north. Then two clicks back (east). Then one more click north, one final click back (east). Conrad's seventh redesignation put the automatic landing algorithm targeted very near where Surveyor III had landed in 1967, but he would not use the auto mode to land.

At about four hundred feet he called, "I got it." He then took over control, in P66, canceling the LPD mode (which would have allowed him to redesignate for forty-nine more seconds). Despite three northerly LPD designations, Conrad steered further to the right (north) when he took over. He also slowed the LM down to land closer up range than his last designation.

At about three hundred feet, Conrad banked toward the crater again and began to descend. Later he noted, "I saw a suitable landing area between the Surveyor crater and head crater, which now meant I had to maneuver to my left and sort of fly around the side of the crater, which I started to do."[10]

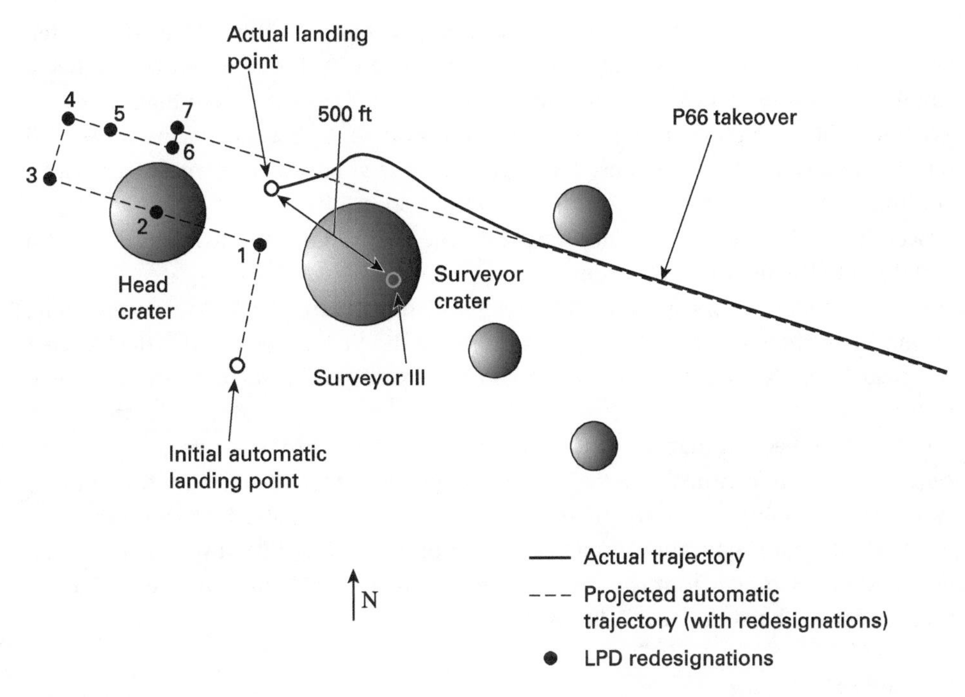

Figure 10.1
Conrad's use of LPD redesignations during the Apollo 12 landing. (Redrawn by the author from Floyd V. Bennett, "Apollo Experience Report: Mission Planning for Lunar Module Descent and Ascent," 340.)

The dust began at nearly two hundred feet. Conrad lost his visual reference. He could see how fast he was moving by looking at the ground, "but I couldn't tell what was underneath me. I knew I was in a generally good area and I was just going to have to bite the bullet and land, because I couldn't tell whether there was a crater down there or not."[11]

Conrad began to focus inside the LM, on instruments. But he kept glancing out the window. He thought that the "crosspointers," the needles that indicated lateral velocity, were not working. Actually, they were working, but he was nearly stopped, moving so slowly that the needles had essentially no deflection. So Conrad tried to sense his velocity by looking out the window. If he'd trusted the needles, Conrad later said, "I probably would never have looked at the ground in the last 50 or 100 feet."[12] As they drew nearer the surface, the dust cloud overwhelmed the view.

The LM descended three and then two feet per second. The probes hit the moon. "Contact light." Conrad killed the engine and the LM dropped the last few feet.

Then Conrad vented his own relief valve: "Okay. Man, oh man, Houston. I'll tell you. I think we're in a place that's a lot dustier than Neil's. It's a good thing we had a simulator, because that was an IFR landing." IFR stands for "instrument flight rules" in aviation, when the pilot cannot see due to clouds or fog. "It's a good thing we leveled off high and came down, because I sure couldn't see what was underneath us once I got into that dust. . . . That stuff was going to the horizon."

Bean hailed Conrad. "It's a real pleasure to ride with a number 1 aviator." Pride still tied them to the piloting profession.

When Conrad looked around, he later recalled, "it turned out there were more craters there than we realized, either because we didn't look before the dust started or because the dust obscured them."[13] Fortunately, they did not land in an unseen crater.

Conrad had been in manual control for one minute and fifty seconds, about 30 percent less time than Armstrong had. Sixty seconds had remained before the land-or-abort decision point—twice the time that remained on the Apollo 11 mission.[14]

Despite the confusion, they'd made a good shot. Conrad and Bean were able to walk over and take samples from the craft, to see how its materials had survived for a few years under the harsh lunar conditions.

Instruments and Trust

NASA hailed Apollo 12 as proof of precision-landing capability. Floyd Bennett got the NASA Exceptional Service Medal for his role in the pinpoint operation.[15] Indeed Apollo 12 came down right next to Surveyor III, which had landed in exactly the same spot, with no human involvement at all. By its very triumph, landing near a robotic spacecraft, Apollo 12 implicitly compared automated and the human-controlled lunar landing, implying that the human landing was precisely controlled, whereas the machine had just plunked down in any old spot.

Yet how much did the human hand on the stick improve accuracy over an automated system? Conrad later described "my slight overflying and taking over to stop what I thought was a relatively high horizontal velocity." His comment is ambiguous whether "overflying" means "flying too much" or "overshooting." Conrad said later that his initial two redesignations were "short" because he felt he was tracking a little high: "I didn't like the size of the area short, where we had normally been trying to land, and I looked for a more suitable place."

This explanation is puzzling, because it does not match the data. Conrad's first two redesignations were not short; the first moved to the right (north) and the second and third actually went long (west). Then he moved right, and only then began to move back east (short). NASA's mission report describes this move this way: "The manual descent program was entered at approximately four hundred feet altitude to prevent an apparent downrange miss and to maneuver to the left." It doesn't mention that the "downrange miss" was caused by Conrad's initial redesignations past the target. His final "accurate" landing was exactly as far down range as the original spot. A post-flight analysis with the Grumman team reported that Conrad "related that an examination of the post-flight data indicated that these redesignations were not required but he felt that his real-time view using the LPD showed that crater as the targeted site."[16]

Other analyses confirmed that Conrad's maneuvering put the LM pretty close to its original spot. "If no redesignations had been made," ran the mission planning and analysis division (MPAD) post-flight analysis, "then it (LM) would have landed approximately 600 feet southwest of Surveyor III. If the LM had landed automatically after the seventh redesignation, then it would have landed approximately 1,000 feet northwest of Surveyor III."[17] Either one would have qualified as a "precision" landing under the one-kilometer criteria. Conrad's intervention improved the proximity to Surveyor, but only by a few hundred feet.

Apollo 12 also proved the feasibility of an instrument landing on the moon. Conrad thought that he landed in a dustier spot than Armstrong had, because his vision was totally obscured, whereas the Apollo 11 crew reported they could see through the dust (on their moonwalks Conrad and Bean did seem to kick up a little more dust than on the previous flight). The sun angle was also lower (five degrees compared to more than ten degrees on Apollo 11), which may have made the dust cloud more attenuating of vision. Nevertheless, post-flight analysis showed that the Apollo 12 LM had higher thrust levels lower down than did Apollo 11, kicking up more dust. Conrad's approach came in steeper than Armstrong's.

"I took over manually at about 700 feet and immediately killed the rate of descent. It looked like we were going at the ground like a bullet. I had plenty of gas and I wanted enough time to look around," Conrad said. He reported that he then overcompensated, killed the rate of descent to 3 feet per second at 500 feet, essentially hovering the spacecraft. "I got a little nervous." His recollections were in error: he actually took over below 400 feet and slowed the rate of descent to one foot per second, then came down steep.

Just a few weeks after Conrad's landing, he and Armstrong compared their experiences at a meeting of the Flight Readiness Review Board for the LLTV. Conrad noted that because of the dust cloud, he felt more comfortable on instruments during the final seconds but he continued to look outside because he didn't believe the crosspointers were working, as they were indicating zero horizontal velocity. In retrospect, he stated, it was a mistake not to verify proper operation of that indicator during the initial checkout because he couldn't bring himself to trust it during descent.

In fact, the velocity indicator was working, but Conrad entered his vertical descent so high (300 feet) that he didn't see the very slight deviations of the needles (under one foot per second for much of the final seconds).[18] Nevertheless, in Conrad's words, "I was perfectly satisfied that we were in a clear enough spot that I didn't need to look out anymore." Overall, Conrad said he felt comfortable with his training, "and I had all the confidence in the world in the inertial guidance system, which made it very easy to put my head in the cockpit when I thought I had to do it."

Conrad's experience, and the way he related it, raised a question about the human role in the landings. At the meeting, Chris Kraft asked whether, during an IFR landing such as Apollo 12, when the pilot is "heads down" during the final descent and touchdown, it would be better to have him just flick a switch and have the computer "null the rates." It could automatically bring the LM down in a hover, at a particular rate controlled by the commander's left-hand switch. Put another way, if the pilot's major asset is his visual recognition and selection of the landing site, then why continue to fly by hand?

Armstrong's account of his own landing argued for unique human involvement. He declared himself "probably a little more reluctant to accept an instrument landing than Pete was." The reason, he explained, was some experience flying with a Doppler radar on a helicopter, which demonstrated to him that the indications could acquire a bias, as much as six feet per second, as the vehicle approached the ground. In other words, if one was flying IFR using the radar and touched down reading zero velocity, the vehicle might actually be moving as much as six feet per second, presenting a danger of tipping over on landing ("stubbing your toe," as Armstrong put it). A reasonable enough objection, although Armstrong realized such a bias in the radar was "probably an effect of rotor interference by the helicopter.... It probably would not exist in the LM" (he still presented this velocity bias in public as a source of his objection to automated landing). Unlike Conrad, Armstrong did believe his radar was working during his vertical descent, but he spent most of his time looking out the window for visual cues (although he became comfortable with his landing spot much later than Conrad did).[19] The two men distrusted the radar for opposite reasons: Armstrong didn't trust the needles because he thought they might have a large bias in a hover; Conrad didn't trust them because they indicated zero when he was hovering.

The Apollo 11 and 12 landings occurred just a few months apart and before an unplanned gap of more than a year before the next landing, so they formed something of an analytical pair. Floyd Bennett wrote a detailed, insightful technical analysis of the two flights, based on his Lunar Landing Branch's post-flight analyses. Where Armstrong's focus was on accomplishing the first landing, Conrad sought to land with precision. The addition of Noun 69 to update the landing target during the descent indeed prevented an Apollo 11-type overshoot on Apollo 12. Where Armstrong fought the distraction of program alarms, Conrad struggled with the LPD, redesignating it close to its original spot, and then taking over. Where Armstrong landed visually, Conrad landed using instruments, with some uncertainty as to what was beneath him. Neither wished to trust their instruments entirely, even though for numerous other phases of flight they had staked their lives on instruments alone, with no visual or manual backups.

Hardware Failure and Software Fix

Before his lunar landing flight, Jim Lovell said that he planned to land the LM in fully automatic mode. He never got a chance to try. The Apollo 13 story has become among the best-known aspects of the program, thanks to a number of books and a popular movie. Of course, that flight made no attempt at a lunar landing, so it will not be part of our analysis here. The crises during that mission generated plenty of interesting human-machine interactions, however, and the guidance system proved both redundant and flexible in ways that were critical to saving the lives of the crew.

Apollo 14 came ten months after the near disaster, more than a year after the previous landing. In LMP Ed Mitchell's words, "Apollo 14 had to be a flawless mission."[20] Mitchell had originally been on the Apollo 13 prime crew and the Apollo 10 backup crew. Hence he had a great deal of training, having been through numerous LM procedures, and having worked with Grumman at Bethpage, New York, building the spacecraft. Apollo 14 Commander Alan Shepard, now forty-seven years old and the only one of the original Mercury Seven astronauts to fly to the moon, was a pilot's pilot—old school, hands on, stick and rudder, and an ego to match.

About four hours before the planned PDI burn, Shepard and Mitchell checked out the LM. Everything was normal, until about fifteen minutes before the LM, nicknamed Antares, was scheduled to disappear behind the moon and lose contact with Houston.

For all of the Apollo missions, ground controllers spent the long hours of the flights staring at lists of numbers on computer screens as they downloaded the telemetry in real-time. The controllers became exquisitely sensitive to the binary bits that made up the numbers, which appeared as numerical hash to an untrained observer. Lights on their consoles reflected these bits, signaling "flags" that represented various conditions on board the spacecraft. Now, as PDI burn time approached for Apollo 14, they noticed something strange. Amid the hundreds of bytes of data, a bit was set at 1 that should have been a 0. It was the abort bit, indicating that somewhere along the line the computer had gotten an indication to abort the landing. The erroneous bit was harmless at this point, because the LGC in the LM was not running the descent program, so it was not looking for an abort input. But once PDI occurred and P63, the braking phase program, began, an uncommanded abort would be a serious problem.

In the real world of electromechanical devices, things can be "flaky." Houston treated the problem this way, asking Mitchell to press the ABORT button, and then the STOP button. He pressed, and reset both buttons. The faulty indication went away.

The crew twiddled around with antennas and gain controls, trying to get the communications links solid and noise-free. Never an easy job.

About forty-five minutes later, Antares reappeared from behind the moon. Everything looked good. But half an hour later, the abort bit set again. Houston mentioned the discrepancy to the astronauts, noting that they were working on a fix to get rid of the problem after the PDI. A little more than two hours remained before the critical burn.

Houston: "And, while we've got that display up, Ed, could you tap on the panel around the ABORT pushbutton and see if we can shake something loose?"

Mitchell complied: Tap, tap. Again the error went away.

Houston: "OK. Antares, we'd like to kind of sit here a minute and watch it." Ground controllers uploaded a navigation vector into the computer. The astronauts spotted their landing site on the moon as they passed over, on this the last orbit before landing.

The radio conversation was calm and matter of fact, but everyone knew the story: the flaky abort button was a showstopper. Without some kind of fix, a patch, even a "kludge," the landing could not continue.

In Houston, and in Cambridge, engineers like Don Eyles were working furiously. Jack Garman, who had been in the back room during the Apollo 11 program alarms, remembered the intense pressure of only having two hours to solve the problem: "The worst nightmare of all."[21]

Unaware of the feverish activity on the ground, Mitchell became concerned. "Do you think we're going to come up with something on this problem with the ABORT button?" he asked Houston.

"We're working it right now," EECOM Fred Haise replied, "and also MIT's working it. Needless to say, we're busy here, but we think we got a solution."

"Good enough. Something—is it something like a solder ball?" Mitchell's early guess was exactly right; according to the post-flight analysis, it was indeed a ball of solder in the switch, jarred loose by the flight, floating around in zero gravity and causing electrical mischief. Several similar switches were later x-rayed and found to have similar problems.[22]

"Well, we don't know yet," Houston replied. "We got about 19 minutes until loss of signal here, so we'll have something to you before then, and we'll have some time to pick it up on the other side."

The bit came back again. Tap went Mitchell. Out again.

Ground now had some ominous commentary: "The implications of that bit being set, I guess you also realize, means that in 63 we're going to find ourselves in P70."

That was Apollo computer lingo for disaster. Translation: *when we're in the braking phase (P63), we could instantaneously go into an uncommanded abort (P70)*. If this were to happen when the LM was on its way down, it would incorrectly abort the mission—at best a massive disappointment, at worst a dangerous surprise.

Shepard was nervous and felt powerless to fix anything. "You can't climb into a computer buried deep inside a spaceship with a screwdriver or a wrench," he later wrote. The familiar mechanical world of nuts and bolts was long gone.[23]

Finally Eyles came up with a clever solution. "As soon as he identified it," remembered Garman, "everybody went, 'Yep, that's it.'" Houston tested the procedure and then called it up to Apollo 14.[24]

Mitchell wrote the procedure down on his timeline for the PDI: "Verb 25 Noun 7 Enter; 105 Enter; 400 enter; 0 enter." The procedure was complex, and touchy. If the bit set before a certain command, there would be problems.[25]

That was the last Houston heard from the LM before it went around the moon for the final time. About forty-five minutes later, the LM emerged. As soon as it came in on the radio, Houston sent another procedure. It turned out the one they'd prepared before had some time-critical moments, so they scratched it and came up with a new one.

The ground controllers could have uploaded the procedure automatically, through the digital telemetry link. But instead they called this complex series of commands up to the astronauts by voice, had them write it down, and then key it in by hand. To this day, the computer's designer Eldon Hall believes this meant that the flight crews didn't really trust the computer. While potentially more error-prone, the manual entry gave the people in the loop the confidence that they knew what was being entered.

This fix has often been described as a program change, which is incorrect. Actually, it was a kind of bit-level hack right into the main registers of the computer. The procedure involved about seven steps. An initial DSKY command fooled the computer into thinking it was already executing an abort, so it would not respond to the stuck button. This prevented the computer from checking the abort bit, but it also inhibited the automatic commands for throttle up, the automatic guidance steering, and processing of the landing radar data. Thus the astronauts would manually throttle up for PDI, enter some DSKY commands to start the computer's descent guidance, restore some of the mode flags in the computer, and then command the computer to decrease the throttle when the burn ended.[26] Because of these changes, the computer could no longer accomplish an abort with the button, so the crew would have to initiate it themselves by key presses on the computer, separating the stages and going through a number of other steps.[27] As with other Apollo procedures, the fix involved a relatively simple, automated ride if everything went well, but high-workload critical manual procedures if there was a problem.

Shepard and Mitchell said later that they felt confident about the procedure. They felt the ground had handled the problem well—not burdening them with the details of developing the new procedure, just calling up one that was tested and accurate. Of course, Shepard admitted, "we didn't have much of a choice. It was either try that or give it up."[28] It will always remain a hypothetical question whether the crew of Apollo 14 could have aborted safely with the new procedures (although the same could be said of the original procedures as well).

Mitchell didn't feel he was tempting fate by typing in such strange mode changes at the last minute. "I was so darn familiar with the systems that I could play games with them, you know, like a computer hacker." He was comfortable with the computer; it was something he'd worked with in many different modes. "I could do things with the AGS and with the main guidance system that really weren't in there, that weren't even in the checklist."[29] For all its complexity, the human-machine system of the crew and their two computers was flexible, adaptable to new tasks on the spur of the moment. But the new procedures also introduced unwelcome uncertainty.

Relieved to get the procedure into the computer, Shepard said he "felt like we were home free." He realized later that he was being too optimistic—plenty of things could still go wrong.[30]

Preparing for the PDI, the crew was tense. "We were trying to be a little light-hearted," Mitchell later recalled, "about the fact that the abort switch problem was really a disaster. I mean, it had us pretty ginchy and gun-shy. God only knows what could happen next." He added, "We suspected, in the back of our minds but didn't know for sure, that when you re-write a system and re-engineer a system . . . you could create side effects."[31]

The astronauts went through the checklist for the PDI burn. The RCS engines automatically fired for ullage. Mitchell hit PROceed. Next: DPS ignition.

Shepard: "Okay. And the Master Arm is off." This switch would normally have been on—enabling the PNGS to signal an automatic abort. From now on all such procedures would be manual. About half a minute after ignition, the commander manually throttled it up.

Some of the workload was shifted to Shepard at this point to enable Mitchell to work the computers.[32]

Mitchell: "001, Enter. Should have guidance. And you have Command and Throttle."

Shepard: "Okay. We have guidance."

At this point Shepard began to feel the procedure was working, as the computer took the craft down the trajectory. "PNGS was happy with itself. So that gave us a little more confidence that things were going along well."[33]

One minute into the burn. Now Mitchell: "400, Enter; 0, Enter. Okay. Landing radar enable, Verb 21 Noun 1, Enter; 1010, Enter; 77, Enter. The landing radar is there. Al, you can reduce your throttle to Minimum."

Shepard: "Okay. It's coming down."

Mitchell: "You have Command and Thrust. Okay, Houston. The procedure is complete." What would have been accomplished by the computer now took tens of keystrokes.

About two and a half minutes after ignition, Shepard keyed in Noun 69, plus 2800, the navigation correction.

They were now into a normal descent. A little bit fast, about ten feet per second, a little bit low. But all within normal bounds. Mitchell reported: "It looks good, it looks good."

About four minutes into the burn, Mitchell started looking for the landing radar to lock in. "Down to 32,000, we should be getting landing radar in very soon."

No radar. Forty-five seconds later he checked it again. Mitchell fiddled: "On radar, set the lock ON on radar." Then: "Can't get the radar in."

The radar was supposed to be coming in (on Apollo 11 it had come in at 37,000 thousand feet; on Apollo 12 at 40,000). "At 20 thousand feet, that's when were frantically trying to get it to come in," Mitchell later recalled, "because, at 10 thousand feet, there was an automatic (meaning 'mandatory') abort without the landing radar."[34]

"We still have ALTITUDE, VELOCITY lights," lights that should go out when the radar locked on the surface.

Shepard added, nervously, "I'll bet they know that."

Houston issued instructions: "We'd like you to cycle the LANDING RADAR breaker."

The radar, it turned out, had somehow switched to "low scale," designed for close-in measurement, before the burn even began, which prevented it from locking on the moon at long distances. Recycling the circuit breaker switched it back.[35] This was the electrical engineer's oldest and crudest troubleshooting technique, but one frequently effective with balky electronics. Cycle the power: turn it off and turn it back on again. "Just a body performing the job," Mitchell said of himself later, "searching for a solution to a dilemma, functioning like an extension of the computers at my fingertips."[36]

A tense silence followed, as all watched the lights for twenty seconds.

Mitchell: "Come on in . . . How's it look Houston?"

Mitchell: (To the radar) "Come on in! . . . Okay!"

It had popped in. Mitchell went to work: "VERB 57 ENTER," reviewing the radar data. He asked Houston, "Can we accept?"

Haise: "Okay. We'd like to accept the radar."

Shepard: "Okay. PRO(ceed). Converging. PRO." Relieved, he continued, "Okay, go . . . go great. Great. Oh, that was close."

About 11 minutes after PDI, Apollo 14 passed the high gate. Said Mitchell, "Okay, there's pitchover."

He looked out the window and saw his landmark. "There's Cone Crater."

He had no trouble recognizing the landing spot. It looked just like the plaster-of-paris model they had used for training. For Shepard, the "determining factor" was the "high fidelity of the simulator visual display" and the time he spent training with it.[37]

Tension broke in the thrill of recognition.

Mitchell: "And there it is."

Shepard: "Hot damn! Right on the money!"

Unlike Buzz Aldrin, the LMP who had concentrated wholly on the computer displays, Mitchell was looking out, "stealing glances out the window with virtually every sweep" of the instruments, he said, backing up Shepard's own senses, giving guidance, suggesting moves.[38]

Shepard redesignated once, moving the landing point approximately three hundred and fifty feet to the left. As the LM descended, past about one thousand feet, Shepard saw that his redesignated point was no good—the crater there was too large. He entered P66 and flew back to the original spot.[39] The commander took over control at about 360 feet altitude, about 2,200 feet short of the target.[40] He also flew 2,000 feet down range, because the coordinates of the landing site were in error by 1,800 feet.[41]

The spacecraft landed on a slight slope, with what Shepard described as "about a 7-degree right-wing down attitude." If he'd come in faster, he thought, the vehicle would have ended up more level.[42]

Pilot Skills and Backup Systems

Apollo 14's mission report recounted the events of the descent to justify human presence: "The advantages of manned spaceflight were again clearly demonstrated on this mission by the crew's ability to diagnose and work around hardware problems and malfunctions which otherwise might have resulted in mission termination."

Similarly, the *Boston Globe* reported, "Because of the computer problem, Shepard took over manual control of the landing craft."[43] The Apollo 14 crew had experienced a systems problem. First, an ordinary mechanical failure, caused by contamination of the abort button, had necessitated a computer workaround. For all the difficulty in software development, and the complexity introduced by the digital computer, in this case it had saved the day—few other control systems could have been flexible enough to route around at the last minute. The work of solving the problem was offloaded from the crew in the LM to the NASA ground controllers, and then yet again to the IL programmers in Cambridge. When they did come up with a fix, they were able to test and verify it on the ground, all while the crew members were busy with other tasks. A triumph of human problem solving in space? Or of the crews on the ground?

After the clever solution to the abort button dilemma, the system presented another problem, the landing radar. Even many years later, the crew thought that the two were somehow related. "As is often the case," Mitchell wrote in his memoir, "errors in a system tend to propagate."[44] The landing radar failed to lock on to the ground because it had tripped into the wrong mode setting, likely the result of some noise in the system and unrelated to the computer or the abort system. The astronauts believed the system was acting up, that their changes to the computer and procedures had somehow induced a problem elsewhere.

When Shepard retold the story afterward, he described it as his own struggle against the balky equipment, against the stultifying mission rules, "rebelling against the glitches and all this crap about wiring and circuitry screwing up." He recalled telling his partner, "Ed, if the radar doesn't kick in, we're going to turn her over and fly her down."

Of course, the radar did kick in, and Shepard's memoir describes his landing as the triumph of his manual skill: "Using thirty years of pilot skills, threaded a needle between the hills and ridges along their approach path and dropped his ship down into a narrow valley, craters and boulders everywhere."[45] After landing, the book says Mitchell asked him "Would you really have flown us down without the radar?" The book reports Shepard's answer: "The Tom Sawyer grin was never so wide on Alan's face: 'You'll never know, Ed. You'll never know.'"[46] (None of the recordings, air to ground, or within the LM, or the technical debriefs afterward, recorded such a conversation. Ed Mitchell does not recall it either.)

Flight director Gene Kranz, in his memoir *Failure Is Not an Option,* recalled that Shepard had confided later to Flight Director Gerry Griffin, "'I had come too far to abandon the Moon. I would have continued the approach even without the radar.'" Kranz did not doubt that Shepard was serious, but also was sure he would have had to abort, because he would not have known his altitude accurately and, as Kranz noted, "The fuel budget was just too tight."[47] Whether Shepard or any other astronaut could have landed without the radar data will never be known, but the systems problem on Apollo 14 once again stressed the interactions between humans and machine, and among the pilots, ground controllers, and engineers. In the later flights, these relationships would mature and evolve.

The J-Missions

Just prior to PDI on Apollo 15, LMP Jim Irwin commented: "I'm going to write me a joke—Astronauts come back from the Moon; said it's great, but no atmosphere."[48] The moment of lightheartedness reflected a relaxation typical of the later landings as successful flights built confidence and refined techniques. First of the so-called "J-missions," Apollo 15 incorporated a number of changes in design and operations that allowed a heavier LM to descend to the surface, stocked with consumables for a longer surface stay.

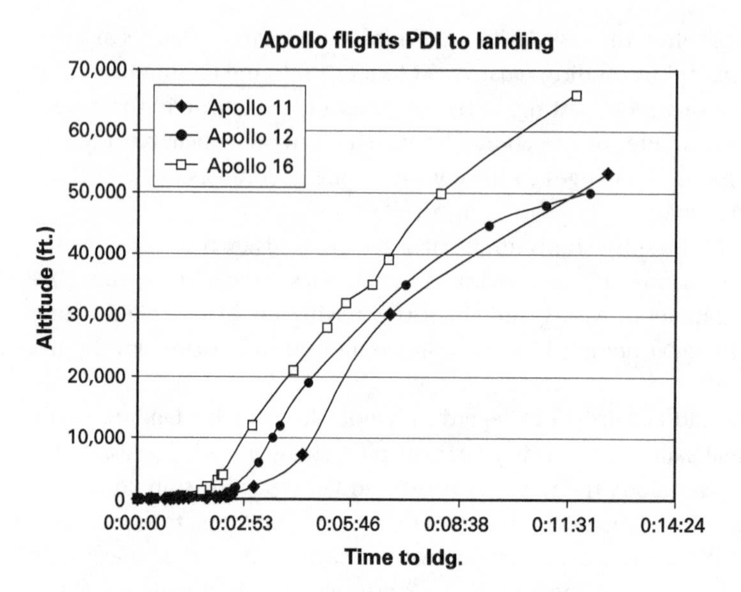

Figure 10.2
Comparison of radar altitudes for PDI to landing for Apollos 11, 12, and 16. Note the steeper descent for the later, J-mission (16). (Drawn by the author from Apollo mission transcripts.)

The J-missions also carried the lunar roving vehicle, the small buggy the crews used to cover significantly greater distances on the moon. The rover had implications for landings, because the ability to drive around extended the radius of operations from a few hundred feet to several miles. This extended range in turn relaxed the requirements for precision landing. If the LM ended up far away from a landing spot, the astronauts could just drive there in the rover. At least one commander mentioned that the presence of the rover eased his concern for accuracy at touchdown.[49] For the J-missions the angle of descent also increased from fourteen degrees to twenty-five degrees, and the vertical descent phase began at 200 rather than 100 feet. The steeper, longer descent would provide improved accuracy, better visibility at pitch over, and better control of the LPD redesignations.[50] Figure 10.2 illustrates the different trajectories for three of the six landings. Note the higher angles for the J-mission, Apollo 16.

By Apollo 15, the landings themselves were considered proven, although hardly routine. In training the astronauts began to focus more on new techniques and scientific work on the lunar surface.[51] For them, the simulators were more regularly available than for the earlier flights, allowing their training to become more reliable and routine. Now the astronauts spent as much as 40 percent of their time training for lunar surface science work, whereas the earlier missions concentrated on systems and procedures.[52] "Everybody else had done it," Charlie Duke said of the Apollo 16 landing, "and so we felt real confident, when we were there, that we were going to have plenty of gas."[53]

Landing Variations

Despite the increased confidence, each landing still had unique demands, problems, and anomalies. Each was shaped by its place within the Apollo program and events on the national and international stage. By Apollo 16, the final three missions (Apollos 18, 19, and 20) had been canceled, and the program was winding down. Developing new technology and methods for lunar exploration took a back seat to science, exploration, and the safe completion of the program.

By Apollo 17, though the missions had developed a sophisticated, reliable system, Gene Cernan was acutely aware of his status as "the last man on the moon." Even though the landings spanned a mere three and a half years, a great deal had changed by 1972. Federal budgets were tight, war in Vietnam was draining the nation's confidence, and a new consciousness in public discourse was proving skeptical of technology. Moreover, President Nixon had made a conscious decision to reverse the Kennedy-era focus on well-funded spaceflight that had given rise to the Apollo program.[54]

In this changed world, Apollo soldiered on. Each landing varied in everything from lunar topography to crew training. Increased confidence in proven systems led to choosing more difficult landing sites. By no coincidence did Apollo 11 land in the Sea of Tranquility; the name evokes the flat topography chosen for the mission with the greatest technical uncertainty. As the program progressed, however, experience grew in navigation and landing operations and NASA selected ever more challenging terrain. Here engineering and science reinforced each other, for "difficult" sites—more radical terrain, more changes in topography, higher elevations, deeper craters—made for interesting technical challenges and tended to correspond to "interesting" science. Apollo 15 headed toward Mount Hadley (11,000 feet) at the edge of a larger range, to land near Hadley Rille, a mile-wide, v-shaped trench that runs parallel to the mountains, 80 miles long and 1,000 feet deep. Apollo 17 chose a spot between two mountains so narrow that Floyd Bennett's mission planners had to shrink their specified uncertainty or landing from an ellipse one-by-three kilometers in size to a one-kilometer circle (Bennett rejected other sites he considered too risky for flying).

Pilots enjoyed the difficulty, which supported their sense of skilled flying. "For me, the challenge of flying into an unexplored box canyon was a space pilot's dream," Cernan wrote in his memoirs, "pushing the LM's envelope of performance and testing everything I knew as a pilot."[55]

Confusion and Simulation

The most interesting moments in the final landings came at pitch over, when the commander saw the terrain and searched for his landing spot for the first time. On Apollo 14 this moment engendered expressions of joy and recognition. On Apollo 15, the

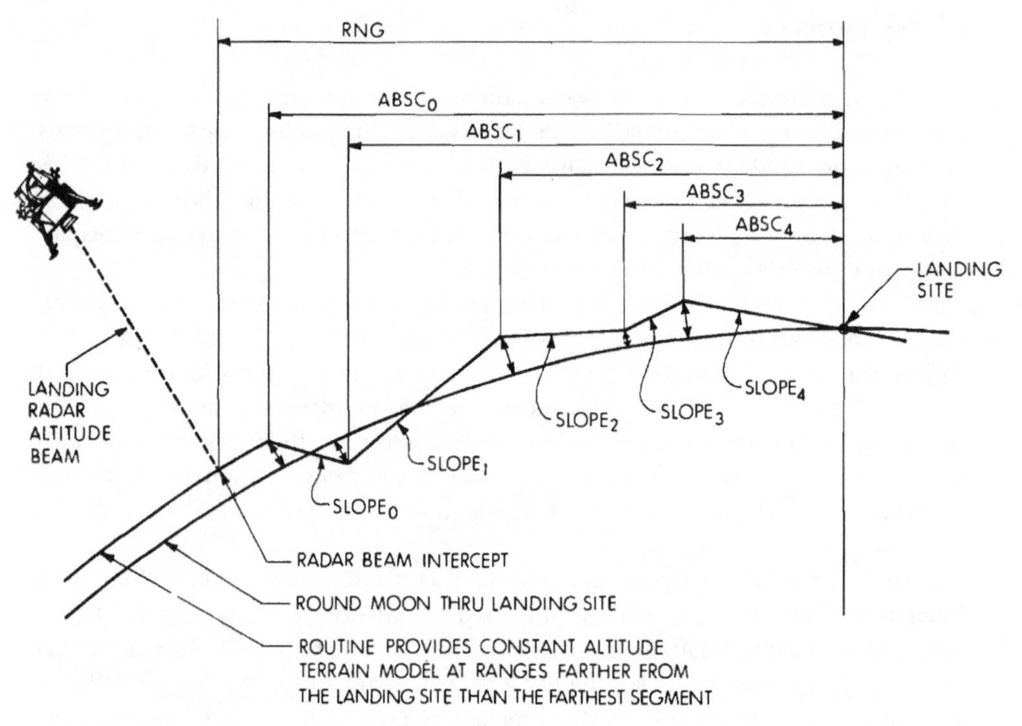

Figure 10.3
Lunar terrain model, stored in the Apollo computer as five line segments in later missions. (MIT Charles Stark Draper Laboratory, "Guidance System Operations Plan for Manned LM Earth Orbital and Lunar Missions Uising Program Luminary 1E: Section 5, Guidance Equations," R-567, December, 1971, 5.3-73.)

crew experienced miscommunication and confusion, the effects of errors in the simulator.

During Apollo 15's landing, the landing radar seemed in particularly good order and began updating at 50,000 feet. "Look at that! ALTITUDE and VELOCITY lights are out at 50 k!" Commander David Scott exclaimed, pleased that this tricky system, which almost aborted the Apollo 14 landing, seemed to be working better than expected. Soon the computer converged to a solution within 150 feet of the radar, working with the five-segment model of the moon's terrain now incorporated into the software (figure 10.3).[56]

Things proceeded normally as the guidance solutions converged, the computer throttled down, and the LM, dubbed Falcon, descended. The computer was expecting to see a one-degree slope of the surface below, but it turned out flatter than expected, so the trajectory flattened out for a minute or so, hardly enough to be a problem.[57]

Scott went into ATT HOLD mode, checking roll, pitch, and yaw controls. Everything seemed okay, so he returned to automatic.

At about 8,000 feet, Mission Control told Scott, "We expect you may be a little south of the site, maybe 3,000 feet." This information, according to Scott, "biased my estimation of where we were going to land." Indeed the ground controller was in error, for the slight southerly velocity that would have caused this error had already been corrected before PDI ("a double correction was made for this single error").[58] "My interpretation [Scott] was that our landing point had been moved . . . that was a confusing call."[59] Compounding this problem, Scott looked out the window and could see the 11,000-foot summit of Mount Hadley Delta. "We seemed to be floating across Hadley Delta and my impression at the time was that we were way long because I could see the mountain out of the window." They were supposed to land just past the mountain, but they were still more than 10,000 feet high. Scott couldn't see Hadley Rille outside the window, but from training he was used to seeing it in the left forward corner of his vision. So he concluded he was heading long and south.[60]

At 7,000 feet, P64 occurred, the LM began to pitch over.

Then Scott's disorientation began. "The problem was, when we pitched over and began to look out the window, there was nothing there!" He was looking for a series of craters dubbed Matthew, Mark, Luke, and Index (the last one located next to the landing spot). But if he could see them at all, they were extremely subtle.[68] From his long experience in the LM simulator, Scott expected to look out and see numerous craters on the surface, defined clearly by their shadows. Instead he saw very few, "not nearly as many as we were accustomed to seeing." He could not get oriented toward the craters he was expecting.[62]

Another problem, Scott later concluded, was the simulator, and the plaster of paris model it used of Mount Hadley, produced by the U.S. Geological Survey. Modelers worked from the best photographs they could get of the area from lunar orbiters, but they were relatively low resolution (sixty-five feet) and taken at a different sun angle than Apollo 15 was now experiencing.[63] The modelers "made themselves think the terrain had more topographic relief than it really did," overestimating the true relief of the site. The crew trained on the Lunar Mission Simulator, flying a little camera from just below the plaster of paris model. Once on the moon, features from the model either looked entirely different or seemed to be missing entirely. "I was very surprised," Scott stated, "that the general terrain was as smooth and flat as it was . . . there were very few craters that had any shadows at all, and very little definition."

Now Scott could see Hadley Rille, and "we were in fairly good shape, relative to the Rille, but we were south."[64] Irwin, meanwhile, was head-down inside the cockpit, concentrating on the instruments, reading aloud as much information as Scott could use.

Scott found the LPD to be excellent, and had confidence the LM was heading to the points he designated. "We were south, and I redesignated immediately four clicks to the right, and then very shortly thereafter . . . I redesignated two more right and three up range." Scott finally saw what he thought was the area of his intended landing spot, Salyut Crater (although afterward he concluded he had really seen another crater, Last Crater). He redesignated to that point. "I'm not sure how many other redesignations I put in heading for the target as Jim called the numbers. I may have put in a couple more." He put in nine more, making a total of eighteen, the most redesignations of any mission.[65] The new landing spot was 1,110 feet up range and 1,341 feet to the north of the original spot.

Scott's plan was that if he was on target, he would try to land exactly there. If he was not, he would try to find a place within the circle of acceptable error and land as soon as possible to conserve fuel. Now he realized that not only wasn't he heading for the original landing spot, he really didn't know where he was relative to any familiar landmarks. So he picked a smooth area nearby and headed right for it.

At about four hundred feet, Scott switched into P66 with a click of the left-hand rate-of-descent switch, and the LM began to descend. His intention was to stay on a constant flight path all the way down. Most of the previous flights had followed a "stair-step" pattern, where they leveled out too high due to the difficulty of perceiving altitude. "When you get to the Moon, there's no runway." Scott said, "There's nothing there to tell you how high you are; and I think the trend had been for people to start slowing up their rate of descent too soon."[66]

Scott found the LM's controls responsive, a familiar experience after working out in the LLTV. "Comfort and confidence existed throughout this phase," he recounted in his pilot's report.[67] "From the LLTV training . . . I knew I could land the machine if it would stay upright and the engine kept burning."[68]

About fifty feet high, the dust came up, completely obscuring the view outside. "It was completely IFR . . . and [I] flew with the instruments from there on down," Scott said. He kept his eyes outside, stole a few glances at the "eight ball" attitude indictors inside, and relied on Irwin's callouts.[69] Irwin read off the rates from the needles, and Scott slowed it down to one foot per second in the last fifteen feet.

One of the LM's modifications for the J-missions extended the length of the bell on the descent engine, and there was some concern about it burying into the lunar surface. So as soon as Irwin called "contact light," Scott pushed the STOP button on the engine (whereas other commanders had waited a second or two).

The LM dropped dead the last few feet and made a hard landing (6.8 feet per second, the hardest of the six landings).

"Bam!" yelled a startled Irwin, as the equipment inside the LM rumbled and rattled around ("like you would shake the vehicle").[70]

Scott: "Okay, Houston. The Falcon is on the Plain at Hadley."

Irwin: "No denying that. We had contact (that is, it was a hard bump)." As the heaviest LM yet, the Falcon hit harder than Irwin expected.[71]

Indeed, there had been a bump: the LM was overlapping the edge of a small crater, with the bell of the descent engine dropping directly on its rim, buckling it.

Downslope

On Apollo 16, pitch over proved less critical. A pre-mission study had suggested that the commander might be able to see the landing site out the window before the pitch over, so at 20,000 feet Commander John Young looked out the window. He had already seen the site during the two previous orbits so he was not surprised. Young saw Stone Mountain and South Ray Crater, indicating to him that the LM was right on target.[72]

At 14,000 feet Young saw the entire landing site: "We're right in there."[73]

At pitch over the LPD kicked in, allowing the commanders to redesignate the landing site. Young paused. "I was just letting the LM float in there until I could see where it was going," he said. Between 4,000 and 3,000 feet in altitude, Young made five redesignations to the south. At about four hundred feet, he noticed they were coming in a bit long. Young redesignated an additional five times, more to the south, and to "back up a hair." They were still going to land north, but having the rover on board eased the need for precision.

After all the redesignations, Young was tempted to let the LPD land automatically, "to let it do the thing all by itself," but saw as he got closer that it was going into a "pothole."[74] He took over manual control at 257 feet, pitching forward a bit to move along. From then on he never looked back into the cockpit, but concentrated on the rocks. "It was just like flying the LLTV. Your reference is on the ground outside."[75]

When the contact light came on, the astronauts could tell they were still a few feet up. Young counted "one potato," before hitting the button to stop the engine. Duke experienced "a definite sensation of falling," as they dropped for the last three feet.[76]

Apollo 16 landed about seven hundred feet from the original landing position.[77] It turned out they were only a few feet from the lip of a crater. Duke didn't notice it until he went around the back of the LM to retrieve some instruments. "If we'd have landed like 3 meters back to the east, we'd have been—had one—back leg in the crater."

In fact, astronauts on the moon had a great deal of difficult judging shallow slopes, and if they had landed a hundred feet in any direction they would have been on a slope of six to ten degrees. Young said afterward that he "just lucked into" a flat landing spot. Ironically, the spot had been chosen because the lack of contour lines indicated it was particularly flat, but that data gave no indication of craters or rocks. Young later said that he would have preferred to land in a place they saw on the traverse, near the Palmetto Crater. It didn't look great on the contour map, but when they got there it looked acceptable and had no rocks at all.[78]

Pilots or Scientists?

What sorts of skills were required to fly these landings? Of the six commanders, five were navy pilots, or had some training for landing on aircraft carrier landings (all except David Scott). Only two had graduate degrees, and only one a Ph.D. All commanders had been to space before, some as many as three times. Of the LMPs, the men in the right-hand position, only Aldrin had been into space before. Though named "lunar module pilots," they performed as systems managers, a task reflected in their educations: five out of the six had graduate degrees, three of them Ph.Ds. Some had detailed experience with the computer, such as David Scott who had studied for his master's degree at the MIT IL. He had been deeply involved in engineering decisions regarding the computer, particularly the user interface. The difference in education levels partly derives from the fact that the commanders were more senior and tended to have been picked earlier when piloting was emphasized over engineering, but it also points to the nature of the tasks. The commanders' tasks were engineered to be hand-eye and decision-making tasks at critical moments, drawing on more traditional pilot skills. The LMPs, by contrast, were systems engineers or systems monitors. As Charlie Duke recalled of his job during descent, "I was just keeping the LM running."[79]

Apollo 17 also saw a radical change in crew when the first non-test pilot, geologist Jack Schmidt, from a group of "scientist-astronauts," landed as LMP. As early as 1962, the National Academy of Sciences backed manned spaceflight on the condition that scientific observations would be made by "trained scientists, not pilots trained as astronauts." The academy proposed that on each flight one of the three Apollo crewmen be a scientist. They made the analogy with Darwin: it took a scientist, and not a bevy of mariners aboard the *Beagle*, to make his critical observations. NASA, of course, opposed the plan, arguing that "a test pilot must be present to deal with any emergency," and emphasized the flight-testing role of the missions.[80]

Still, in 1964, under continuing pressure from the National Academy, NASA began a selection process for scientist-astronauts, reluctantly admitting "there is not all that much distinction between a scientist and an astronaut."[81] The move raised the suspicions of the pilots who felt "NASA was caving in to the scientific community, bargaining for dollars and support by promising a ride for some guy toting test tubes."[82] Gene Cernan, who had already flown on Apollo 10 as lunar module pilot, considered science "a parasite" on the moon program, noting, "Science is not the reason we learned to fly."[83]

Nevertheless, NASA selected six scientists in 1965, including Harrison "Jack" Schmidt, the geologist who would fly on Apollo 17. Upon selection, NASA immediately sent the four who were not pilots for a year of air force jet training.[84] At least one scientist-astronaut thought the jet training irrelevant and objectionable, found the pilot-astronauts' scientific training to be superficial, and quit the program due to the clash of cultures between test pilots and scientists. Disputes arose about how much time was required for a scientist to stay "proficient" at research. "Jet jockeys versus absent-minded professors. What two groups could be more different?"[85] For Deke Slayton, head of the astronaut office, it came down to an issue of skill: "The lunar landing is purely a piloting technique.... You'd feel pretty bad to have to abort a mission because a guy on the right (the co-pilot) [LMP] wasn't competent, and a dead scientist won't do anybody good."[86]

A number of senior NASA scientists resigned after Apollo 11, at least in part because of dissatisfaction with the role of scientists in the upcoming missions. Several of the LMPs on those flights were test pilots with less seniority than the first batch of scientist-astronauts. Astronomer Eugene Shoemaker, who had been a vigorous supporter of Apollo science, told the press that unmanned flights could have accomplished the same amount of science, at much lower cost, years before.[87] These pressures produced Schmidt's crew assignment on Apollo 17.

Cernan eventually came to terms with his geologist LMP, calling him an "adequate" pilot, despite his lack of a career-long love for aviation. "Quite frankly, I don't think he had the aptitude or the desire to be a great pilot. Jack flew airplanes in NASA because he had to fly. That was one of the squares to fill to get on the Moon." Schmidt himself agreed with the assessment, calling himself a "pretty good" pilot, never putting himself within the class of fighter and test pilots.[88] Cernan qualified his judgment, however, regarding Schmidt's abilities in space: "When it came to flying the lunar module, 'adequate' does not describe Jack's ability to fly in the right-hand seat of the lunar module. He was outstanding." So, for Cernan, it was possible to be a merely "adequate" aviator but an "outstanding" LM pilot.

By contrast, Cernan called test pilot Joe Engle, who had been bumped from the roster for Schmidt's role in Apollo 17, an "outstanding aviator," but "only an adequate lunar module pilot." We should interpret these statements with some historical perspective, as they tend to justify the events of the mission: Engle being bumped becomes palatable if his skills were not of the highest quality, and Schmidt's presence is justified by his performance on the flight.

Once on the moon, Cernan bristled at being thought of as "Dr. Rock's taxi driver," and believed that Schmidt would not be able to get out of a tough spot on his own. "Dr. Rock was a product of Mission Control and the scientific laboratory environment," Cernan wrote in his memoir, "while I was an aviator."[89] Nevertheless, Apollo 17 was "the most productive and trouble-free manned mission," according to the mission report, which concluded that the flight "demonstrated the practicality of training scientists to become qualified astronauts and yet retain their expertise and knowledge in the scientific field."[90]

"Computer, I am now smarter than you"

As he approached the moon on Apollo 17, Cernan scanned the lunar surface, looking for his landing site, glancing inside at the instruments. Before pitching over, at about 8,000 feet, Cernan leaned forward toward the window and saw the tops of the mountains out the bottom of his window.[91] The crew could also see the earth, straight ahead outside the window. "You're allowed two quick looks out the window," Commander Cernan told his non-aviator LMP, "one now and one when we pitch over."

In his memoir, Cernan describes the moment of P64: "I pitched smoothly upright," suggesting he was controlling, when actually it was automatic.[92]

Then "all of a sudden, bam, the lunar surface filled up almost the entire window," and Cernan instantly recognized his landmarks. Using the LPD, Cernan removed the exact correction he had just entered with the Noun 69 command, "which means their [Houston's] targeting was essentially perfect."[93]

Still, Cernan decided that he could find better areas to land than the selected spot. "It was evident that boulders and craters were going to be determining factors in the selection of the final landing point," he later recalled. "I used LPD frequently . . . several clicks back, a couple left, a couple right. I just flew it where I wanted to fly it." Cernan redesignated several hundred meters south of the targeted point, to the right of the Poppie Crater (named after what Cernan's daughter called his father). "I just sort of tumbled in on that area and did some more LPDs to finally what I'd call a suitable landing site.[94]

Now Cernan took over in P66, at just under three hundred feet, to find an area level and free of boulders. He described the moment personally, in terms of his relationship to the machine: "In effect, I was saying, 'Computer, I am now smarter than you. You think you know where the target is, but I'm looking at it out the window. I know where it is, and I'm going to tell you.'"[95] In his memoir, he remembers it as an intimate connection with the spacecraft, "the LM had become part of me, responding to my wishes as well as my touch on the controls as we lowered closer to the surface."[96]

Cernan felt "extremely comfortable flying the bird," in both LPD and manual, which he attributed to his experience practicing in the LLTV.[97] In debrief, he concluded, "I can't say enough for what I consider the accuracy of the guidance. Manual control of the spacecraft was hard and firm, different certainly than the command module operation but exactly what I expected the LM to be."[98]

Vision, Skill, and Automation

Despite the differences between the landings, all had one thing in common: the commander did not allow the computer to land in automatic mode and instead operated the attitude "manually" in P66 (table 10.1). David Scott proved most reflexive about

Table 10.1
P66/ATT HOLD takeover altitude for six flights (compiled by the author from Apollo mission transcripts).

Flight	P66 height (ft.)
11	550
12	400
14	370
15	400
16	240
17	240

his use of the automation, perhaps because of his education at MIT and his comfort with automatic controls. He described a single, integrated system: "we—as a group flying the machine, the PNGS, AGS, Irwin, and me, we were all flying that thing." This "collective entity" was organized around the pilot to make it "safer and more efficient if there was a focal point. And I was the focal point. Jim fed things into my ears. The Moon fed things into my eyes, and I could feel the machine operating." Irwin read Scott as much information as possible so the commander could keep his eyes focused on the outside. "I wasn't going to do any talking." Scott said, "I was doing the flying. I was going to look outside the window, and he [Irwin] was going to tell me what was going on inside."[99]

As commonly told in newspapers, astronaut memoirs, and popular books, stories of Apollo landings usually recount how "the computer was taking us into a bad area," necessitating the commander's manual override of the mindless automation. Such accounts fail to acknowledge the heavily mediated nature of the "manual" intervention, which was really semiautomatic. Software created the very semblance of "manual" control of a complex beast like the LM and maintained a variety of feedback loops to make it flyable. Moreover, until a relatively low altitude of a few hundred feet, the commander could use the LPD to redesignate to a different area. Only below that level would "manual" takeover be required. Consider the comments of each of the commanders on their interventions:

Neil Armstrong, Apollo 11:
"The LPD was pointing . . . just short (and slightly north) a large rocky crater surrounded with the large boulder field with very large rocks covering a high percentage of the surface. . . . Continuing to monitor the LPD, it became obvious that I could not stop short enough to find a safe landing area."[100]

Pete Conrad, Apollo 12:
"I didn't like the size of the area short, where we had normally been trying to land, and I looked for a more suitable place."[101]

Alan Shepard, Apollo 14:

"Using thirty years of pilot skills, [he] threaded a needle between the hills and ridges along their approach path and dropped his ship down into a narrow valley, craters and boulders everywhere."[102]

David Scott, Apollo 15:

"I knew I could land the machine if it would stay upright and the engine kept burning. And all the other things I had was icing on the cake."[103]

"I was very surprised that the general terrain was as smooth and flat as it was.[104]

John Young, Apollo 16:

"It was working so well I was tempted to let it (the PNGS) do the thing all by itself; but the trouble is, we got down low and I could see that we were going to land in that pothole down there."[105]

Gene Cernan, Apollo 17:

"The reason I took over is that I wanted to slow our forward velocity down. I did not want to go any farther west, because there were more blocks and more hummocky terrain."[106]

"Finding a place to land wasn't as easy as anticipated. A boulder the size of a house—that wasn't supposed to be there—loomed right in front of me... I scanned for an empty space in a parking lot of boulders as big as automobiles."[107]

Each commander describes human control of the mission in the last few moments. Nearly all mention that the LPD was carrying the LM into a rocky, cratered area. Of course, given the rocky nature of the moon, that's not too surprising. But it did happen on every single flight. Fortunately, according to the commanders, on every single flight there was a smooth area suitable for landing a short distance away.

The exception in the comments above is from David Scott, who explained his manual control as preparation for reacting quickly in case something went wrong. "If I'm in the loop mentally I can respond more quickly than if I have to watch an automatic system and take over and then get my mind into the loop. So I was more comfortable in flying a manual descent than an automatic descent... my mental computer is already running at full speed." As he indicated in the quote at the beginning of this chapter, this apparent safety might have been an illusion—a computer problem would likely have affected the manual control as much as the automatic. But a hand on the stick gave the commander a feeling of confidence and control. Scott admitted that "mainly, it's a challenge" to do the landing manually, and being in the loop for emergency response was secondary. Scott was thrilled by the difficulty of the landing. For him, and for the other pilots, it represented the ultimate in flight. "I don't care what anybody says, that's damn hard. I mean, that takes real aviation. That's flying!"

The astronauts' own comments tend to emphasize visual recognition of the landing spot as the unique human capability, demanding hands-on piloting to land safely. "You could fly automatically, but it's not likely," Armstrong told the SETP, noting, "one reason is that the auto system doesn't know how to pick an area and it can't change its mind."[108] Armstrong conflated two issues in a way that was common in discussions of the landings: the perception and judgment required for selecting a landing site and manual versus automatic control of the LM.

NASA management, and at least some of the astronauts, acknowledged this conflation. Comparisons of the Apollo 11 and 12 landings raised the question of why, once the landing spot had been chosen, the computer should not gently land the LM, given that an instrument landing might be required anyway. Indeed, after Apollo 12 the IL team added software to make an automatic touchdown easier for the commander. Now anywhere below 500 feet the commander could enter could enter "P66 Auto" mode (velocity-nulling guidance), in which the computer would bring the LM to a gentle stop, hover, and descend at a constant rate. P66 Auto could land a LM blind, through a dust cloud like the one Conrad had encountered.[109] The feature was included on every flight from Apollo 13 to 17. It was never used.

Even P66 Auto did not represent the limit of the computer's capability. Astronaut John Young suggested an additional feature. In P66 Auto mode, the astronaut could "mark," or designate a landing spot as it moved past a particular point across the LPD marker on his window. P66 Auto would then fly the LM to that spot, null the velocities, and land. Young's suggestion (which became known as "P66 LPD") would allow the commander to request specific horizontal velocities by jogging his control stick, much as he jogged the stick to move the LPD point around. A left deflection of the stick, for example, would increment the horizontal velocity two feet per second to the left, and a right deflection could cancel it out. Similar motions could move forward and backwards.[110] Don Eyles implemented this mode in simulations, demonstrating precise landings, but it was never included in an Apollo flight.[111] Given more time, and perhaps more flights, the landings would likely have evolved further computer control, decreasing pilots' workload and the attendant risk during the final descent.

None of this is to say that the astronauts, their perceptions, judgments, or flying skills were irrelevant to the landings. Quite the contrary. The sociotechnical system, including the engineers, flight controllers, and programmers on the ground, as well as pieces of machinery, was impressive, precise, even wondrous, achieving a successful landing on all six attempts. But it was not perfect. Programs alarmed, guidance overshot, boulders appeared, people misspoke, and buttons failed. In each case, human abilities intervened in unplanned ways, made decisions, and landed the spacecraft on the moon.

11 Human, Machine, and the Future of Spaceflight

For a while I was afraid that Apollo might be one of the last battlefields on which the human race took up arms against machines. Catch phrases such as "man in the loop," "man out of the loop," the middle ground of "man across the loop" and, I suppose, even man just "looped" have purported to represent the proper solution to one of the more subtle system problems facing the program. . . . From this point of view, the terms "manual" and "automatic" carry more emotional than technical content.

—Joe Shea, Deputy Director of Manned Space Flight (Systems), NASA

In 1963 Joe Shea addressed the second Manned Space Flight Meeting of the American Institute of Aeronautics and Astronautics (AIAA). He explained the "systems view" that had originated in Bell Telephone Laboratories and in the U.S. Air Force's ballistic missile programs and was now pervading NASA's upper administration. As the paramount example of this philosophy, Shea offered NASA's approach to the human-machine relationship, "one area in danger of being overwhelmed by dogma." Systems engineering, he argued, could help engineers move beyond that dogma.[1]

Shea granted that both humans and machines had critical roles to play in space missions. While human pilots could control docking, "selection of the landing site from the hover point and, perhaps, lunar touchdown," their most important function would be monitoring systems and selecting alternate modes in case the systems failed; hence his evaluation of the "emotional" nature of the manual versus automatic dichotomy. To call a control manual, when the output of an inertial platform is sent through a digital computer and then displayed to a human who operates a hand controller also connected to that computer, "is stretching things a bit"; under these conditions, the distinction between humans "in" or "out" of the loop, Shea concluded, became primarily one of "semantics."

Shea presciently recognized how traditional understandings of "automatic" and "manual" no longer applied in the world of programmable, digital systems. Traditional social roles of the operators would change as well (what the IL's Jim Nevins called "a transition in the art of piloting"). Moreover, Shea speculated on the broader implications of these changes, leading some to see the Apollo program as a "battlefield," on which humans fought for their cause against the encroachment of machines. Shea's speech to the AIAA came early in 1963, when he was still at NASA headquarters, months before he would head to Houston and bring his approach to bear on its pilot-oriented culture, years before Apollo hardware actually flew or men landed on the moon. Nonetheless, he asked questions about the relationships between systems approaches and pilot-oriented design that played out in Apollo, from Apollo 11's program alarms to Apollo 17's professional explorations.

"Wings and wheels at last"

Were these questions unique to Apollo? Did they arise from some special set of social and technological circumstances during the early history of spaceflight and computer? Absolutely not: the questions raised by Apollo's history remain fundamental to human spaceflight and other technological endeavors.

Immediately after Apollo (indeed during Apollo), NASA chose to send an aircraft into orbit. The space shuttle put pilots at the center of American manned spaceflight for a generation. Chauffeurs versus airmen? The pilots won.

Robert Chilton, the man responsible for the digital computer on Apollo, saw the shuttle's genesis this way: "I was very unhappy with the way the program was going away from the Moon, and that was all largely an astronaut influenced decision. They were pilots and they wanted to—nobody had ever built a spaceship that had wings or landed, and they thought that was a deficiency. . . . They wanted something they could fly and land. So I didn't believe any of the justification for Shuttle."[2] In 1970, during the shuttle's formative phase, one engineer complained, "They're [NASA] talking about automatic landings and the pilots are yelling, 'Keerist, no!' [But] you can design this thing so that the pilot still makes the decision with respect to the landing and yet it's done automatically."[3] Like the LM, the shuttle relies on a fly-by-wire control system (running software created in part by Intermetrics, a Cambridge, Massachusetts, firm founded by five MIT programmers from Apollo).[4] Like the LM, the space shuttle has an automatic landing feature that is never used.

In his memoir, astronaut Walter Cunningham made the point in terms of the pilots' professional dignity. Whereas the Apollo command module fell into the sea, the crew recovered "by helicopter like a bag of cats saved from a watery grave," the space shuttle orbiter, by contrast, "makes a smooth landing at the destination airport and the flight crew steps down from the spacecraft in front of a waiting throng in a dignified and properly heroic manner."[5]

The shuttle first flew in April 1981, culminating with Columbia's landing at Edwards Air Force Base, Apollo 16 Commander John Young at the controls. The moment represented America's return to space after a seven-year hiatus, but also the return of spaceflight to California's high desert, which some saw as its proper home, the test pilots reigning supreme. Michael Collins perhaps put it best in his book on the history of spaceflight: "Wings and wheels . . . a dignified flying machine at last, no more awkward capsules."[6]

In the 1950s, advocates for the X-15 project argued that reentry would require human skill. Yet humans have manually flown only one shuttle reentry: flight number two of Columbia, when former X-15 pilot Joe Engle flew from Mach 25 down to the ground to test the stability and control system. Subsequent shuttle reentries have flown automatically, though pilots still control the landing: despite the presence of automated systems, every shuttle flight has ended with a human flying final approach and touchdown "manually."

During the shuttle's development phase, voices within NASA called for including the ability to fly fully automatic missions. NASA neither planned such missions nor designed options for them into the system, arguing that only human presence would maximize the chances for successful return of the orbiter.[7] In recent comments, Chris Kraft, who was director of the Manned Spacecraft Center at the time, regretted this decision because an automatic system would have allowed a quicker and easier return to flight after the Columbia and Challenger disasters.[8] (The Russian copy of the shuttle, named Buran, flew once in 1988 with no crew and landed automatically on a runway in a stiff crosswind.)

Strangely, the only aspect of a shuttle reentry that cannot be automated is the landing gear, which the crew must deploy manually with a lever. After the Columbia accident, shuttle flights do include a crude automatic landing feature as a last resort. If the shuttle's reentry protection was crippled (as happened with Columbia), the crew could evacuate to the International Space Station. They would then rewire the shuttle to install a basic capability to separate the orbiter from the station, reenter, deploy the gear, and land under computer control. Experiences of disaster and evolving attitudes toward risk prompted changes in the balance between human and machine in the shuttle system.

This short survey of pilots, professional identity, and automation in the space shuttle's history is merely suggestive. The shuttle emerged from a complex mix of engineering, politics, and institutions (the physical dimensions of the vehicle literally reflect NASA's political alliances to secure the program's funding) combined with impossibly optimistic estimates of cost savings for a reusable vehicle. Still, a scholarly, document-based history of the role of astronauts and NASA's astronaut office in the conceptualization and development of the shuttle would elucidate the role of human-machine questions in setting NASA's agenda in manned spaceflight for the past four decades.

As if to comment on that history, NASA's next manned spacecraft, designed for travel to the space station and possible return to the moon, resembles an Apollo-like command module. NASA Administrator Michael Griffin calls it "Apollo on steroids," suggesting a spacecraft that can still be muscular without wings or wheels. The new Crew Exploration Vehicle (roughly equivalent to the command module in Apollo) will have fully automated rendezvous and docking, with manual control as a backup.

For lunar landings, at least some NASA engineers are still contemplating pilots' hands on the stick in the final moments (although several Apollo veterans I spoke to suggested future landings be fully automated). Current plans involve automated landers delivering supplies and returning to the earth as engineering tests before any human landings.[9] High-resolution maps, precision navigational sensors, computer imaging, advanced user interfaces, and modern computer power will surely change the equation from the 1960s. Will future lunar landings be fully automated? If so, who will be in command? Chauffeurs or airmen? Scientists or pilots? Or others?

"Automated Cockpits: Who's in Charge?"

Apollo's connections to human-machine issues are not limited to spaceflight but also extend into aviation. In the early 1970s, engineers at NASA's Dryden Research Center (formerly the Flight Research Center at Edwards Air Force Base) began building on their experience with the Lunar Landing Research Vehicle (LLRV) and its analog fly-by-wire control system. Neil Armstrong suggested they use a digital computer, an Apollo computer in fact, which had proven its reliability. The cancellation of the later Apollo flights meant extra hardware would be available for research.

The Dryden group traveled to the Instrumentation Laboratory for meetings and learned that developing a digital flight control system now involved writing software.[10] From an unused Apollo spacecraft they acquired a computer, an inertial platform, and a DSKY and installed it in an F-8 fighter jet, the first digital fly-by-wire installation in an aircraft. Perhaps most important, the Dryden group also employed the MIT group's software development tools, program validation techniques, and testing processes. Sixty percent of the software for the first digital fly-by-wire aircraft consisted of Apollo code.[11]

In NASA's fly-by-wire aircraft, as in the LM, a computer accepted the pilots' inputs and actuated the control surfaces according to a software program, allowing a great deal of flexibility in controlling the vehicle (ideally improving reliability and reducing weight). The pilot merely had to move a control stick that issued commands to the computer. Unlike in the LM, the DSKY was buried in a lower maintenance bay, available only to ground crews.[12] NASA's test pilots would not have an LMP available to punch keys.

In the 1970s fly-by-wire systems made their way into fighter aircraft like the F-16. Continuing the trend from the 1950s, this technology meant that new aircraft (like the space shuttle) no longer needed to be inherently stable; now computer control could provide stability for them. Exotic-looking aircraft like the B-2 stealth bomber, for example, dispensed with traditional tail surfaces altogether as multiple, redundant computers kept the aircraft flying. European aircraft manufacturer Airbus began building fly-by-wire commercial airliners in the late 1980s with the A-320 passenger jet. For the pilots, "glass cockpits" began to replace traditional flight instruments with screens connected to flight computers.

As in previous generations, the new technologies fueled debates over pilots' roles. A philosophical difference emerged between Airbus and its American counterpart, the Boeing company. The former tended to include more automation and side-stick controllers (not unlike those on the X-15 or the shuttle) instead of traditional hands-on control yokes. Under certain circumstances Airbus software would prevent pilots from making dangerous maneuvers with the aircraft. By contrast, Boeing retained traditional flight controls and allowed pilots ultimate authority, even when the company introduced fly-by-wire in the 1990s.

In the late 1980s and early 1990s, pilots' interaction with flight computers caused a series of high-profile accidents in both Airbus and Boeing aircraft, drawing attention to cockpit automation as a public safety issue. An ominous cover story in *Aviation Week and Space Technology* asked, "Automated Cockpits: Who's in Charge?" Inside, an editorial pressed, "Preserve pilot's roles in automated cockpits. There are justified concerns that computers are taking control authority out of the hands of the pilots."[13]

The lay press immediately picked up on the social implications of these changes in the airborne workplace: "today, having the right stuff means the byte stuff... when aviator becomes computer monitor."[14] Pilots grew concerned about losing their touch, what one pilot called the "I-can't-fly-anymore-but-I-can-type-80-words-a-minute syndrome" (indeed today's newest Airbus airliner places keyboards where control yokes used to be). One trade publication warned, in language reminiscent of the SETP's concerns in the 1950s: "Men and women who expected to participate in a great adventure of individualism and personal control are memorizing tedious procedures manuals.... It is a great irony called... automation. Rather suddenly, the profession of aviation—that vocation of leather jackets and exciting experiences—has become a career of computer monitoring and equipment management. You once thought you would dance on silver wings, and now you are merely tapping on a keypad."[15] Government and industry responded with further human-factors research, novel simulators, and improved training methods. The automation-induced accident rate decreased.

Now remote-controlled robot aircraft, called Unmanned Aerial Vehicles (UAVs), are beginning to pervade military operations. Some ask whether automated or remote-controlled airliners can be far behind. Will people be willing to risk their lives on pilots sitting on the ground hundreds of miles away?

In addition, digital and automated cockpits have come to "general aviation" aircraft, right down to the small Cessnas flown by private pilots. These new technologies raise familiar questions about skill, training, and whether inexperienced pilots become too reliant on automation and prone to distraction by computers and display screens. Yet the new control systems also offer safe access to new realms for amateur pilots. Human lives depend on striking the right balance.

Aviation is not alone; professional identities in a broad range of fields face similar challenges from automation, simulation, remote operation, and telepresence. Architects now build prototype buildings entirely inside computers, leading some to question whether they are becoming mere computer operators and losing touch with the "reality" of their craft. Scientists conduct experiments entirely in simulated environments, leading some to question whether they lose the "feel" of the physical world. Archeologists use remote robots to explore shipwrecks in the very deep ocean, leading some to question whether one can do "real" archaeology without physically touching a site. Oceanographers debate whether to build new generations of human-occupied submersibles or fleets of unmanned robots, leading some to question whether "real" oceanographers must personally visit the sea floor. Surgeons use precise robots to augment their manual skills and conduct novel surgeries, pressing their hands into computer interfaces rather than physical bodies. What do we make of a military UAV pilot who flies a robotic bombing mission from a remote control station, thousands of miles from an actual war?

Agendas for Research

Space history offers one approach to studying these thorny, fundamental issues. It is, of course, impossible to explore their full spectrum within the scope of this book. Even within the Apollo program, questions were not limited to the landings. From the astronauts' relative passivity during launch, to their systems monitoring during other phases of flight, to their operations as geologists on the lunar surface, each phase of flight entailed a series of trade-offs between human and machine. Nevertheless, a close analysis of the Apollo landings illustrates an approach to the history and sociology of human spaceflight that will have broad impact.

Questions about human-machine interaction, when addressed from primary documentation, open up new avenues into seemingly well-worn topics. As a meeting point for the social and technical aspects of a system, the human-machine relationship connects a variety of dimensions of space history that are otherwise difficult to integrate into a coherent narrative. For instance, the iconic role of astronauts as American heroes was critically dependent on their roles (real and perceived) in actual piloting of the missions. Questions of control unify Apollo's cultural and political dimensions with the evolution of its hardware and software.

As a final example, consider the lunar EVAs ("extra vehicular activities," colloquially known as "moonwalks"). Here the astronauts sought new professional roles, seeking to become like scientists, or, as they put it, "explorers." Where did they exercise scientific judgment in collecting samples and data? Which actions, specifically, constituted exploration (versus science, data collection, instrument deployment, etc.)? How did these actions necessitate human presence?

Beginning with Apollo and continuing during the 1970s (and onward), the professional identity of astronauts began to expand—from the exclusive focus on test pilots to scientists and engineers to new job titles like "mission specialist" and "payload specialist," coupled with demographic expansion beyond white men. What did these new professionals do? I recently asked an astronomer-astronaut how much he used his scientific judgment while in orbit. "Not at all," he quickly replied. Most of his time had been spent following well-established procedures to deploy and operate other scientists' automated experiments. Under such conditions, what is the necessity for scientific training? Or for human presence at all? Still, that same astronaut acknowledged that being able to "speak the same language" as the scientists on the ground proved an important part of his job. Clearly, tacit knowledge, social interaction, and shared vocabulary play an important role in space operations.

Scholars who can deeply analyze scientific practice can explore these issues further, across a broad range of spaceflight. Like this study, an anthropology of space operations would examine skill, training, professional identity, automation, risk, organizations, divisions of power, and other aspects of human-machine relationships. Sources exist for Skylab, the space shuttle, and the International Space Station, as well as for similar endeavors in other countries. What, *exactly*, are engineers, pilots, and scientists doing in orbit? When are they using judgment, skill, experience, and expertise, and when are they following scripts? What kinds of contingency operations do they perform? How are skills and judgment divided between those in orbit and those on the ground? What kinds of "repair" do humans do (including rework and workarounds) when operating complex systems? Numerous spaceflights have been recorded and transcribed in detail (all of the Apollo flights and at least some of the shuttle flights), allowing such real-time ethnography. Mission transcripts, combined with deep analyses of operations, provide an empirical basis for exploring such questions.

Similarly detailed studies of unmanned projects will help us see the degree to which remote systems provide scientists and engineers on earth a sense of "being there," as well as the value of that remote presence for missions' success. How do operators of remote vehicles—in air, in space, or even undersea—explore unknown places? What skills do they require? What sensors, imagery, or data presentations can enhance their experience? What are the limitations of remote presence?

Such studies, if rigorously done by disinterested scholars, would have implications for engineering design, training, mission planning, and safety in spaceflight. Moreover, comparing such studies of human- and remote- spaceflight will provide much-needed insight into the appropriate mix of these technologies in a national space policy. Only by conducting such research will we be able to understand the evolution and operation of human (and remote) spaceflight and make informed engineering and policy decisions for future endeavors. This work will also likely generate insights into other complex technical systems whose operations are rarely as well documented or as accessible as human spaceflight.

We can also help clarify the relationship between science and exploration. The documents leading up to Kennedy's 1961 Apollo decision assumed that that "exploration" is done by humans, and "science" is remote or automated, assumptions that broke down during the Apollo missions. The distinctions remain unclear today. George Bush's January 2004 speech kicking off a new American space policy used the word "exploration" more than twenty-five times, while mentioning "science" only once (regarding elementary education).[16]

When must exploration entail physical human presence? Exploration has a long history, but when that history has been brought to bear on spaceflight it has tended to serve advocacy and hagiography more than critical analysis. The history of science and exploration has a great deal to offer current debates.[17]

Exploration often includes science, but usually as one component of a broader agenda, and not usually the most important one. We can make this oversimplified distinction: science involves collecting observations to learn about the natural world, whereas exploration expands the realm of human experience. Sometimes the two overlap, but not always. Exploration has always had significant components of state interest, international competition, technological display, public presentation, national and professional identity, and personal risk—components ideally suited to human spaceflight (under the right political and cultural conditions). Hence the prominence of these nonscientific elements in Apollo seems less anomalous than sensible in a historical context.

In this light, justifying human spaceflight by emphasizing unique human abilities to collect scientific data seems a distraction, always to be overtaken by the next generation of technology. Better machines will always come along to choose landing sites, collect images, map terrain, and so on. Human judgment will, by definition, always be required to interpret the data, but perhaps in the comfort of earthly offices or control rooms.

By contrast, justifying human spaceflight as an expansion of human experience is independent of technology, ultimately a human rather than a technological aspiration. Perhaps NASA's technical and political orientation has prevented it from appealing for public support for so humanistic an ambition. Perhaps the agency lacks the language to make the case—although NASA administrator Michael Griffin has recently spoken of "acceptable reasons" for human spaceflight, favored by the policy process, and the "real reasons," that motivate those involved. Only a precise and rigorous framing of the case for (or against) human spaceflight can generate informed, productive, and engaged public discussion of NASA's envisioned futures.

My goal here is not to take a side on whether we should be sending people or robots into space. Rather, I argue that a historical understanding of the human-machine relationship helps to clarify the terms of this important debate, a clarification critical for those seeking to intelligently commit public resources for future projects.

The Virtual Explorer

In the spring of 2004, the Explorers Club of New York City held its 100th annual dinner at the Waldorf Astoria. At this glitzy, black-tie affair, a few thousand people consumed hors d'oeuvres of roasted tarantulas, braised earthworms, and poached bovine brains, and then crowded into the grand ballroom. The Explorer's Club has always included scientists but also welcomes a panoply of mountain climbers, navy captains, pilots, sailors, divers, trekkers, photographers, astronauts, and wannabe adventurers.

Above the crowd, on the ballroom's stage, sat a number of the greatest living explorers. Each rose in turn to give inspiring speeches about their own experiences and the importance of exploration. Bertrand Piccard, heir of the great Swiss exploring family, recounted his circumnavigation of the earth in a hot-air balloon. Sir Edmund Hillary described the feeling of his first steps atop Mount Everest. Buzz Aldrin spoke about his Apollo journey and advocated for a human return to the moon and a venture to Mars.

The evening's last speaker was Dr. Stephen Squyres of Cornell University, the chief scientist of the Mars Exploration Rover Project that had recently landed two robotic vehicles on the surface of the red planet. I leaned over to my friend sitting next to me and whispered, "This ought to be interesting because the rest of those guys have actually gone places, where Squyres has done all of his work remotely." A moment later, on the heels of the great explorers, Squyres got up, in front of thousands of people, and said (I paraphrase): "I must say I'm a little intimidated, because all of these people have actually gone somewhere, whereas all I've done is rack up frequent flyer miles between Ithaca and Pasadena."

But nothing in the speech that followed lacked the excitement of the others. "We build fiendishly complex robots. Our techniques lack the glamour and glory, but if you really like exotic destinations, we've got you covered," Squyres said, and went to give an account of his group's remote robotic exploration of Mars that easily matched the others' in energy and interest. He explained how the scientists and engineers in his group at the Jet Propulsion Laboratory in Pasadena "live" on Mars, from darkened control rooms, for months at a time, through technologies of remote, virtual presence.

For Squyres, as for the Apollo astronauts, exploration entangled humans and machines, defining new forms of heroic action in a technological world.

Notes

1 Human and Machine in the Race to the Moon

1. The term "mythology" in this book does not imply a false story. Rather, I use the term in the anthropological sense: a mythology is a story or collection of stories told within a culture to convey something about its origins, goals, or character, regardless of its truth-value.

2. Lindbergh and Green, *We*.

3. Smith, "Selling the Moon," 189. See also Ward, "The Meaning of Lindbergh's Flight," and Launius, "Heroes in a Vacuum."

4. Webb, quoted in Logsdon, *The Decision to Go to the Moon*, 90, 125.

5. Wiesner Committee, "Report to the President-Elect of the Ad Hoc Committee on Space."

6. John F. Kennedy, "Special Message to the Congress on Urgent National Needs."

7. Beattie, *Taking Science to the Moon*; Benson and Faherty, *Moonport*; Brooks, Grimwood and Swenson, *Chariots for Apollo*.

8. Aldrin and Warga, *Return to Earth*; Armstrong et al., *First on the Moon*; Bean and Fraknoi, *My Life as an Astronaut*; Cernan and Davis, *The Last Man on the Moon*; Conrad and Klausner, *Rocket Man*; Duke and Duke, *Moonwalker*; Irwin and Emerson, *To Rule the Night*; Mitchell and Williams, *The Way of the Explorer*; Shepard and Slayton, *Moon Shot*; Thompson, *Light This Candle*; Schirra and Billings, *Schirra's Space*; Borman and Serling, *Countdown*; Cunningham and Herskowitz, *The All-American Boys*; Lovell and Kluger, *Lost Moon*; Stafford and Cassutt, *We Have Capture*.

9. Kraft, *Flight*; Kranz, *Failure Is Not an Option*; Liebergot, *Apollo EECOM*.

10. Murray and Cox, *Apollo*.

11. Kelly, *Moon Lander*.

12. Chaikin, *A Man on the Moon* was made into the HBO series *From the Earth to the Moon*.

13. Atwill, *Fire and Power*; Carter, *The Final Frontier*; McCurdy, *Space and the American Imagination*; Kauffman, *Selling Outer Space*; McDougall, *The Heavens and the Earth*.

14. Campbell, *The Hero with a Thousand Faces.*

15. Siddiqi, "American Space History."

16. Collins, *Carrying the Fire*, 16–17.

17. Aldrin, quoted in Smith, *Moondust*, 116.

18. Glines, *Roscoe Turner.*

19. Kauffman, *Selling Outer Space*, 36–37, 56–67.

20. Braverman, *Labor and Monopoly Capital*; Noble, *Forces of Production*; Bix, *Inventing Ourselves Out of Jobs?*

21. Hong, "Man and Machine in the 1960s"; Turner, *From Counterculture to Cyberculture*; Dick, *Do Androids Dream of Electric Sheep?*; Ellul, *The Technological Society*; Mumford, *The Myth of the Machine*; Pynchon, *Gravity's Rainbow.*

22. Ackmann, *The Mercury 13*; Weitekamp, *Right Stuff, Wrong Sex*; also see Weitekamp, "Critical Theory as a Toolbox."

23. Faludi, *Stiffed*, chapter 9.

24. McCurdy, *Inside NASA.*

25. Clynes and Kline, "Cyborgs and Space."

26. For an idiosyncratic but useful review of the debate, see McCurdy, "Observations on the Robotic versus Human Issue in Spaceflight." Also see Roland, "Barnstorming in Space."

2 Chauffeurs and Airmen in the Age of Systems

1. Society of Experimental Test Pilots, *History of the First Twenty Years*, 39–43.

2. Horner, "Banquet Address," 3–4.

3. Ibid., 7.

4. Ibid., 8.

5. Ibid., 9.

6. Quoted in Vincenti, *What Engineers Know*, 57.

7. Gibbs-Smith, *Aviation*, 58.

8. Ibid., 96.

9. Wright, "Some Aeronautical Experiments," 100; Crouch, *The Bishop's Boys*, 167–169, 212–213.

10. Wilbur Wright, quoted in Gibbs-Smith, *Aviation*, 222.

11. *The Oxford English Dictionary*, etymology for "skill."

12. Abbott, *The System of Professions.*

13. Wohl, *A Passion for Wings*; Corn, *The Winged Gospel*, 41.

14. Kennett, *The First Air War*, 156–157.

15. Fritzsche, *A Nation of Fliers.*

16. Quoted in Vincenti, *What Engineers Know*, 63.

17. Ibid., 62.

18. Gann, *Fate Is the Hunter*, 7, 15.

19. Hallion, *Test Pilots*, 101.

20. Conway, *Blind Landings.*

21. Doolittle and Glines, *I Could Never Be So Lucky Again*, 73.

22. Ibid., 104.

23. Doolittle, "The Effect of the Wind Velocity Gradient on Airplane Performance"; "Wing Loads as Determined by the Accelerometer."

24. Leary, "The Search for an Instrument Landing System"; Hughes, *Elmer Sperry*; Conway, *Blind Landings.*

25. Hallion, *Test Pilots*, 103. Doolittle and Glines, *I Could Never Be So Lucky Again*, 129–131.

26. Dennis, "A Change of State," 58, 67–69; Draper, quoted in Dennis, "A Change of State," 111.

27. Kelly and Parke, *The Pilot Maker*; Conway, *Blind Landings*, 264–267; Cameron, *Training to Fly.*

28. Vincenti, *What Engineers Know*, 79.

29. Roland, *Model Research*; Bilstein and Anderson, *Orders of Magnitude.*

30. Hansen, *Engineer in Charge.*

31. Gilruth, "Requirements for Satisfactory Flying Qualities of Airplanes"; Gilruth and White, "Analysis and Prediction of Longitudinal Stability of Airplanes."

32. Phillips, *Journey in Aeronautical Research*, 32.

33. Barthes, "The Jet-Man"; Constant, *The Origins of the Turbojet Revolution*; English, "Jet Pilots Are Different."

34. Hallion, *Test Pilots*, 143.

35. Armstrong quoted in Hansen, *First Man*, 188. On Walker, Rathert interviewed by Greenwood and Swenson.

36. Cooper, "Understanding and Interpreting Pilot Opinion," 19.

37. Ibid., 21.

38. Ibid., 23.

39. Abzug and Larrabee, *Airplane Stability and Control*, 33.

40. Armstrong, "Where Do We Go from Here?"

41. Mindell, *Between Human and Machine*, chapter 5.

42. Evans, "Graphical Analysis of Control Systems."

43. McRuer, Ashkenas, and Graham, *Aircraft Dynamics and Automatic Control.*

44. Abzug and Larrabee, *Airplane Stability and Control*, chapter 20.

45. Blackburn, "Flight Testing Stability Augmentation Systems for High Performance Fighters," 2.

46. Armstrong, email to author, April 20, 2004.

47. Roberts, "The Case against Automation in Manned Fighter Aircraft," 10.

48. *The Oxford English Dictionary* entry for "system."

49. Ridenour, *Radar System Engineering*; Goode and Machol, *Systems Engineering.*

50. Wiener, *Cybernetics*; Heims, *Constructing a Social Science for Postwar America*; Gerovitch, *From Newspeak to Cyberspeak.*

51. Clynes and Kline, "Cyborgs and Space."

52. Hughes, *Rescuing Prometheus*, 99; Hughes, and Hughes, *Systems, Experts, and Computers.*

53. Sapolsky, *The Polaris System Development*; Pinney, "Projects, Management, and Protean Times"; Walker and Powell, *Atlas: The Ultimate Weapon.*

54. For a history of systems thinking in the Atlas project, see Hughes, *Rescuing Prometheus*, chapter 3. Simon Ramo is quoted on p. 67. Ramo, "ICBM: Giant Step into Space," 83.

55. Armstrong, "Where Do We Go from Here?"

56. Blackburn, "Flight Testing in the Space Age," 10–11.

57. Ibid., 10.

58. Ibid.

59. Quesada, "A Pilot's Philosophy for the Space Age." Emphasis added.

60. Blair, "Automation and the Space Pilot."

3 Flying Reentry: The X-15

1. Thompson, *Flight Research*, 45.

2. Gorn, *Expanding the Envelope*, 199.

3. Becker et al., "NACA Views on a New Research Airplane," quoted in meeting minutes, NACA Committee on Aerodynamics, NASA HQ, October 4–5, 1954.

4. Ibid.

5. Meeting minutes, NACA Committee on Aerodynamics, NASA HQ, October 4–5, 1954.

6. Ibid.

7. Johnson, "Minority Opinion on High Altitude Research Airplane," attached to meeting minutes, NACA Committee on Aerodynamics, NASA HQ, October 4–5, 1954.

8. Hallion and Gorn, *On the Frontier*, 106–107.

9. Crossfield, quoted in Jenkins and Landis, *Hypersonic*, 35.

10. "IAS Chanute Award to Armstrong," *X-Press, NASA Flight Research Center Newsletter*, June 22, 1962, NASA Dryden Archives.

11. Thompson, *At the Edge of Space.*

12. Crossfield, quoted in Jenkins, *Hypersonic*, 99. Also see Thompson, *At the Edge of Space*, 91.

13. Thompson, *Flight Research*, 27–28.

14. Crossfield, "The Way to the Stars."

15. Holliday and Hoffman, "Systems Approach to Flight Controls."

16. Waltman, *Black Magic and Gremlins*, 29–32.

17. Ibid., 16.

18. Milton Thompson, "General Review of Piloting Problems Encountered during Simulation and Flights of the X-15," reprinted in Waltman, *Black Magic and Gremlins.*

19. Thompson, *At the Edge of Space*, 69.

20. Waltman, *Black Magic and Gremlins*, 9.

21. Thompson, *At the Edge of Space*, 70.

22. Stanley Butchard, quoted in Waltman, *Black Magic and Gremlins*, 154.

23. Thompson, *At the Edge of Space*, 166.

24. Thompson, "General Review of Piloting Problems," reprinted in Waltman, *Black Magic and Gremlins*, 3.

25. Pilot memoirs printed in Waltman, *Black Magic and Gremlins.*

26. Thompson, *At the Edge of Space*, 58; idem., "Flight Research," 33.

27. Richard Day, "Training Considerations during the X-15 Development," reprinted in Waltman, *Black Magic and Gremlins.*

28. Thompson, *At the Edge of Space*, 123.

29. Ibid., 109.

30. *X-15 Research Results*, NASA SP-60, 1964.

31. Walker and Weil, "The X-15 Program." For more on the X-15 SAS and the ventral fin, see Hoey, "X-15 Ventral Off."

32. Armstrong, email to author, April 20, 2004.

33. Bailey, "Development and Flight Test of Adaptive Controls for the X-15."

34. Armstrong, "Pilot Notes," X-15 Flight 3-1-2, 3-2-3, 3-3-7, December 1961–April 1962, NASA Dryden Archives.

35. Armstrong, "Pilot Notes," X-15 Flight 3-4-8, April 20, 1962, NASA Dryden Archives. Thompson, *At the Edge of Space*, 100–102. Also see Hansen, *First Man*, 179–183.

36. Thompson and Welsh, "Flight Test Experience with Adaptive Control Systems," 141.

37. Thompson, *At the Edge of Space*, 188.

38. Walker and Weil, "The X-15 Program."

39. Thompson and Welsh, "Flight Test Experience with Adaptive Control Systems," 142.

40. Adams accident report reprinted in Thompson, *At the Edge of Space.*

41. Thompson and Welsh, "Flight Test Experience with Adaptive Control Systems," 145.

42. Ibid., 143, 145.

43. Armstrong, "Electronics and the Pilot."

44. R. G. Nagel and R. E. Smith, "An Evaluation of the Role of the Pilot and Redundant Emergency Systems in the X-15 Research Airplane." The SETP publication is a summary of the full study, idem., "X-15 Pilot-in-the-Loop and Redundant/Emergency Systems Evaluation," Technical Documentary Report No. 62-20, Air Force Flight Test Center, Edwards Air Force Base, Calif., October 1962, NASA Dryden Archives L2-5-1D-3.

45. Ibid., 1.

46. "X-15 The Movie: Correspondence." Reprinted in Godwin, *X-15: The NASA Mission Reports*, 384–391.

47. *X-15*, directed by Richard Donner.

48. Paul Bikle, "Foreword," in *X-15 Research Results.*

49. *X-15 Research Results*, 14.

50. X-15 press release, Edwards Flight Research Center, April 27, 1969. Reprinted in Godwin, *X-15 Mission Reports*, 393–394.

51. Jenkins, *Hypersonics Before the Shuttle*, 74.

4 Airmen in Space

1. Joachim Kuettner, a former Luftwaffe test pilot and member of von Braun's Huntsville group, played a significant role in Mercury. Kuettner had been the first person to fly a manned version of the V-1 cruise missile during World War II, and led the effort to man-rate the Redstone rocket. Quoted in Swenson, Grimwood, and Alexander, *This New Ocean*, 172.

2. Joe Shea, "The Goddard Lecture," March 14, 1967, NASA HQ Folder 013363; also see Shea, quoted in McCurdy, *Inside NASA*, 92, 97.

3. Von Braun, "Address to the Society of Experimental Test Pilots."

4. Von Braun and Ryan, *Conquest of the Moon*, 36–37, 63. For a discussion of early visions of the human role in space, see McCurdy, "Observations on the Robotic versus Human Issue in Spaceflight."

5. Blackburn, *Aces Wild: The Race for Mach 1*. Blackburn's book explores whether other pilots reached Mach 1 before Chuck Yeager, but it is full enough of personal reminiscences and opinions to be considered a memoir. Blackburn remembers von Braun proposing to anesthetize the pilots for space travel until they reached their final destination, 226–228.

6. Society of Experimental Test Pilots, *History of the First Twenty Years*, 60.

7. Von Braun, "Address to the Society of Experimental Test Pilots."

8. Godwin, *Dyna-Soar*; Jenkins, *Space Shuttle*, 22–31.

9. Dornberger, "The Rocket Propelled Commercial Airliner," reprinted in Godwin, *Dyna-Soar*, 19–37.

10. U.S. Air Force, Air Photographic and Charting Service, "The Story of Dyna-Soar," CD included with Godwin, *Dyna-Soar*. Emphasis in original.

11. Clark and Hardy, "Preparing Man for Space Flight."

12. Waldman, *Black Magic and Gremlins*, 176.

13. Armstrong and Holleman, "A Review of In-Flight Simulation Pertinent to Piloted Space Vehicles"; Holleman, Armstrong, and Andrews, "Utilization of the Pilot in the Launch and Injection of a Multistage Vehicle"; Matranga, Dana, and Armstrong, "Flight Simulated Off-the-Pad Escape and Landing Maneuver for a Vertically Launched Hypersonic Glider."

14. Murray, "Pilot-Oriented Dyna-Soar Designs"; Wood, "Pilot Control of the X-20/Titan II Boost Profile."

15. Thompson, *At the Edge of Space*, 146.

16. Gordon, "Concepts for Piloted, Maneuvering, Reentry Vehicles."

17. Crossfield, "Pilot Contributions to Mission Success."

18. Walker, "Some Concepts of Pilot's Presentation."

19. Kauffman, *Selling Outer Space*, chapter 4.

20. Wolfe, *The Right Stuff*, 186.

21. Grimwood, *Project Mercury: A Chronology*, 30–31; Murray and Cox, *Apollo*, 32–33.

22. Faget and Buglia, "Preliminary Studies of Manned Satellites."

23. Johnson, quoted in Grimwood, *Project Mercury*, 6.

24. Kraft, *Flight*, 30–35.

25. Ibid.

26. Chilton interview with Swenson, 8.

27. Ibid.

28. Ibid., 23.

29. Astronaut Press Conference, September 16, 1960, Cape Canaveral, Fla, NASA press release 60-276, 8.

30. Pitts, *The Human Factor*, 18. Voas interview with Sherrod, 131.

31. Voas interview with Sherrod, 131.

32. Ibid. Voas felt the second group of astronauts was better qualified than the first. Voas interview with Sherrod, 4, 131.

33. Swenson, Grimwood, and Alexander, *This New Ocean*, 131. Also see Goldstein, *Reaching for the Stars*, 42–44.

34. Faber interview with Swenson; Kraft, *Flight*, 84.

35. Voas, "Manual Control of the Mercury Spacecraft," 18.

36. Ibid., 38.

37. Voas, "A Description of the Astronaut's Task in Project Mercury."

38. Goldstein, *Reaching for the Stars*, 44–77.

39. Swenson, Grimwood, and Alexander, *This New Ocean*, 194–195.

40. Ibid., 194.

41. Ibid., 196–197; Voas, "Manual Control of the Mercury Spacecraft"; Gainor, *Arrows to the Moon*, 48–49.

42. Kraft, *Flight*, 91; Thompson, *Light This Candle*, 239.

43. Kraft, *Flight*, 93; idem. "Some Operational Aspects of Project Mercury."

44. Slayton, "Operation Plan and Pilot Aspects of Project Mercury."

45. *New York Times*, October 9, 1959, p. 12.

46. Swenson, Grimwood, and Alexander, *This New Ocean*, 355.

47. Ibid., 429.

48. Voas interview with Sherrod, 58.

49. Kauffman, *Selling Outer Space*, 85.

50. Collins, *Carrying the Fire*, 76.

51. Schirra, "A Real Breakthrough—The Capsule Was All Mine," 44–47, 87.

52. Heinlein, "All Aboard the Gemini," 116; "Let Man Take Over," *New York Times*, February 25, 1962, p. D10. Both are quoted in Kauffman, 63–64, which contains an excellent summary of press responses to Mercury on the issue of control.

53. Hacker and Grimwood, *On the Shoulders of Titans*, 22, 40. See also Gainor, *Arrows to the Moon*, 97–103.

54. "Talk Delivered by Major Virgil Grissom at an SETP East Coast Section Meeting," November 9, 1962, *SETP Newsletter* (November–December 1962): 5–12.

55. Ibid., 10.

56. Finney, "Pilots Will Control Gemini Spacecraft."

57. Schirra and Billings, *Schirra's Space*, 164.

58. Kramer, Aldrin, and Hayes, "Onboard Operations for Rendezvous."

59. Box et al., "Controlled Reentry."

60. Hacker and Grimwood, *On the Shoulders of Titans*, 283; Schirra and Billings, *Schirra's Space*, 161; Harland, *How NASA Learned to Fly in Space*, 121.

61. Aldrin, "Line-of-Sight Guidance Techniques for Manned Orbital Rendezvous"; McElheny, "Aldrin Simplified Moon Procedure."

62. McDivitt and Armstrong, "Gemini Manned Flight Program to Date."

63. McDivitt interview with Ward, 12–49; Slayton and Cassutt, *Deke!*, 15; Hacker and Grimwood, *On the Shoulders of Titans*, 246; Harland, *How NASA Learned to Fly in Space*, 54.

64. Cooper and Chow, "Development of On-Board Space Computer Systems."

65. See, "Engineering and Operational Approaches for Projects Gemini and Apollo."

66. McDonnell Corporation, *NASA Project Gemini Familiarization Manual*; Tomayko, *Computers Take Flight*, 13.

67. Box et al., "Controlled Reentry." Hacker and Grimwood, *On the Shoulders of Titans*, 4.

68. Funk interview with Swenson, 15.

69. Cohen interview with Swenson.

70. Cunningham and Herskowitz, *The All-American Boys*, 335–336.

71. Gerovitch, "The New Soviet Man in a Man-Machine System." Siddiqi, *Challenge to Apollo*, 264.

72. Gerovitch, "Human Machine Issues in the Soviet Space Program."

73. Ibid.

74. Ibid.

75. Paul Bikle to NASA Headquarters, May 11, 1960, "Review of Areas of Flight Research Center Competence in Manned Lunar Program," NASA Dryden Archives, L1-6-7A-1.

76. John Gibbons, Milton O. Thompson, Victor Horton, "Flight Research Center Manned Rocket Flight Study Summary," August 1960. Included in packet "Flight Research Center Rocket Flight Study," Paul Bikle to Commander Air Research and Development Command, Wright Patterson Air Force Base, August 9, 1960, NASA Dryden Archives, L1-6-7A-2. Also see Milt Thompson, "Report to Apollo Guidance and Control Technical Liaison Group," n.d., NASA Dryden Archives, Milton Thompson Papers, L1-5-4-14.

77. Joseph Walker and John B. McKay, "Pilot's Role in Flight Research," included in packet "Flight Research Center Rocket Flight Study," Paul Bikle to Commander Air Research and Development Command, Wright Patterson Air Force Base, August 9, 1960, NASA Dryden Archives, L1-6-7A-2.

78. Hubert Drake, "Pilot's Role in the Rendezvous Mission," February 24, 1961. NASA Dryden Archives, L1-6-7A-1.

79. Chilton interview with Bergen, 28–29.

80. Chilton, "Command and Communications."

81. Chilton interview with Swenson, 8–9.

82. Apollo feasibility studies and proposals are found in UHCL Chrono 83-42/45.

83. Collins, *Carrying the Fire*, 257.

5 "Braincase on the tip of a firecracker": Apollo Guidance

1. Webb, quoted in Logsdon, *The Decision to Go to the Moon*, 90, 125.

2. John F. Kennedy, "Memorandum for Vice President," April 20, 1961. Presidential Files, John F. Kennedy Library, Boston, Mass.

3. Logsdon, *The Decision to Go to the Moon*.

4. MacKenzie, *Inventing Accuracy*.

5. Mindell, *Between Human and Machine*, 22.

6. Seamans, *Aiming at Targets*.

7. MacKenzie, *Inventing Accuracy*; Dennis, "A Change of State."

8. Battin, *Astronautical Guidance*, 1.

9. For an assessment of the state of the art, see Farrior, "Guidance and Navigation: State of the Art—1960."

10. MacKenzie, *Inventing Accuracy*, chapter 3.

11. Sapolsky, *The Polaris System Development*.

12. Battin, "Space Guidance Evolution—A Personal Narrative."

13. Hall, *Journey to the Moon*, 38–46. Hall and Jansson, "Miniature Packaging of Electronics in Three Dimensional Form."

14. Lecuyer, *Making Silicon Valley*, chapters 5–6.

15. Hall, "From the Farm to Pioneering with Digital Control Computers," 22–31.

16. Laning interview with Brown, quoted in Brown, "Probing Mars, Probing the Market," 18.

17. Mudgway, *Uplink-Downlink*.

18. Laning et al., "Preliminary Considerations," quoted in Brown, "Probing Mars, Probing the Market," 19.

19. Battin, "Space Guidance Evolution—A Personal Narrative."

20. Battin, *Astronautical Guidance*, 17.

21. Battin uses his numerology in regular lectures; see, for example, "Some Funny Things Happened on the Way to the Moon," lecture at MIT, January 17, 2007. Cohen interview with Swenson, 3.

22. Brown, "Probing Mars, Probing the Market"; Battin interview with Ertel.

23. Brown, ibid.; Hoag, "The History of Apollo On-Board Guidance, Navigation, and Control," 276.

24. Battin, "A Statistical Optimizing Navigation Procedure for Space Flight."

25. Battin, "Space Guidance Evolution"; Kalman, "A New Approach to Linear Filtering and Prediction Problems"; Battin interview with Mindell and Brown.

26. Robert G. Chilton, "Memorandum for Associate Director: Meeting with MIT IL to Discuss Navigation and Guidance Support for Project Apollo," November 28, 1960, UHCL Chrono 62-34. Italics added.

27. Chilton interview with Bergen, 12–30.

28. Robert G. Chilton, "Memorandum for Associate Director: Visit to MIT Instrumentation Laboratory, March 23, 24, 1961," April 3, 1961, UHCL Chrono 62-34.

29. Hoag, "The History of Apollo On-Board Guidance, Navigation, and Control," 272.

30. Chilton interview with Bergen, 12–29.

31. Hoag interview, with Ertel 2.

32. MIT IL, "Bimonthly Progress Report No. 1, Project Apollo Guidance and Navigation Study," February 7–May 3, 1961, UHCL Chrono.

33. Trageser interview with Ertel.

34. "Project Apollo Guidance and Navigation: A Proposal for a Research, Development, and Space Flight Program," MIT Instrumentation Laboratory, August 4, 1961. Courtesy of Eldon C. Hall.

35. Hall, "MIT's Role in Project Apollo," 11.

36. Miller interview with Ertel, 6.

37. Trageser interview with Ertel.

38. Chilton interview with Swenson, 7; Chilton interview with Bergen, 30–31.

39. Ertel, *The Apollo Spacecraft*, vol. 1, 106.

40. Cohen interview with Swenson, 10.

41. Chilton interview with Bergen.

42. Welch interview with Ertel, 1.

43. Klass, "Apollo Guidance Bidders Protest NASA Choice of Non-Profit Firm."

44. Draper to Seamans, November 21, 1961; Seamans to Draper, November 27, 1961. NASA HQ, Biography file—Draper.

45. Seamans comments to MIT class, "Engineering Apollo," April 2005.

46. "Project Apollo: Navigation and Guidance System Development Statement of Work, Space Task Group, Langley Field, Virginia, August 10, 1961." Also see a similarly titled document dated December 4, 1961, UHCL Chrono 62-54/55.

47. Arthur Ferraro, NASA MSC News Conference, MIT, Cambridge, Mass., September 24, 1963, UHCL Chrono 63-64, 21.

48. Chris Kraft to manager, Apollo Spacecraft Program Office, December 1, 1964. UHCL Chrono 64-63/4. Also see D. Slayton to manager, Apollo Spacecraft Program Office, "Hard Copy Printer for Block II CM," November 16, 1964. UHCL Chrono 64-62/3. The Slayton memo reports a study of the teleprinter utility and feasibility, and recommends against it.

49. Ertel, *The Apollo Spacecraft*, vol. 1, 110.

50. Ibid., 113.

51. Ibid., 121.

52. Ibid., 122.

53. Ibid., 137.

54. James Webb, "Statement of the Administrator, NASA, on Selection of Contractors for Apollo Spacecraft Navigation and Guidance System MIT Industrial Support," n.d., UHCL Chrono 62-66.

55. "AC's Role in the Aerospace Industry," press release, AC Electronics, Milwaukee, Wis., February 13, 1967, MIT Museum.

56. James Webb, op. cit.

57. Ertel, *The Apollo Spacecraft*, vol. 1, 160.

58. Hoag, "The History of Apollo On-Board Guidance, Navigation, and Control," 273.

59. For good accounts of the LOR decision, see Seamans, *Project Apollo: The Tough Decisions*, chapter 10; Murray and Cox, *Apollo*; Hansen, *Enchanted Rendezvous*.

60. Lickly, HRST1.

61. Cohen interview with Swenson, 12–14.

62. Gavin, HRST1.

63. Hoag, Gavin, HRST1.

64. Hall, "MIT's Role in Project Apollo," 96–97.

65. Collins, *Carrying the Fire*, 410.

66. Gilbert interview with Ertel.

67. Hoag interview with Ertel, 19.

68. Ibid.

69. Hall, "MIT's Role in Project Apollo," 89.

70. Hoag interview with Ertel.

71. Kelly, *Moon Lander*, 77–79.

72. Armstrong, quoted in Hansen, *First Man*, 544; Collins, *Carrying the Fire*, 406.

73. Gilbert interview with Ertel.

74. Trageser to IL distribution, "Proposed Agenda for the Astronaut's Review of the AGE System," November 26, 1962. Courtesy of J. Nevins.

75. Langone, "Astronauts Get 'Moon'"; "Noisy Welcome for Astronauts."

6 Reliability or Repair? The Apollo Computer

1. Hall, *Journey to the Moon*, chapter 5. Alonso, Blair-Smith, and Hopkins, "Some Aspects of the Logical Design of a Control Computer."

2. Lecuyer, *Making Silicon Valley*, 159–162. Berlin, *The Man Behind the Microchip*.

3. Alonso, Blair-Smith, and Hopkins, "Some Aspects of the Logical Design of a Control Computer."

4. Hall, *Journey to the Moon*, chapter 6.

5. Ibid.

6. Gavin, HRST1.

7. "Cite Wide IC Use in Apollo Guidance Unit."

8. "Fairchild Div. Ships 110,000 Integrateds for Apollo Project."

9. Noyce, "Integrated Circuits in Military Equipment."

10. Hall, "MIT's Role in Project Apollo," 14–15.

11. NASA MSC News Conference, MIT, Cambridge, Mass., September 24, 1963, UHCL Chrono 63-64, 37.

12. David Gilbert, "A Summary of Apollo Spacecraft Navigation and Guidance System Reliability and Quality Assurance Program," January 23, 1963, UHCL Chrono 63-65; Hall, "A Case History of the AGC Integrated Logic Circuits."

13. Chilton interview with Bergen.

14. Faget interview with Ertel, 7.

15. Hall, "General Design Characteristics of the Apollo Guidance Computer," appendix III.

16. Frasier, HRST1.

17. John French to C. W. Frick, "Apollo Navigation and Guidance System Trip Report," March 30, 1962, UHCL Chrono 62-63.

18. For a good description of the different approaches to this problem, see Murray and Cox, *Apollo*, 141–142. On systems accidents, see Perrow, *Normal Accidents*; Leveson, *Safeware*.

19. Karth to Webb, and Webb to Karth, March 2, 1965. NASA HQ Archives administrator's correspondence.

20. Partridge, Hanley, and Hall, "Progress Report on Attainable Reliability of Integrated Circuits for Systems Applications," 7.

21. Hall, "General Design Characteristics of the Apollo Guidance Computer."

22. Sato, "Local Engineering in the Early American and Japanese Space Programs," 87–88.

23. C. S. Draper, E. C. Hall, G. W. Mayo, J. E. Miller, and E. G. Schwarm, "Engineering and Reliability Techniques for Apollo Guidance, Navigation, and Control at MIT Instrumentation Laboratory." Unpublished manuscript, July 28, 1963. MIT Museum. Also see Hall, "General Design Characteristics of the Apollo Guidance Computer," section III.

24. Holley et al. "Apollo Experience Report," 33.

25. Speer, "Strict Control Kept Out Semiconductor Flaws," 29.

26. Hall, "Case History of the Apollo Guidance Computer," 26. On issuing blank drawings, see Kupfer interview with Ertel.

27. Hall, "From the Farm to Pioneering with Digital Computers," 28.

28. Hall, "MIT's Role in Project Apollo," 267.

29. Mueller interview with Ertel, 7, 11. Also see Logsdon, *Managing the Moon Program,* 14–17.

30. Johnson, "Samuel Philips and the Taming of Apollo," 695.

31. Philips, quoted in ibid., 699.

32. Shea interview with Kelley, 5; Brooks, Grimwood, and Swenson, *Chariots for Apollo,* 120–121.

33. Frick interview with Ertel, 6.

34. Brooks, Grimwood, and Swenson, *Chariots for Apollo,* 133–135.

35. Shea interview with Kelley, 10.

36. Ibid., 12.

37. Shea memo to Mueller, August 5, 1964, UHCL Chrono 64-43/44.

38. Johnson interview with Grimwood, 8.

39. Kraft, *Flight,* 197.

40. Ibid., 196.

41. Sato, "Local Engineering in the Early American and Japanese Space Programs."

42. Shea memo to Mueller, August 5, 1964, UHCL Chrono 64-43/44.

43. Cohen interview with Swenson, 23.

44. Meeting minutes on contract realignment, April 6, 1964, UHCL Chrono 64-43/44.

45. Shea memo to Mueller, August 5, 1964, UHCL Chrono 64-43/44.

46. Apollo quarterly report, period ending September 30, 1962, UHCL Chrono. Shepard comment remembered by Miller, HRST2. Nevins to author, personal communication, September 2006.

47. Apollo monthly progress report, October 16–November 15, 1963, SID 62-300-19. UHCL Chrono; Ertel, *The Apollo Spacecraft,* volume 2, 25.

48. Brooks, Grimwood, and Swenson, *Chariots for Apollo*, 135.

49. Vonbun, "Ground Tracking of Apollo."

50. Frasier, HRST1.

51. Klass, "First Apollo Control Prototype Is Readied"; Littleton, "Apollo Experience Report—Guidance and Control Systems."

52. "More Apollo Guidance Flexibility Sought."

53. Frasier, HRST1.

54. Littleton, "Apollo Experience Report—Guidance and Control Systems."

55. Shea memo to Trageser, November 30, 1964, UHCL Chrono 64-64.

56. Scott, "The Apollo Guidance Computer,"

57. Frasier, HRST1.

58. David Hoag interview by NASA historian, May 15, 1967 (not transcribed), MSC oral histories, audio at NASA UHCL.

59. Martin and Battin, "Computer-Controlled Steering of the Apollo Spacecraft," 400–407; Crisp and Keene, "Apollo Command and Service Module Reaction Control by the Digital Autopilot."

60. David Hoag interview by NASA historian, May 15, 1967 (not transcribed), MSC oral histories, audio at NASA UHCL.

61. Frasier, HRST1; Cohen interview with Swenson, 5.

62. Minutes of meeting, North American MIT/IL, NASA/MSC, June 23, 1964, UHCL Chrono 6L-41. Johnson and Giller, "MIT's Role in Project Apollo, Vol. V," 19–21.

63. Apollo quarterly reports, periods ending June 30 and September 30, 1964, and March 31, 1965, UHCL Chrono.

64. Hall, "MIT's Role in Project Apollo, Vol. III" 27.

7 Programs and People

1. *The Oxford English Dictionary*; MacKenzie, *Mechanizing Proof*, 26; Ceruzzi, *Beyond the Limits*, 269.

2. Project Apollo, "Navigation and Guidance System Development Statement of Work," August 10, 1961, Space Task Group, Langley, V., NASA HQ, 8.

3. Battin, "A Funny Thing Happened on the Way to the Moon," 3.

4. Johnson and Giller, "MIT's Role in Project Apollo, Vol. V," 17.

5. Garman interviewed by Rusnak.

6. Lickly and Kosmala, HRST2.

7. Lickly, HRST1.

8. Hamilton, HRST2.

9. Martin, HRST2.

10. Poundstone, HRST3.

11. Martin, Lickly, HRST2.

12. Copps, HRST4.

13. Johnson and Giller, "MIT's Role in Project Apollo, Vol. V" 17.

14. Felleman, "Hybrid Simulation of the Apollo Guidance Navigation and Control System"; Sullivan, "Hybrid Simulation of the Apollo Guidance Navigation and Control System"; Glick and Femino, "A Comprehensive Digital Simulation for the Verification of Apollo Flight Software."

15. Miller, HRST2.

16. Ibid.

17. Ibid.

18. Johnson and Giller, "MIT's Role in Project Apollo, Vol. V," 13.

19. Eyles, "Tales from the Lunar Module Guidance Computer," 5; Hall, "MIT's Role in Project Apollo," 155–159.

20. Hamilton, HRST2.

21. Martin and Battin, "Computer Controlled Steering of the Apollo Spacecraft," 400–406.

22. E. M. Copps, Jr., "Recovery from Transient Failures of the Apollo Guidance Computer"; Eyles, "Tales from the Lunar Module Guidance Computer," 9.

23. Copps, HRST4.

24. Martin, HRST2.

25. Johnson and Giller, "MIT's Role in Project Apollo, Vol. V," 19.

26. Ibid., 15–16.

27. Ibid., 7. Hoag, "The History of Apollo On-Board Guidance, Navigation, and Control," 287.

28. Martin and Kosmala, HRST2.

29. Funk interview with Swenson, 15.

30. NASA MSC News Conference, MIT, Cambridge, Mass., September 24, 1963, UHCL Chrono 63-64, 13.

31. Poundstone, HRST3.

32. Ibid.

33. Miller and Lickly, HRST2.

34. Johnson and Giller, "MIT'S Role in Project Apollo," 82.

35. Miller, HRST2.

36. Hoag, "The Eagle Has Returned," 280.

37. Alonso, HRST1.

38. Frasier, HRST1.

39. Alonso, HRST1.

40. Gavin, HRST1.

41. Lickly, HRST1.

42. Battin interview with Mindell and Brown.

43. Collins, *Carrying the Fire*, 75. Also see "Apollo Astronaut's Guidance and Navigation Course Notes, Prepared by MIT Instrumentation Laboratory," MIT IL/E-1250 November 1962–February 1963, sections I–VII, MIT Museum.

44. Hoag, "History of Apollo On-Board Guidance, Navigation, and Control," 289.

45. Hoag interview with Ertel, 16.

46. Miller, HRST2.

47. Hamilton, HRST1.

48. Lickly, HRST2.

49. NASA MSC News Conference at MIT, September 24, 1963, UHCL Chrono 63-64; Hoag, "History of Apollo On-Board Guidance, Navigation, and Control."

50. R. Metzinger to J. Nevins, "Definition of C/M OP SIM," March 1965. Courtesy of J. Nevins.

51. Copps, HRST4.

52. Kosmala, HRST3.

53. Miller, HRST3.

54. Draper, Whitaker, and Young, "Roles of Men and Instruments in Control and Guidance Systems for Spacecraft."

55. Copps, HRST2.

56. Ibid.

57. Conway, *Blind Landings*.

58. J. L. Nevins, "Man-Machine Design for the Apollo Navigation, Guidance, and Control System—Revisited." Also see Nevins, Johnson, and Sheridan, "Man/Machine Allocation in

the Apollo Navigation, Guidance, and Control System." The first of these papers (1970) has a detailed analysis of crew activities during a lunar landing; the second (1968) uses the example of rendezvous.

59. Alonso and Frasier, "Evolutionary Dead Ends."

60. Green and Filene, "Keyboard and Display Program and Operation."

61. Alonso, HRST1.

62. Nevins interview with Mindell, 11.

63. Hoag, "History of Apollo On-Board Guidance, Navigation, and Control," 281.

64. Scott, "The Apollo Guidance Computer, A User's View."

65. Nevins, Woodin, and Metzinger, "Man-Machine Simulations for the Apollo Navigation, Guidance, and Control System."

66. Nevins to Distribution, "The Space Navigator, its Capabilities and Uses," January 12, 1966, MIT Museum; Nevins interview with Mindell, 5–10.

67. Nevins interview with Mindell, 27.

68. Martin, Alonso HRST1.

69. David Hoag, "Apollo Guidance and Navigation Program at MIT Instrumentation Lab, Material in Support of a $31 Million 30 Month Proposal to NASA to Continue Work from 1 Jan 1968 to 30 June 1970," October 4, 1967, MIT Museum.

70. Shea to Draper, May 9, 1966. Draper to Shea, June 8, 1966, MIT Museum.

71. Murray and Cox, *Apollo*, 292–297.

72. Martin, HRST1.

73. Garman interview with Rusnak, 9.

74. Tindall to distribution, May 31, 1966, MIT Museum.

75. Johnson and Giller, "MIT's Role in Project Apollo," 15.

76. Tindall to distribution, June 13, 1966, MIT Museum.

77. Rankin, "A Model of the Cost of Software Development for the Apollo Spacecraft Computer."

78. Lickly, HRST2.

79. NASA, "Postlaunch report for mission AS-202," Manned Spacecraft Center, Houston, Tex., October 12, 1966, 7-185, 7-192.

80. Low to Gilruth, September 19, 1967, UHCL Chrono, Apollo Note #100.

81. Martin, HRST1.

82. Battin interview with Wright.

83. Copps, HRST4.

84. Tindall to distribution, March 24, 1967, MIT Museum.

85. Tindall to distribution, May 17, 1967, MIT Museum.

86. Guidance Software Validation Committee, "Apollo Guidance Software Development and Validation Plan," October 4, 1967, NASA MSC. Available at http://klabs.org/history/history_docs/mit_docs/sw.htm, accessed on January 10, 2007.

87. Hoag memo to Draper, October 9, 1967, MIT Museum. The memo was titled "Another Schedule Emergency on Software (Another Wolf Call)" and shows how the IL team, while worried about schedules, was becoming skeptical of Kraft's warnings that they were holding up the programs. "Perhaps the call of wolf will deliver a wolf this time. The Apollo program needs it!" was Hoag's conclusion to his boss.

88. Ken Young, "My Head Was Full of Space and Other Ditties," NASA MSC, Houston, Tex., 1987, 10–33, MPAD History Files CD-ROM, UHCL.

89. Miller, HRST2.

90. Ibid.

91. Apollo quarterly report, "Summary," March 30, 1968, UHCL Chrono 83-12, 23.

92. McElheny, "Space Test Pinpoint Lunar Module Bugs." Samuel C. Phillips, "The Shakedown Cruises," in Cortright, *Apollo Expeditions to the Moon*, 167.

93. Eyles, "Tales from the Lunar Module Guidance Computer," 2.

94. Hoag to Apollo personnel, "G&N performance in Apollo 5 and Apollo 6 flights." April 15, 1968, MIT Museum. Also see LM-1 Flight Performance Evaluation Team, "LM-1 Flight Performance Evaluation Report," Grumman Aircraft Engineering Corporation LED-541-1, n.d., courtesy of Paul Fjeld. For a good summary of the GNC performance on the early flights, through Apollo 9, see Hoag, "Apollo Navigation, Guidance, and Control Systems: A Progress Report."

95. TRW Systems, "Mission Requirements for Apollo 7 CSM Development Mission," NAS 9-3810, May 1, 1967, MIT Museum.

96. Hoag to Apollo personnel, "G&N Performance in Apollo 7," MIT Museum.

97. Cunningham and Herskowitz, *The All-American Boys*, 192.

98. Low's decision-making process is documented in detail in Low, "Special Notes for August 9, 1968 and Subsequent," UHCL Chrono.

99. McElheny, "MIT Unit to Set Orbit."

100. NASA "Apollo 8 Mission Report."

101. "Apollo 8 Transcript," 4:10:29, 70–71.

102. Hoag to IL, January 10, 1969, "Report on Apollo 8," MIT Museum. Hoag, "Apollo Navigation, Guidance, and Control Systems: A Progress Report."

103. Cohen interview with Swenson.

104. Herfort, "MIT Computers 'Miraculous.'" Sewall, "Happy Day for MIT Experts."

105. *Aviation Week and Space Technology*, January 20, 1969, 40–46.

106. Battin interview with Wright.

8 Designing a Landing

Epigraph: Minutes of meeting, Flight Readiness Review Board, Lunar Landing Training Vehicle, Houston, Texas, January 12, 1970, UHCL Chrono.

1. Armstrong interview with Ambrose and Brinkley, 85–86.

2. Kelly, *Moon Lander*; Gavin, "Engineering Development of the Apollo Lunar Module."

3. Sherman interview with Ertel, 21.

4. Gavin, HRST1.

5. Kelly, *Moon Lander*, chapter 8.

6. Ibid., 94–95.

7. NASA, "Apollo 9 Mission Report," 10–26.

8. Hoag memo to IL, "How Did We Do on Apollo 9?" April 10, 1969, MIT Museum; "Apollo 9 Mission Report," 10–21.

9. Cheatham and Hackler, "Handling Qualities for Pilot Control of Apollo Lunar-Landing Spacecraft."

10. For one description of MPAD's approach to trajectories, see Howard W. Tindall, Jr., "Techniques of Controlling the Trajectory," in *What Made Apollo a Success*, chapter 7.

11. Bennett, "Apollo Experience Report," 34. For another perspective on the design of the landing trajectory, see Funk interview with Ross-Nazzal, 29–35.

12. Klumpp interview with Mindell.

13. Cheng, "Lunar Terminal Guidance," in Leondes and Vance, eds., *Lunar Missions and Exploration*; Eyles, "Tales from the Lunar Module Guidance Computer," 16.

14. Bennett interview with Ross-Nazzal, 11, 20.

15. Von Braun and Ryan, *Conquest of the Moon*, 63–64.

16. "Preliminary Report on Automatic LEM Mission Study," Grumman Aircraft Engineering Corporation, LED-54-02, April 29, 1963. Available at NASA Technical Report Server, http://ntrs.nasa.gov (accessed January 12, 2007).

17. Bennett, interview with Ross-Nazzal, 10; Bennett interview with Mindell.

18. Bennett interview with Ross-Nazzal, 7, 13.

19. Manned Space Flight Center, "Apollo Lunar Landing Mission Symposium"; Gainor, *Arrows to the Moon*, 129.

20. Widnall, "Lunar Module Digital Autopilot"; Johnson and Giller, "MIT's Role in Project Apollo," 240–279, also has a good description of the LM digital autopilot. The LM digital autopilot was like a Kalman filter, but not exactly, because it did not update its statistical weighting coefficients in real-time the way a Kalman filter does. Rather, these coefficients were derived analytically by engineers ahead of time.

21. Nevins, "Man-Machine Design for the Apollo Navigation, Guidance, and Control System—Revisited"; MIT Instrumentation Laboratory, "Guidance Systems Operations Plan, AS-278."

22. Stengel, "Manual Attitude Control of the Lunar Module."

23. Also see NASA, "Apollo 12 Technical Debrief," 12-11/12-12.

24. Cheatham and Bennett, "Apollo Lunar Module Landing Strategy," 177.

25. Charles Stark Draper Laboratories, "Guidance Systems Operations Plan," 5.3–5.50.

26. Hoag, "Guidance, Navigation, and Control of Manned Lunar Landing."

27. Actually, the landing radar only measured 1-d velocity, so the cross-x-pointers on the x-y display reflected the state vector rather than the radar.

28. Norman Sears memo to Apollo Spacecraft Program Office, March 26, 1963, UHCL Chrono.

29. Cheatham and Bennett, "Apollo Lunar Module Landing Strategy," 178.

30. Johnson and Gillers, "MIT's Role in Project Apollo, Vol. V," 180.

31. Cheatham and Bennett, "Apollo Lunar Module Landing Strategy," 189.

32. Ibid., 193.

33. Klumpp, "A Manually Retargeted Automatic Descent and Landing System for the Lunar Module (LM)"; Jay D. Montgomery, "LM Landing Point Designator Procedures and Capability," NASA MSC Internal Note No. 67-EG-24, Houston, Tex., August 1, 1967. Collection of Paul Fjeld.

34. Klumpp, "A Manually Retargeted Automatic Descent and Landing System for the Lunar Module (LM)," 130.

35. "Guidance System Operations Plan for Manned LM Earth Orbital and Lunar Missions Using Program Luminary 1E, Section 5 Guidance Equations Revision 11," MIT Charles Stark Draper Laboratories, R-567, December 1971.

36. Bennett, "Apollo Experience Report," 10.

37. Cheatham and Bennett, "Apollo Lunar Module Landing Strategy," 201.

38. Eyles, "Apollo LM Guidance Computer Software for the Final Lunar Descent."

39. Goldstein, *Reaching for the Stars*, 128.

40. The "great train wreck" was John Young's term for the CM Mission Simulator in Houston. Cortright, *Apollo Expeditions to the Moon*, chapter 8.3.

41. Woodling et al., "Apollo Experience Report: Simulation of Manned Space Flight for Crew Training."

42. Collins, *Carrying the Fire*, 191.

43. Goldstein, *Reaching for the Stars*, 172.

44. Actually, the LM had a few RCS thrusters pointing horizontally, but all horizontal moves were made by swiveling the descent engine, which was much more fuel-efficient. Keller, "Study of Spacecraft Hover and Translation Modes above the Lunar Surface."

45. For Armstrong's account of the LLRV, see Armstrong, "Wingless on Luna."

46. Rusnak, "Avionics Aspects of the Lunar Landing Research Vehicle."

47. Dean Grimm, quoted in Matranga, Ottinger, and Jarvis, *Unconventional, Contrary, and Ugly*; Kleuver et al., "Flight Results with a Non-Aerodynamic, Variable Stability, Flying Platform."

48. Matranga and Walker, "Investigation of Terminal Lunar Landing with the Lunar Landing Research Vehicle"; Duda et al., "Human Interfaces and Pilot-Vehicle Interactions of the Lunar Landing Research Vehicle," paper submitted for MIT course "Engineering Apollo: The Moon Project as a Complex System," May 16, 2005.

49. Scott interview with Jones.

50. Minutes of meeting, Flight Readiness Review Board, Lunar Landing Training Vehicle, Houston, Tex., January 12, 1970, UHCL Chrono.

51. Strickland, "Series of Lunar Landings Simulated."

52. Minutes of meeting, Flight Readiness Review Board, Lunar Landing Training Vehicle, Houston, Tex., January 12, 1970, UHCL Chrono.

53. Ibid. Armstrong didn't trust the landing radar velocity at low speed, because he had seen the data go bad close to the ground when operating from a helicopter because of radar interference with the helicopter blades. He admitted this was not a good reason not to trust the radar on the moon, there being no helicopter blades to interfere. In any case, the P66 Auto function relied primarily on the inertial system and not on radar to measure velocity.

9 "Pregnant with alarm": Apollo 11

1. Numerous versions of the Apollo 11 landing exist in the literature but I have compiled this account from a few primary sources, and will cite the major ones here: NASA's "Apollo 11 Mission Report"; "Apollo 11 Technical Air-to-Ground Voice Transcription"; "Apollo 11 On-Board Voice

Transcription"; and "Apollo 11 Technical Debrief." The best technical account is Floyd V. Bennett, "Apollo Experience Report: Mission Planning for Lunar Module Descent and Ascent," by the man who planned the descent, for both Apollo 11 and 12 landings. A version of this paper was published as Floyd V. Bennett, "Apollo Lunar Descent and Ascent Trajectories." These papers incorporate data from Floyd Bennett, "Apollo 11 LM Approach and Landing Phase Groundtrack," NASA MSC Memo 69-RM21-265, October 1, 1969, and idem. "Apollo 11 LM Descent Postflight Analysis," NASA MSC Memo 69-FM22-220, August 13, 1969, UHCL Chrono 71-64 and 79-21, respectively. Also see the Grumman technical evaluation of the LM performance, Grumman Engineering Corporation LED-541-10 (no title, no date), which has a detailed discussion of systems performance, generously provided from the personal collection of Paul Fjeld.

2. Bennett interview with Mindell.

3. Kranz email to Eric Jones, *Apollo Lunar Surface Journal*, available at http://www.hq.nasa.gov/alsj/a11/ (accessed January 5, 2007). For other possible reasons, see McElheny, "Little Errors Added Up to 4-Mi. Apollo Mistake."

4. NASA, "Apollo 11 Technical Debrief," 63.

5. Duke interview with Ward, 21.

6. Mailer, *Of a Fire on the Moon*, 377.

7. Eyles, "Tales from the Lunar Module Guidance Computer," 7.

8. Armstrong interview with Ambrose and Brinkley, 21. For Armstrong's account, also see Hansen, *First Man*, 465.

9. Neil Armstrong, "Apollo 11 Postflight Crew Press Conference," August 12, 1969, Houston, Tex., 21.

10. Neil Armstrong, "Apollo: Past, Present, and Future," Proceedings of the 13th SETP Symposium, Beverly Hills, Calif., September, 1969.

11. Inadvertent LPD in Bennett, "Apollo Experience Report."

12. Bennett interview with Ross-Nazzal, 24; Bennett, "Apollo 11 LM Descent Postflight Analysis"; figure 12 seems to show that Armstrong was well outside the abort boundary between 200 and 100 feet.

13. Mailer, *Of a Fire on the Moon*, 380.

14. Bennett, "Apollo 11 LM Approach and Landing Phase Groundtrack"; and idem. "Apollo 11 LM Descent Postflight Analysis," 6.

15. Bennett, "Apollo 11 LM Descent Postflight Analysis," 9.

16. Hansen, *First Man*, 467.

17. Martin, "Apollo 11: 25 Years Later."

18. My explanation of the program alarms is based on: George Cherry, "Exegesis of the 1201 and 1202 Alarms Which Occurred During the Mission G Lunar Landing," MIT Instrumentation Laboratory memo AG#370-69, August 4, 1969; Clint Tillman, "Program Alarms in Powered Descent, Apollo 11," Grumman Aircraft Engineering Corporation Memo #LAV-500-940, July 31, 1969; Clint Tillman, "Simulating the RR-CDU Interfaces When the RR Is in the SLEW or AUTO (not LGC) Mode in the FMES/FCI Laboratory," Grumman Aircraft Engineering Corporation Memo #LMO-500-723, August 13, 1969. Documents generously provided from personal collection of Paul Fjeld. Also see Eyles, "Tales from the Lunar Module Guidance Computer."

19. Donald Arabian, manager, Mission Evaluation Team, "Apollo 11 Problem and Discrepancy List," November 5, 1969; "Apollo 11 Mission Report," MSC-00171, Apollo Mission File 074-63, UHCL. For Robert Chilton's version of this story, see Chilton interview with Bergen, 46.

20. Garman interview with Rusnak, 21–22; Garman, "Computer Overload and the Apollo 11 Landing: An Insider's View," presentation to Military and Aerospace Applications of Programmable Logic Design conference, Washington, D.C., September 2005. Available at http://www.klabs.org/mapld05/abstracts/108_garman_a.html (accessed January 24, 2007). Also see "Steve Bales: Guidance Officer, Apollo 11," in Watkins, *Apollo Moon Missions: The Unsung Heroes*, chapter 1; Hansen, *First Man*, 461–463, for an account of the simulation.

21. Aldrin quoted in Eric Jones, *Apollo Lunar Surface Journal*, available at http://www.hq.nasa.gov/alsj/a11/ (accessed January 12, 2006).

22. For a detailed, technical account of this issue, see Eyles, "Tales from the Lunar Module Guidance Computer."

23. Cherry, "Exegesis of the 1201 and 1202 Alarms," 9.

24. Quoted in Hansen, *First Man*, 568.

25. "Lunar Landing Had Its Earthbound Heroes."

26. Kaye, "The Indispensable Man."

27. Eyles, "Apollo LM Guidance Computer Software for the Final Lunar Descent," 243–250.

28. Martin, HRST2.

29. Battin interview with Wright, 35.

30. Larsen memo to Moore, Elyes, and Klumpp, April 22, 1970; Chilton to distribution, June 29, 1970. Documents courtesy of Allen Klumpp. Also see Eyles, "Tales from the Lunar Module Guidance Computer," for an explanation of this problem.

31. Hugh Blair-Smith, oral comments on Jack Garman presentation cited in note 21.

10 Five More Hands On

1. "Mission Requirements: SA-507/CSM-108/LM-6 H-1 Type Mission," July 18, 1969, MSC Houston, UHCL Chrono.

2. For details of the Apollo 12 descent, see NASA, "Apollo 12 Mission Report," 4-27; Landing Analysis Branch, "Apollo 12 LM Descent Postflight Analysis," NASA MSC, 69-FM22-324. Houston, Tex., December 10, 1969.

3. NASA, "Apollo 12 Mission Report," 4-25.

4. For one account of this change, see Bean interview with Compton, 23–24; Eyles, "LGC-Astronaut Interfaces during Landing (revised for Luminary 1E)," n.d., ca. 1970. Courtesy of Don Eyles.

5. NASA, "Apollo 12 Mission Report," 4-1, 4-4, 5-1.

6. Bennett, "Apollo Experience Report," 26–28, has a detailed comparison of mission planning for Apollo 11 and Apollo 12 landings.

7. NASA, "Apollo 12 Mission Report," 4-12.

8. NASA, "Apollo 12 Mission Report," 8-7.

9. Bean, quoted in Smith, *Moondust*, 196–197.

10. NASA, "Apollo 12 Technical Debrief," 9-10.

11. NASA, "Apollo 11 Technical Debrief."

12. NASA, "Apollo 12 Technical Debrief," 9-14.

13. Ibid.

14. Bennett, "Apollo Experience Report," 31.

15. Bennett interview with Mindell.

16. LM-6 Flight Performance Evaluation Team, "LM-6 Flight Performance Evaluation Report," Grumman Aircraft Engineering Corporation LED-541-12, March 19, 1970, 5.16-4.

17. Landing Analysis Branch, "Apollo 12 LM Descent Postflight Analysis," 6–7.

18. For authoritative data on this issue, see LM-6 Flight Performance Evaluation Team, "LM-6 Flight Performance Evaluation Report." Grumman Aircraft Engineering Corporation LED-541-12. March 19, 1970, 5.1-10–5.1-11. 5.16-4.

19. Minutes of meeting, Flight Readiness Review Board, Lunar Landing Training Vehicle, Houston, Tex., January 12, 1970, UHCL Chrono.

20. Mitchell and Williams, *The Way of the Explorer*, 39.

21. Garman interview with Rusnak, 33.

22. NASA, "Apollo 14 Mission Report," 14-29; "Apollo 14 30-Day Failure and Anomaly Listing Report," NASA MSC, Houston, Tex., 24. UHCL Chrono 79-53.

23. Shepherd and Slayton, *Moon Shot*, 299.

24. Garman interview with Rusnak, 34. At the IL, David Hoag gave credit to Eyles for the fix, writing to Dale Myers that "I realize there must be a single hero for this mission-saving work-around," but he also gave credit to Bruce McCoy, Russel Larson, Samuel Drake, Philip Fellman, Steven Copps, and others for the Apollo 14 fix. Hoag memo to Myers, February 16, 1971, MIT Museum.

25. "Grumman Flight Performance Evaluation Report for LM-8," 5.1-10–5.1-12 explains the procedure in detail. 5.1-24 lists the precise commands and their sequence. Courtesy personal collection of Paul Fjeld. Also see Landing Analysis Branch, "Preliminary Postflight Analysis of Apollo 14 LM Descent," Mission Planning and Analysis Division, Houston, Tex., February 12, 1971. JSC Archives.

26. Ibid.

27. NASA, "Apollo 14 Mission Report," 8-11.

28. NASA, "Apollo 14 Technical Debrief," 151.

29. Mitchell, quoted in Eric Jones, *Apollo Lunar Surface Journal*, at http://www.hq.nasa.gov/alsj/a14/ (accessed January 5, 2007). NASA, "Apollo 14 Technical Debrief," 151.

30. Shepard, in NASA, "Apollo 14 Technical Debrief," 151.

31. Mitchell, quoted in Eric Jones, *Apollo Lunar Surface Journal*, at http://www.hq.nasa.gov/alsj/a14/ (accessed January 5, 2007).

32. Ibid.

33. NASA, "Apollo 14 Technical Debrief," 151.

34. Mitchell, quoted in Eric Jones, *Apollo Lunar Surface Journal*, available at http://www.hq.nasa.gov/alsj/a14/ (accessed January 5, 2007).

35. NASA, "Apollo 14 Mission Report," 8-5.

36. Mitchell and Williams, *The Way of the Explorer*, 51.

37. NASA, "Apollo 14 Technical Debrief," 152.

38. Mitchell, quoted in Eric Jones, *Apollo Lunar Surface Journal*, available at http://www.hq.nasa.gov/alsj/a14/ (accessed January 5, 2007).

39. NASA, "Apollo 14 Technical Debrief," 152.

40. NASA, "Apollo 14 Mission Report," 9-7.

41. Ibid., 6-7, 9-7.

42. NASA, "Apollo 14 Technical Debrief," 152.

43. NASA, "Apollo 14 Mission Report," 15-1, 2, 11-2; McElheny, "Man Returns to the Moon."

44. Mitchell and Williams, *The Way of the Explorer*, 51.

45. Shepard and Slayton, *Moon Shot*, 304.

46. Ibid., 305.

47. Kranz, *Failure Is Not an Option*, 351. In response to Apollo 14 post-flight analysis, Floyd Bennett's group in MPAD conducted an analytical study of whether it would be possible for a crew to complete a landing in the event of a complete guidance failure (but with the computer's control systems working). The report, which did not consider any human factors or human-in-the-loop simulation data, concluded that it would be possible to control the descent if the crew stayed very close to the planned trajectory, and controlled the thrust/weight ratio of the vehicle to within 10 percent and the pitch attitude within plus or minus two degrees; see Landing Analysis Branch, "Manually Controlled Lunar Descent Approach Phase," MPAD NASA MSC, June 8, 1970, UHCL Chrono 72-51.

48. NASA, "Apollo 14 Technical Air-to-Ground Voice Transcription," 4:07:58:51.

49. NASA, "Apollo 15 Mission Report," 245–246.

50. Ibid., 93. Landing Analysis Branch, "Comparison of a 16° and a 30° Glide Angle Powered Descent Trajectory," MPAD NASA MSC, September 24, 1970, UHCL Chrono 72-64.

51. NASA, "Apollo 12 Mission Report," 89.

52. NASA, "Apollo Program Summary Report," chapter 6.

53. Duke interview with Ward, 12–27.

54. Logsdon, "A Failure of National Leadership," 296–297.

55. Cernan and Davis, *The Last Man on the Moon*, 282.

56. Landing Analysis Branch, "Use of the A Priori Terrain over Littrow," MPAD NASA MSC, April 8, 1970, UHCL Chrono 72-41.

57. NASA, "Apollo 15 Mission Report," 68.

58. Landing Analysis Branch, "Out-of-Plane Velocity during Apollo 15 Descent," MPAD NASA MSC, December 9, 1971, UHCL Chrono 73-56.

59. NASA, "Apollo 15 Technical Debrief," 9-13–9-17.

60. NASA, "Apollo 15 Mission Report," 9-12.

61. Scott, quoted in *Apollo Lunar Surface Journal*, available at http://www.hq.nasa.gov/alsj (accessed January 24, 2007).

62. Irwin, quoted in *Apollo Lunar Surface Journal*, available at http://www.hq.nasa.gov/alsj (accessed January 24, 2007).

63. Scott, quoted in *Apollo Lunar Surface Journal*, available at http://www.hq.nasa.gov/alsj (accessed January 24, 2007).

64. NASA, "Apollo 15 Mission Report," 94; NASA, "Apollo 15 Technical Debrief."

65. NASA, "Apollo 15 Technical Debrief," 9-14.

66. Scott, quoted in *Apollo Lunar Surface Journal*, at http://www.hq.nasa.gov/alsj (accessed January 24, 2007).

67. NASA, "Apollo 15 Technical Debrief," 9-15; NASA, "Apollo 15 Mission Report," 95.

68. NASA, "Apollo 15 Mission Debrief," 9-15; NASA, "Apollo 15 Mission Report," 95; Scott interview with Jones.

69. Scott, quoted in *Apollo Lunar Surface Journal*, available at http://www.hq.nasa.gov/alsj (accessed January 24, 2007).

70. NASA, "Apollo 15 Technical Debrief," 9-15.

71. Scott, quoted in *Apollo Lunar Surface Journal*, available at http://www.hq.nasa.gov/alsj (accessed January 24, 2007).

72. Duke interview with Ward, 27.

73. NASA, "Apollo 16 Mission Report," 9-18. Also see Flight Performance Evaluation Team, "LM-11 Flight Performance Evaluation Report," Grumman Aircraft Engineering Corporation 5-1.3–5-2.4.

74. NASA, "Apollo 16 Technical Debrief," 9-8; NASA, "Apollo 16 Mission Report," 9-18.

75. NASA, "Apollo 16 Technical Debrief," 9-9.

76. Ibid., 9-16.

77. NASA, "Apollo 16 Technical Debrief," 9-19.

78. Duke interview with Ward, 33; NASA, "Apollo 16 Mission Report," 9-19; NASA, "Apollo 16 Technical Debrief," 9-10.

79. Charlie Duke, class presentation, MIT, April 25, 2005.

80. Finney, "NASA Considering New Space School."

81. "Scientists, Engineers, Seek Roles in Space"; Beattie, *Taking Science to the Moon*, 173–174; Compton, *Where No Man Has Gone Before*, has an excellent analysis of the test-pilot versus scientist tensions in Apollo.

82. Cernan and Davis, *The Last Man on the Moon*, 84.

83. Cernan interview with Compton.

84. Schmidt interview with Butler.

85. O'Leary, *Making of an Ex-Astronaut*, 66.

86. Recer, "They Feud Over Moon Flights."

87. "Geologist to Quit Apollo Project; Weak Scientific Effort Charged"; Compton, chapter 10. Even NASA's life scientists felt the agency ignored scientific study of humans in space, despite all of the rhetoric and technology focused on human participants. McElheny, "Report Says NASA Concentrates on Machine, Ignores Man"; idem., "Some Fear Showy Space Trips Will Stifle Progress"; idem., "More Luna Probes Urged by Group of U.S. Scientists."

88. See Cernan and Schmidt comments in Eric Jones, *Apollo Lunar Surface Journal*, available at http://www.hq.nasa.gov/alsj/a17/ (accessed January 5, 2007).

89. Cernan and Davis, *The Last Man on the Moon*, 331.

90. NASA, "Apollo 17 Mission Report," 16-11.

91. Cernan, in NASA, "Apollo 17 Technical Debrief," 10-5.

92. Cernan, *The Last Man on the Moon*, 316.

93. NASA, "Apollo 17 Technical Debrief," 9-4.

94. NASA, "Apollo 17 Mission Report," 10-11; NASA, "Apollo 17 Technical Debrief," 9-6.

95. Cernan, quoted in Eric Jones, *Apollo Lunar Surface Journal*, available at http://www.hq.nasa.gov/alsj/a17/ (accessed January 5, 2007).

96. Cernan and Davis, *The Last Man on the Moon*, 317.

97. NASA, "Apollo 17 Technical Debrief," 9-17.

98. Cernan, in NASA, "Apollo 17 Technical Debrief," 9-7.

99. Scott, quoted on Eric Jones, *Apollo Lunar Surface Journal*, available at http://www.hq.nasa.gov/alsj/a15/ (accessed January 5, 2007).

100. NASA, "Apollo 11 Technical Debrief."

101. NASA, "Apollo 12 Technical Debrief."

102. Shepard and Slayton, *Moon Shot*, 304.

103. Scott, quoted in Eric Jones, *Apollo Lunar Surface Journal*, available at http://www.hq.nasa.gov/alsj/a15/ (accessed January 5, 2007).

104. NASA, "Apollo 15 Technical Debrief."

105. NASA, "Apollo 16 Technical Debrief."

106. NASA, "Apollo 17 Technical Debrief."

107. Cernan and Davis, *Last Man on the Moon*, 317–318.

108. Armstrong, "Apollo: Present," 122. The second reason Armstrong gave was that he did not trust the velocity indicators, which he thought would bring the LM down with some residual velocity and possibly tip it over.

109. Floyd Bennett, "Use of P-66 Auto in landing phase," MPAD NASA MSC, March 16, 1970. UHCL Chrono 72-34; Bennett, "Apollo Experience Report," 31; NASA, "Apollo 12 Mission Report," 6-5.

110. Astronauts already had "rate control" of the LM's attitude in ATTITUDE HOLD, but attitude controlled horizontal *acceleration* (actually, horizontal acceleration was proportional to the sine of the angle of the LM's attitude off the vertical times the descent engine's thrust). This made the LM difficult to stop. P66 LPD amounted to rate control in *velocity*, not attitude, which was arguably easier to control.

111. Eyles, "Apollo LM Guidance Computer Software for the Final Lunar Descent," 243–250; Eyles, "Tales from the Lunar Module Guidance Computer."

11 Human, Machine, and the Future of Spaceflight

1. "Address by Dr. Joseph Shea, Deputy Director of Manned Space Flight (Systems), National Aeronautics and Space Administration, at the American Institute of Aeronautics and Astronautics, 2nd Manned Spaceflight Meeting," Dallas, Tex., April 22, 1963, 12–13, NASA History Office, Joe Shea file.

2. Chilton interview with Bergen, 12–30, 34.

3. Sherman interview with Ertel.

4. Apollo alums who founded Intermetrics: John Miller, Jim Flanders, Jim Miller, Dan Lickly, and Ed Copps, joined soon after by Fred Martin. Martin, "Apollo 11: 25 Years Later."

5. Cunningham and Herskowitz, *The All-American Boys*, 181. For another pilot's view, see Fred W. Haise, Jr., "Space Transportation System as Seen by an Astronaut," NASA JSC, November 1, 1974, UHCL.

6. Collins, *Liftoff*, 202.

7. Heppenheimer, *History of the Space Shuttle*, vol. 2, 380–381.

8. Chris Kraft, comments to MIT students on the space shuttle system, November 8, 2005.

9. See comments by John Connolly in Reichardt, "Son of Apollo," 26. For the Crew Exploration Vehicle NASA is keeping the cockpit design in-house rather than farming it out to the contractor building the spacecraft. See Covault, "Piloting the CEV." On automated landings to deliver supplies, see Morring, "A Base to Build On."

10. Szalai and Jarvis interview with Wallace.

11. For a good account of the Apollo/F-8 relationship, see Tomayko, *Computers Take Flight*, chapter 3. Also see NASA, "Proceedings of the F-8 Digital Fly-By-Wire and Supercritical Wing First Flight's 20th Anniversary Celebration."

12. Dwain Deets (for Cal Jarvis), "The Digital Fly-By-Wire Program," in NASA, "Proceedings of the F-8 Digital Fly-By-Wire," 32.

13. "Preserve Pilots' Roles in Automated Cockpits," 86.

14. Roughton, "Jets with Byte Stuff Hovering on Horizon for Airline Industry."

15. Manningham, "The Cockpit."

16. Bush, "Remarks on U.S. Space Policy."

17. Pyne, "Seeking Newer Worlds: An Historical Context for Space Exploration."

Glossary

AGC Apollo guidance computer: main digital computer guiding the CSM

AGS Abort guidance system backup digital computer for lunar landings

AOS Acquisition of signal: the moment the LM reappears from behind the moon during an orbit, allowing earth-based tracking to receive radio signals

apogee High point of an orbit

apolune High point of a lunar orbit

attitude Orientation of a spacecraft or aircraft

CAPCOM Capsule communicator: astronaut on the ground who speaks with crew during a mission

CSM Command and service module

delta v budget Means of keeping track of velocity changes (delta v), roughly equivalent to fuel, during spacecraft maneuvers

DELTAH Quantity describing the difference between the on-board inertial estimate of height above the moon and that measured by the radar altimeter

DPS Descent propulsion system: rocket engine on the bottom of the Lunar Module

DSKY Display/keyboard interface to Apollo computer, pronounced "dis-key"

gimbal lock Condition where three-gyro inertial system in CSM and LM loses track of attitude due to unusual orientation of spacecraft

LGC Lunar module guidance computer: identical to AGC but installed in LM

LLRV Lunar landing research vehicle: "flying simulator" developed to research lunar-landing trajectories and control techniques

LLTV Training version of LLRV

LM Lunar module: pronounced "lem"

MIT IL Massachusetts Institute of Technology Instrumentation Laboratory, today known as the Charles Stark Draper Laboratory Inc. (spun out of MIT in the early 1970s)

MPAD Mission Planning and Analysis Division: group at NASA Houston that developed landing trajectories

PDI Powered descent initiation: moment at ten miles above the moon when descent engine (DPS) fired to begin descent to lunar surface

perigee Low point of an orbit

perilune Low point of lunar orbit

PNGS Primary navigation and guidance system ("pings"): the digital computer plus the inertial and optical systems

PRO Proceed button on DSKY display; astronauts had to press this button before the computer would initiate critical events

RCS Reaction control system: collection of small thrusters that controlled the attitude of the CSM or the LM

SETP Society of Experimental Test Pilots: professional organization of test pilots

IMU Inertial measurement unit: collection of gyroscopes and accelerometers that measured velocity changes in CSM and LM

DAP Digital autopilot: computer programs that controlled steering of Apollo spacecraft

state vector List of numbers that described position and velocity of spacecraft at a given time

Bibliography

Oral Histories

Abbreviations

JSC   Johnson Space Center Oral History Collection

UHCL   Manned Space Flight Center Oral History Collection, University of Houston, Clear Lake.

UHCL Chrono   Apollo chronological file.

MSFC   NASA Manned Space Flight Center.

All interviews are oral histories unless otherwise noted.

Four group oral history interviews conducted under the History of Recent Science and Technology (HRST) project on the World Wide Web are referenced in the text as follows:

HRST1 (Apollo Guidance Computer)   Ramon Alonso, Dan Lickly, Joe Gavin, David Hoag, Cline Frasier, Eldon Hall, Margaret Hamilton, Fred Martin, group oral history interview by David Mindell, Alexander F. Brown, and Slava Gerovitch, Cambridge, Mass., July 27, 2001.

HRST2 (Software and Simulation)   Alex Kosmala, Jim Miller, Herb Thaler, Ramon Alonso, Margaret Hamilton, Dan Lickly, Fred Martin, group oral history interview by David Mindell, Alexander F. Brown, and Slava Gerovitch, Cambridge, Mass., September 14, 2001.

HRST3 (Manufacturing)   Cline Frasier, Bard Turner, Dave Bates, Jack Poundstone, Ed Blondin, Hugh Blair-Smith, Herb Briss, Eldon Hall, David Hanley, Ed Duggan, group oral history interview by David Mindell and Alexander F. Brown, Cambridge, Mass., November 30, 2001.

HRST4 (Intermetrics)   Alex Kosmala, Ed Copps, Dan Lickly, John Miller, John Green, Fred Martin, group oral history interview by David Mindell, Alexander F. Brown, and Slava Gerovitch, Cambridge, Mass., September 6, 2002.

All HRST interviews are available at digitalapollo.mit.edu.

Armstrong, Neil, interviewed by Stephen E. Ambrose and Douglas Brinkley, Houston, Tex., September 19, 2001. JSC.

Battin, Richard, interviewed by Ivan Ertel, Cambridge, Mass., April 29, 1966. UHCL.

Battin, Richard, interviewed by David Mindell and Alexander F. Brown, Cambridge, Mass., September 30, 2002. Notes and tape in the author's possession.

Battin, Richard, interviewed by Rebecca Wright, Cambridge, Mass., April 18, 2000. JSC.

Bean, Alan, interviewed by W. D. Compton, Houston, Tex., April 10, 1984. JSC.

Bennett, Floyd, interviewed by author (telephone), August 24, 2005. Notes in the author's possession.

Bennett, Floyd, interviewed by Jennifer Ross-Nazzal, Houston, Tex., October 22, 2003. JSC.

Cernan, Eugene, interviewed by W. D. Compton, Houston, Tex., April 6, 1984. JSC.

Chilton, Robert, interviewed by Summer Chick Bergen, Houston, Tex., April 5, 1999. JSC.

Chilton, Robert, interviewed by Lloyd Swenson, Houston, Tex., March 30, 1970. UHCL.

Cohen, Aaron, interviewed by Lloyd Swenson, Houston, Tex., January 14, 1970. UHCL.

Crossfield, A. Scott, interviewed by Peter Merlin, Lancaster, Calif., February 3, 1998. NASA Dryden Oral History Collection.

Duke, Charles M. Jr., interviewed by Doug Ward, Houston, Tex., March 12, 1999. JSC.

Faber, Stanley, interviewed by Lloyd Swenson, Houston, Tex., April 22, 1970. UHCL.

Faget, Max, interviewed by Ivan Ertel, Houston, Tex., December 15, 1969. UHCL.

Frick, Charles W., interviewed by Ivan Ertel, Palo Alto, Calif., June 26, 1968. UHCL.

Funk, Jack, interviewed by Lloyd Swenson, Houston, Tex., June 25, 1970. UHCL.

Funk, Jack, interviewed by Jennifer Ross-Nazzal, Houston, Tex., October 30, 2003. JSC.

Garman, John R., interviewed by Kevin M. Rusnak, Houston, Tex., March 27, 2001. JSC.

Gilbert, David W., interviewed by Ivan Ertel, Houston, Tex., Cambridge, Mass., December 16, 1969. UHCL.

Hamilton, Margaret, interviewed by Paul Ceruzzi, Cambridge, Mass., April 4, 1991. NASA Headquarters Historical Collection.

Hoag, David, interviewed by Ivan Ertel, Cambridge, Mass., May 15, 1969. UHCL.

Johnson, Caldwell, interviewed by James Grimwood, Palo Alto, Calif., June 26, 1968. UHCL.

Klumpp, Allan, interviewed by David Mindell, Cambridge, Mass., August 16, 2005. Notes in the author's possession.

Kupfer, Walter, interviewed by Ivan Ertel, Cambridge, Mass., April 27, 1966. UHCL.

Laning, J. Halcombe, interviewed by Sandy Brown, Newton, Mass., November 21, 2002. Transcript in author's possession.

McDivitt, James, interviewed by Doug Ward, Elk Lake, Mich., June 1999. JSC.

Miller, John, interviewed by Ivan Ertel, Cambridge, Mass., April 28, 1966. UHCL.

Mueller, George, interviewed by Ivan Ertel, Washington, D.C., June 27, 1967. UHCL.

Nevins, James, interviewed by David Mindell, Cambridge, Mass., April 26, 2004. Transcript in the author's possession.

Ragan, Ralph, interview, no location, no interviewer, 1966. UHCL.

Rathert, George, interviewed by James Greenwood and Lloyd Swenson, Ames Research Center, Palo Alto, Calif., June 30, 1971 (not transcribed). UHCL.

Schmidt, Harrison, interviewed by Carol Butler, July 14, 1999, Houston Tex. JSC.

Scott, David, interview with Eric Jones, *Apollo Lunar Surface Journal*, http://www.hq.nasa.gov/office/pao/History/alsj, accessed January 23, 2007.

Shea, Joseph, interviewed by Michelle Kelley, August 26, 1998, Weston, Mass. JSC.

Sherman, Howard, interviewed by Ivan Ertel, Bethpage, N.Y., 1970. UHCL.

Szalai, Kenneth, and Calvin R. Jarvis, interviewed by Lane Wallace, Dryden Flight Research Center, August 30, 1995. NASA Dryden Oral History Collection.

Trageser, Milton, interviewed by Ivan Ertel, Cambridge, Mass., April 27, 1966. UHCL.

Voas, Robert, interviewed by Robert Sherrod, Washington, D.C., January 1970. NASA Headquarters Historical Collection.

Welch, Joseph B., interviewed by Ivan Ertel, Valley Forge, Penn., February 16, 1970. UHCL.

Mission Reports, Technical Debriefs, Technical Transcripts

NASA. "Apollo Program Summary Report." Houston, Tex.: Lyndon B. Johnson Space Center, 1975.

NASA. "Apollo 8 Mission Report." Houston, Tex.: NASA Manned Spacecraft Center, 1969.

NASA. "Apollo 9 Mission Report." Houston, Tex.: NASA Manned Spacecraft Center, 1969.

NASA. "Apollo 11 Mission Report." MSC-00171. Houston, Tex.: NASA Manned Spacecraft Center, 1969.

NASA. "Apollo 12 Mission Report." MSC-01855. Houston, Tex.: NASA Manned Spacecraft Center, 1970.

NASA. "Apollo 14 Mission Report." MSC-04112. Houston, Tex.: NASA Manned Spacecraft Center, 1971.

NASA. "Apollo 15 Mission Report." MSC-05161. Houston, Tex.: NASA Manned Spacecraft Center, 1971.

NASA. "Apollo 16 Mission Report." MSC-07230. Houston, Tex.: NASA Manned Spacecraft Center, 1972.

NASA. "Apollo 17 Mission Report." JSC-07904. Houston, Tex.: Lyndon B. Johnson Space Center, 1973.

NASA. "Apollo 11 Technical Debrief." Houston, Tex.: NASA Manned Spacecraft Center, 1969.

NASA. "Apollo 12 Technical Debrief." Houston, Tex.: NASA Manned Spacecraft Center, 1969.

NASA. "Apollo 14 Technical Debrief." Houston, Tex.: NASA Manned Spacecraft Center, 1971.

NASA. "Apollo 15 Technical Debrief." MSC-04561. Houston, Tex.: NASA Manned Spacecraft Center, 1971.

NASA. "Apollo 16 Technical Debrief." MSC-06805. Houston, Tex.: NASA Manned Spacecraft Center, 1972.

NASA. "Apollo 17 Technical Debrief." MSC-07631. Houston, Tex.: NASA Manned Spacecraft Center, 1973.

NASA. "Apollo 8 Technical Air-to-Ground Voice Transcription." Houston, Tex.: NASA Manned Spacecraft Center, 1968.

NASA. "Apollo 11 Technical Air-to-Ground Voice Transcription." Houston, Tex.: NASA Manned Spacecraft Center, 1969.

NASA. "Apollo 12 Technical Air-to-Ground Voice Transcription." Houston, Tex.: NASA Manned Spacecraft Center, 1970.

NASA. "Apollo 14 Technical Air-to-Ground Voice Transcription." Houston, Tex.: NASA Manned Spacecraft Center, 1971.

NASA. "Apollo 15 Technical Air-to-Ground Voice Transcription." Houston, Tex.: NASA Manned Spacecraft Center, 1971.

NASA. "Apollo 16 Technical Air-to-Ground Voice Transcription." Houston, Tex.: NASA Manned Spacecraft Center, 1972.

NASA. "Apollo 17 Technical Air-to-Ground Voice Transcription." Houston, Tex.: NASA Manned Spacecraft Center, 1973.

NASA. "Apollo 11 On-Board Voice Transcription." Houston, Tex.: NASA Manned Spacecraft Center, 1969.

NASA. "Postlaunch Report for Mission AS-202." Houston, Tex.: Manned Spacecraft Center, 1966.

All voice transcriptions can be accessed at http://www.jsc.nasa.gov/history/mission_trans/apollo11.htm.

Archival Collections

JSC Archives Johnson Space Center Archives, Houston, Texas.

NASA Dryden Archives NASA Dryden Flight Research Center Archival Collection, Edwards, California.

NASA HQ NASA Headquarters Archives, Washington, D.C.

NASA MSC Manned Spacecraft Center Archives, University of Houston, Clear Lake, Tex.

UHCL Chrono Manned Spacecraft Center Archives, University of Houston, Clear Lake, Tex., Chronological File.

MIT Museum MIT Instrumentation Laboratory Collection, MIT Museum, Cambridge, Mass.

Charles Stark Draper Laboratories, Library and Archives, Cambridge, Mass.

Books and Articles

"Cite Wide IC Use in Apollo Guidance Unit." August 31, 1964, Fairchild News Service, New York.

"Fairchild Div. Ships 110,000 Integrateds for Apollo Project." August 24, 1964, Fairchild News Service, New York.

"Geologist to Quit Apollo Project; Weak Scientific Effort Charged." *New York Times*, October 9, 1969: 56.

"Lunar Landing Had Its Earthbound Heroes." *Datamation* (September 1969): 145–146.

"Lunar Module Computer Problems." *Datamation* (October 1969): 169–170.

"The Manned Space Flight Network for Apollo." Greenbelt, Md.: Goddard Spaceflight Center, 1968.

"More Apollo Guidance Flexibility Sought." *Aviation Week and Space Technology* (November 16, 1964): 71–74.

"Preserve Pilots' Roles in Automated Cockpits." *Aviation Week and Space Technology* 142, no. 5 (January 30, 1995): 86.

"Scientists, Engineers, Seek Roles in Space." *Aviation Week and Space Technology* (November 23, 1964): 48–52.

"Test Rig Simulates LEM Landings." *Missiles and Rockets* (May 17, 1965): 43–44.

Abbott, Andrew Delano. *The System of Professions: An Essay on the Division of Expert Labor.* Chicago: University of Chicago Press, 1988.

Abernathy, William J., and Bruce A. Wilburn. "Technical Control in the Development Program for the Apollo Guidance Computer." Harvard Business School, Cambridge, Mass. (May 15, 1964): 13.

Abzug, Malcolm J., and E. Eugene Larrabee. *Airplane Stability and Control: A History of the Technologies that Made Aviation Possible*, 2nd ed. Cambridge Aerospace Series 14. Cambridge, UK: Cambridge University Press, 2002.

AC Electronics. "Apollo Guidance and Navigation Lunar Module Student Study Guide." Milwaukee, Wis.: AC Electronics Division of General Motors, 1967.

Ackmann, Martha. *The Mercury 13: The Untold Story of Thirteen American Women and the Dream of Space Flight*, 1st ed. New York: Random House, 2003.

Aldrin, Edwin E. "Line-of-Sight Guidance Techniques for Manned Orbital Rendezvous." Sc.D. dissertation, Massachusetts Institute of Technology, Department of Aeronautics and Astronautics, 1963.

Aldrin, Edwin E., and Wayne Warga. *Return to Earth*, 1st ed. New York: Random House, 1973.

Alonso, Ramon, and Cline Frasier. "Computer Evolutionary Dead Ends: Or How Good Ideas that Successfully Solve Problems Are Not Always the Way of the Future." Presentation at Military and Aerospace Applications of Programmable Logic Devices, Washington, D.C., September 2004.

Alonso, Ramon L., Hugh Blair-Smith, and Albert L. Hopkins. "Some Aspects of the Logical Design of a Control Computer: A Case Study." *IEEE Transactions on Electronics and Computers* (December 1963): 687–697.

Alonso, Ramon L., Albert L. Hopkins, and Herbert A. Thaler. "A Multiprocessing Structure." Cambridge, Mass.: MIT Instrumentation Laboratory, 1967.

Alonso, Ramon L., and J. Hal Laning. "Design Principles for a General Control Computer." Cambridge, Mass.: MIT Instrumentation Laboratory, 1960.

Armstrong, Neil. "Apollo: Present." Paper presented at Society of Experimental Test Pilot's 13th Annual Symposium, September 1969.

Armstrong, Neil. "Apollo 11 Postflight-Crew Press Conference." Houston, Tex., 1969.

Armstrong, Neil. "Wingless on Luna." In *Proceedings of the 25th Wings Club Meeting*. New York: Wings Club General Harold L. Harris "Sight" Lectures, 1988.

Armstrong, Neil. "X-15 Operations: Electronics and the Pilot." *Astronautics* 5 (May 1960): 42–43, 76–78.

Armstrong, Neil, Michael Collins, Edwin E. Aldrin, Gene Farmer, and Dora Jane Hamblin. *First on the Moon. A Voyage with Neil Armstrong, Michael Collins and Edwin E. Aldrin, Jr.*, 1st ed. Boston: Little, Brown, & Co. 1970.

Armstrong, W. T. "Where Do We Go from Here?" *Cockpit* (May 1965): 4–7.

Atwill, William D. *Fire and Power: The American Space Program as Postmodern Narrative*. Athens: University of Georgia Press, 1994.

Bailey, Jr., A. J. "Development and Flight Test of Adaptive Controls for the X-15." *SETP Annual Symposium Proceedings* (1961): 3–23.

Barthes, Roland. *Mythologies*. Paris: Editions du Seuil, 1970.

Barthes, Roland. "The Jet-Man." In Barthes, *Mythologies*, 71–73.

Battin, Richard H. *Astronautical Guidance*. New York: McGraw-Hill and Co., 1964.

Battin, Richard H. "A Statistical Optimizing Navigation Procedure for Space Flight." R-241. Cambridge, Mass.: MIT Instrumentation Laboratory, September 1961, revised May 1962.

Battin, R. H. "Some Funny Things Happened on the Way to the Moon." In *Proceedings of the AIAA 27th Aerospace Sciences Meeting*. Reno, Nev.: AIAA, 1989.

Battin, R. H. "Space Guidance Evolution—A Personal Narrative." *Journal of Guidance, Control, and Dynamics* 5, no. 2 (1982): 97–110.

Bean, Alan, and Andrew Chaikin. *Apollo: An Eyewitness Account by Astronaut/Explorer Artist/Moonwalker Alan Bean*. Shelton, Conn.: Greenwich Workshop Press, 1998.

Bean, Alan, and Beverly Fraknoi. *My Life as an Astronaut*. New York: Pocket Books, 1988.

Beattie, Donald A. *Taking Science to the Moon: Lunar Experiments and the Apollo Program*. New Series in NASA History. Baltimore: Johns Hopkins University Press, 2001.

Benjamin, Marina. *Rocket Dreams: How the Space Age Shaped Our Vision of a World Beyond*. New York: Free Press, 2003.

Bennett, Floyd. "Apollo Experience Report: Mission Planning for Lunar Module Descent and Ascent." Houston, Tex.: NASA MSFC, 1972.

Bennett, Floyd V. "Apollo Lunar Descent and Ascent Trajectories." Paper presented at the AIAA 8th Aerospace Sciences Meeting, New York, N.Y., January 19–21, 1970. NASA TM X-58040.

Benson, Charles D., and William Barnaby Faherty. *Moonport: A History of Apollo Launch Facilities and Operations*. NASA History Series. SP-4204. Washington, D.C.: NASA Scientific and Technical Information Office, 1978.

Berlin, Leslie. *The Man Behind the Microchip: Robert Noyce and the Invention of Silicon Valley*. Oxford, UK: Oxford University Press, 2005.

Bernard, A. V. "Landing Site Selection Criteria." In Manned Space Flight Center, *Apollo Lunar Landing Mission Symposium*, 573–600. Houston, Tex.: NASA, 1966.

Bilstein, Roger E., and Frank Walter Anderson. *Orders of Magnitude: A History of the NACA and NASA, 1915–1990*. NASA History Series. SP-4406. Washington, D.C.: Office of Management, Scientific and Technical Information Division, NASA, 1989.

Bix, Amy Sue. *Inventing Ourselves Out of Jobs?: America's Debate over Technological Unemployment, 1929–1981*. Baltimore: Johns Hopkins University Press, 2000.

Black, Harold. "Stabilized Feedback Amplifiers." *Bell System Technical Journal* 13 (January 1934): 1–18.

Blackburn, A. W. "Flight Testing Stability Augmentation Systems for High Performance Fighters." *SETP Quarterly Review* 6, no. 1 (Summer 1957): 2.

Blackburn, A. W. "Flight Testing in the Space Age." *SETP Quarterly Review* 2, no. 3 (Spring 1958): 3–17.

Blackburn, A. W. *Aces Wild: The Race for Mach 1.* Wilmington, Del.: Scholarly Research Books, 1998.

Blair, Charles F. "Automation and the Space Pilot," *SETP Quarterly Review* 4, no. 2 (Winter 1960): 17–19.

Blair-Smith, Hugh. "AGC4 Memo-Block II Instructions." Cambridge, Mass.: MIT Instrumentation Laboratory, 1966.

Borman, Frank, and Robert J. Serling. *Countdown: An Autobiography*, 1st ed. New York: Dwight W. Morrow & Co., 1988.

Box, David M., Jon C. Harpold, Steven G. Paddock, Neil A. Armstrong, and William H. Hamby, "Controlled Reentry." In Manned Spacecraft Center (U.S.), *Gemini Summary Conference*, 159–166.

Braverman, Harry. *Labor and Monopoly Capital: The Degradation of Work in the Twentieth Century.* New York: Monthly Review Press, 1974.

Brooks, Courtney G., James M. Grimwood, and Lloyd S. Swenson. *Chariots for Apollo: A History of Manned Lunar Spacecraft.* NASA History Series. SP-4205. Washington, D.C.: NASA Scientific and Technical Information Office, 1979.

Brown, Sandy. "Probing Mars, Probing the Market: An Episode in the Early History of Astronautical Guidance and Navigation Systems." Second year paper, Program in Science, Technology and Society. Cambridge: MIT, 2003.

Bush, George W. "Remarks on U.S. Space Policy," January 14, 2004, available at http://www.nasa.gov/pdf/54868main_bush_trans.pdf accessed August 7, 2007.

Byerly, Radford, Jr., ed. *Space Policy Reconsidered.* Boulder, Col.: Westview Press, 1989.

Cameron, Rebecca Hancock. *Training to Fly: Military Flight Training, 1907–1945.* Washington, D.C.: Air Force History and Museum Programs, 1999.

Campbell, Joseph. *The Hero with a Thousand Faces*, commemorative ed., Bollingen Series. Princeton, N.J.: Princeton University Press, 2004.

Carpenter, M. Scott. *We Seven.* New York: Simon & Schuster, 1962.

Carter, Dale. *The Final Frontier: The Rise and Fall of the American Rocket State.* London; New York: Verso, 1988.

Castells, Manuel. *The Power of Identity.* Malden, Mass.: Blackwell Publishers, 1997.

Castells, Manuel. *The Rise of the Network Society*. Cambridge, Mass.: Blackwell Publishers, 1996.

Cernan, Eugene, and Don Davis. *The Last Man on the Moon: Astronaut Eugene Cernan and America's Race in Space*, 1st ed. New York: St. Martin's Press, 1999.

Ceruzzi, Paul. *Beyond the Limits: Flight Enters the Computer Age*. Cambridge, Mass.: MIT Press, 1989.

Chaikin, Andrew. *A Man on the Moon: The Voyages of the Apollo Astronauts*. New York: Penguin Books, 1998.

Chandler, Alfred Dupont. *The Visible Hand: The Managerial Revolution in American Business*. Cambridge, Mass.: Belknap Press, 1977.

Cheatham, Donald C., and Floyd Bennett. "Apollo Lunar Module Landing Strategy." In Manned Space Flight Center, *Apollo Lunar Landing Mission Symposium*, 131–241. Houston, Tex.: NASA, 1966.

Cheatham, Donald C., and Clarke T. Hackler. "Handling Qualities for Pilot Control of Apollo Lunar-Landing Spacecraft." *J. Spacecraft* 3, no. 5 (May 1966): 632–638.

Cheng, Richard K. "Lunar Terminal Guidance." In Leondes and Vance, eds., *Lunar Missions and Exploration*, 308–355.

Chilton, Robert. "Command and Communications." Paper presented at NASA Industry Conference, Goddard Spaceflight Center, Greenbelt, Md., August 30, 1960.

Clark, Carl, and James Hardy. "Preparing Man for Space Flight." *Astronautics* (February 1959): 18–21, 88–90.

Clynes, Manfred E., and Nathan S. Kline. "Cyborgs and Space." *Astronautics* (September 1960): 26–27, 74–76.

Collins, H. M., and Martin Kusch. *The Shape of Actions: What Humans and Machines Can Do*. Cambridge, Mass.: MIT Press, 1998.

Collins, Michael. *Carrying the Fire: An Astronaut's Journeys*. New York: Farrar, Straus & Giroux, 1974.

Collins, Michael. *Liftoff: The Story of America's Adventure in Space*, 1st ed. New York: Grove Press, 1988.

Compton, William David. *Where No Man Has Gone Before: A History of Apollo Lunar Exploration Missions*. NASA History Series. SP-4214. Washington, D.C.: Office of Management, Scientific and Technical Information Division, NASA, 1989.

Conrad, Nancy, and Howard Klausner. *Rocket Man: Astronaut Pete Conrad's Incredible Ride to the Moon and Beyond*. New York: New American Library, 2005.

Constant, Edward W. *The Origins of the Turbojet Revolution*. Johns Hopkins Studies in the History of Technology; New Series, No. 5. Baltimore: Johns Hopkins University Press, 1980.

Conway, Erik M. *Blind Landings: Low-Visibility Operations in American Aviation, 1918–1958*. Baltimore: Johns Hopkins University Press, 2006.

Cooper, A. E., and W. T. Chow. "Development of On-Board Space Computer Systems." *IBM Journal of Research and Development* 20, no. 1 (January 1976): 5–19.

Cooper, George E. "Understanding and Interpreting Pilot Opinion." *SETP Quarterly Review* 1, no. 1 (Summer 1957): 19–30.

Cooper, Henry S. F. *Before Lift-Off: The Making of a Space Shuttle Crew*. New Series in NASA History. Baltimore: Johns Hopkins University Press, 1987.

Copps Jr., Edward M. "Recovery from Transient Failures of the Apollo Guidance Computer." Cambridge, Mass.: MIT Instrumentation Laboratory, 1968.

Corn, Joseph J. *The Winged Gospel: America's Romance with Aviation, 1900–1950*. New York: Oxford University Press, 1983.

Cortright, Edgar M. *Apollo Expeditions to the Moon*. SP-350. Washington, D.C.: NASA Scientific and Technical Information Office, 1975.

Covault, Craig, "Piloting the CEV: Astronauts Define Cockpit Elements for the Crew Exploration Vehicle." *Aviation Week and Space Technology* 164, no. 25 (June 19, 2006): 46–49.

Crisp, Robert, and D. Keene. "Apollo Command and Service Module Reaction Control by the Digital Autopilot." Cambridge, Mass.: MIT Instrumentation Laboratory, 1968.

Cronon, William. *Nature's Metropolis: Chicago and the Great West*. New York: W. W. Norton, 1991.

Crossfield, A. Scott. "Pilot Contributions to Mission Success." *SETP Quarterly Review* 6, no. 3 (Summer 1963): 23–26.

Crossfield, A. Scott. "The Way to the Stars." *Aviation Week and Space Technology* (March 24, 2003): 58–60.

Cunningham, Walter, and Mickey Herskowitz. *The All-American Boys*. New York: Macmillan, 1977.

Dana, William. "A History of the X-15 Program." *SETP Proceedings* (1987): 257–272.

Dennis, Michael Aaron. "A Change of State: The Political Cultures of Technical Practice at the MIT Instrumentation Laboratory and the Johns Hopkins University Applied Physics Laboratory, 1930–1945." Ph.D. dissertation, Johns Hopkins University, 1990.

Dick, Philip K. *Do Androids Dream of Electric Sheep?* New York: Ballantine Books, 1996.

Dick, Steven J., and Roger D. Launius, eds. *Critical Issues in the History of Spaceflight*. Washington, D.C.: NASA, 2006.

Doolittle, James Harold. "The Effect of the Wind Velocity Gradient on Airplane Performance." Sc.D. dissertation, Massachusetts Institute of Technology Department of Aeronautical Engineering, 1925.

Doolittle, James Harold. "Wing Loads as Determined by the Accelerometer." M.S. thesis, Massachusetts Institute of Technology, Department of Aeronautical Engineering, 1924.

Doolittle, James Harold, and Carroll V. Glines. *I Could Never Be So Lucky Again: An Autobiography.* New York: Bantam Books, 1991.

Dornbach, John E. "Lunar Landing Site Data Sources and Analysis." In Manned Space Flight Center, *Apollo Lunar Landing Mission Symposium,* 601–628. Houston, Tex.: NASA, 1966.

Dornberger, Walter. "The Rocket Propelled Commercial Airliner." In Godwin, ed., *Dyna-Soar,* 19–37.

Dotts, Homer W., Roger K. Nolting, Wilburn F. Hoyler, John R. Havey, Thomas F. Carter, Jr., and Robert T. Johnson. "Operational Characteristics of the Docked Configuration," in Manned Spacecraft Center (U.S.). *Gemini Summary Conference,* 41–54.

Drake, Hubert M., Donald R. Bellman, and Joseph A. Walker. "Operational Problems of Manned Orbital Vehicles." NACA Research Memorandum, Washington, D.C., 1958.

Draper, C. S. "The Evolution of Aerospace Guidance Technology at the Massachusetts Institute of Technology 1935–1951: A Memoir." Paper presented at the Fifth History Symposium of the International Academy of Aeronautics, Brussels, Belgium, September 1971.

Draper, C. S., H. P. Whitaker, and L. R. Young. "The Roles of Men and Instruments in Control and Guidance Systems for Spacecraft." Paper presented at XVth International Astronautical Congress, Warsaw, Poland, September 7–12, 1964.

Draper, C. S., W. Wrigley, D. G. Hoag, R. H. Battin, J. E. Miller, D. A. Koso, A. L. Hopkins, and W. E. Vander Velde. "Space Navigation Guidance and Control. Volume I." Cambridge, Mass.: MIT Instrumentation Laboratory, 1965.

Draper, C. S., W. Wrigley, D. G. Hoag, R. H. Battin, J. E. Miller, D. A. Koso, A. L. Hopkins, and W. E. Vander Velde. "Space Navigation Guidance and Control. Volume II." Cambridge, Mass.: MIT Instrumentation Laboratory, 1965.

Duke, Charlie, and Dotty Duke. *Moonwalker.* Nashville, Ten.: Oliver-Nelson Books, 1990.

Duke, Charles M., Jr., and Michael S. Jones. "Human Performance During a Simulated Apollo Mid-Course Navigation Sighting." S.M. thesis, MIT, 1964.

Duncan, Robert. "Apollo Navigation, Guidance and Control." In Manned Space Flight Center, *Apollo Lunar Landing Mission Symposium,* 38–130. Houston, Tex.: NASA, 1966.

Ellul, Jacques. *The Technological Society,* 1st U.S. ed. New York: Knopf, 1964.

English, Richard. "Jet Pilots Are Different." *Saturday Evening Post,* July 9, 1949.

Ertel, Ivan D. *The Apollo Spacecraft: A Chronology,* 4 vols. Washington, D.C.: Scientific and Technical Information Division, NASA, 1969.

Evans, Walter. "Graphical Analysis of Control Systems." *AIEE Transactions* 67 (1948): 547–551.

Eyles, Donald. "Apollo LM Guidance Computer Software for the Final Lunar Descent." *Automatica* 9 (1973): 243–250.

Eyles, Don, "Tales from the Lunar Module Guidance Computer." Paper presented at 27th Annual American Astronautical Society Guidance and Control Conference, Breckenridge, Col., February 2004.

Faget, Maxime A. "The Evolution of Flight Control of the Apollo Mission." In Steinhoff, ed., *The Eagle Has Returned*: 324–330.

Faget, Maxime A., Benjamine J. Garland, and James J. Buglia. "Preliminary Studies of Manned Satellites, Wingless Configuration-Nonlifting." Washington, D.C.: National Advisory Committee for Aeronautics, 1958.

Faludi, Susan. *Stiffed: The Betrayal of the American Man*, 1st ed. New York: W. Morrow and Co., 1999.

Farrior, James S. "Guidance and Navigation: State of the Art—1960." *Astronautics* (November 1960): 150–154.

Felleman, Philip G. "Hybrid Simulation of the Apollo Guidance Navigation and Control System." Cambridge, Mass.: MIT Instrumentation Laboratory, 1966.

Findley, Paul B. "The Systems Development Department." *Bell Laboratories Record* (April 1926): 69–73.

Finney, John W. "Pilots Will Control Gemini Spacecraft." *New York Times*, October 15, 1962: A1, A5.

Fox, Richard J. and T. J. Jackson Lears. *The Culture of Consumption: Critical Essays in American History: 1880–1980*. New York: Pantheon Books, 1983.

Frank, M. P. "Detailed Mission Planning Considerations and Constraints." In Manned Space Flight Center, *Apollo Lunar Landing Mission Symposium*, 131–174. Houston, Tex.: NASA, 1966.

Flight Research Center, NACA. "X-15 Flight Logs." Edwards, Calif.: NACA Flight Research Center, 1959–1967.

Fritzsche, Peter. *A Nation of Fliers: German Aviation and the Popular Imagination*. Cambridge, Mass.: Harvard University Press, 1992.

Gainor, Chris. *Arrows to the Moon: Avro's Engineers and the Space Race*. Burlington, Ont.: Apogee Books, 2001.

Gann, Ernest Kellogg. *Fate Is the Hunter*. New York: Simon & Schuster, 1961.

Garman, John R. "The 'Bug' Heard 'Round the World.'" *Software Engineering Notes* 6, no. 5 (1981): 3–10.

Gavin, Joseph. "Engineering Development of the Apollo Lunar Module." Paper presented at 41st Congress of the International Astronautical Federation, Dresden, GDR, October 6–12, 1990. IAA-90-633.

Geissler, Ernst, and Hausserman, Walter. "Saturn Guidance and Control." *Astronautics* (February 1962): 44–46, 89–92.

Gerovitch, Slava. *From Newspeak to Cyberspeak: A History of Soviet Cybernetics*. Cambridge, Mass.: MIT Press, 2002.

Gerovitch, Slava. "The New Soviet Man in a Man-Machine System: The Technical Intelligentsia, Automatic Control, and the Space Race." Paper presented at conference on Intelligentsia: Russian and Soviet Science on the World Stage, 1860–1960, University of Georgia, October 2004.

Gerovitch, Slava. "Human Machine Issues in the Soviet Space Program." In Dick and Launius, eds., *Critical Issues in the History of Spaceflight*, 107–140.

Gibbs-Smith, Charles Harvard. *Aviation: An Historical Survey from Its Origins to the End of World War II*. London: H.M.S.O., 1970.

Gilruth, Robert. "Requirements for Satisfactory Flying Qualities of Airplanes." NACA Technical Report 755, 1943.

Gilruth, Robert R., and M. D. White. "Analysis and Prediction of Longitudinal Stability of Airplanes." NACA Technical Report 711, 1940.

Gilruth, Robert, Joseph P. Shea, and Samuel C. Philips. "Introductory Remarks." In Manned Space Flight Center, *Apollo Lunar Landing Mission Symposium*. Houston, Tex.: NASA, 1966.

Glick, F. K., and S. R. Femino. "A Comprehensive Digital Simulation for the Verification of Apollo Flight Software." E-2475. Cambridge, Mass.: MIT Instrumentation Laboratory, January 1970.

Glines, Carroll V. *Roscoe Turner: Aviation's Master Showman*. Smithsonian History of Aviation Series. Washington, D.C.: Smithsonian Institution Press, 1995.

Godwin, Robert, ed. *Apollo 7: The NASA Mission Reports*. Burlington, Ont.: Apogee Books, 2000.

Godwin, Robert, ed. *Apollo 8: The NASA Mission Reports*, 2nd ed. Burlington, Ont.: Apogee Books, 2000.

Godwin, Robert, ed. *Apollo 9: The NASA Mission Reports*. Burlington, Ont.: Apogee Books, 1999.

Godwin, Robert, ed. *Apollo 10: The NASA Mission Reports*, 2nd ed. Burlington, Ont.: Apogee Books, 2000.

Godwin, Robert, ed. *Apollo 11: The NASA Mission Reports*, 3 vols. Burlington, Ont.: Apogee Books, 1999.

Godwin, Robert, ed. *Apollo 12: The NASA Mission Reports*. Burlington, Ont.: Apogee Books, 1999.

Godwin, Robert, ed. *Apollo 13: The NASA Mission Reports*. Burlington, Ont.: Apogee Books, 2000.

Godwin, Robert, ed. *Apollo 14: The NASA Mission Reports*. Burlington, Ont.: Apogee Books, 2000.

Godwin, Robert, ed. *Apollo 15: The NASA Mission Reports*. Burlington, Ont.: Apogee Books, 2001.

Godwin, Robert, ed. *Dyna-Soar: Hypersonic Strategic Weapons System*. Burlington, Ont.: Apogee Books, 2003.

Godwin, Robert, ed. *X-15: The NASA Mission Reports, Incorporating Files from the USAF*. Burlington, Ont.: Apogee Books, 2001.

Goldstein, Stanley H. *Reaching for the Stars: The Story of Astronaut Training and the Lunar Landing*. New York: Praeger, 1987.

Goode, Harry H., and Robert Engel Machol. *System Engineering: An Introduction to the Design of Large-Scale Systems*. McGraw-Hill Series in Control Systems Engineering. New York: McGraw-Hill Company, 1957.

Gordon, Henry C. "Concepts for Piloted, Maneuvering, Reentry Vehicles," *SETP Quarterly Review* 6, no. 3 (Summer 1963): 16–19.

Gorn, Michael H. *Expanding the Envelope: Flight Research at NACA and NASA*. Lexington: University Press of Kentucky, 2001.

Graham, Gordon. "The Man-Machine Conflict in High Performance Tactical Aircraft." *SETP Journal* (1965): 180–185.

Green, Alan I., and Robert J. Filene. "Keyboard and Display Program and Operation." E-1095. Cambridge, Mass.: MIT Instrumentation Laboratory, January 1966.

Grimwood, James M. *Project Mercury: A Chronology*. SP–4001. Washington, D.C.: NASA, 1963.

Grimwood, James M., Barton C. Hacker, and Peter J. Vorzimmer. *Project Gemini: Technology and Operations: A Chronology*. NASA History Series. SP-4002. Washington, D.C.: NASA Scientific and Technical Information Office, 1969.

Grumman. "LEM Guidance, Navigation and Control: Apollo News Reference." Bethpage, N.Y.: Grumman Aircraft Engineering Corporation, n.d.

Guidance Software Validation Committee. "Apollo Guidance Software Development and Validation Plan." Houston, Tex.: NASA Manned Spacecraft Center, 1967.

Hacker, Barton C., and James M. Grimwood. *On the Shoulders of Titans: A History of Project Gemini*. NASA History Series. SP-4203. Washington, D.C.: NASA, Scientific and Technical Information Division, Office of Technology Utilization, 1977.

Hall, Eldon C. "From the Farm to Pioneering with Digital Control Computers: An Autobiography." *IEEE Annals of the History of Computing* (April–June 2000): 28.

Hall, Eldon C. *Journey to the Moon: The History of the Apollo Guidance Computer*. Reston, Va: American Institute of Aeronautics and Astronautics, 1996.

Hall, Eldon C. "MIT's Role in Project Apollo, Vol. III: Computer Subsystem." R-700. Cambridge, Mass.: MIT Instrumentation Laboratory, August 1972.

Hall, Eldon C. "Reliability History of the Apollo Guidance Computer." CR-140340. Cambridge, Mass.: Charles Stark Draper Laboratory, 1972.

Hall, Eldon C. "Case History of the Apollo Guidance Computer." E-1970. Cambridge, Mass.: MIT Instrumentation Laboratory, 1966.

Hall, Eldon C. "A Case History of the AGC Integrated Logic Circuits." E-1880. Cambridge, Mass.: MIT Instrumentation Laboratory, December 1965.

Hall, Eldon C. "General Design Characteristics of the Apollo Guidance Computer." R-410. Cambridge, Mass.: MIT Instrumentation Laboratory, May 1963.

Hall, Eldon C. "Computer Displays." E-1105. Cambridge, Mass.: MIT Instrumentation Laboratory, January 1962.

Hall, Eldon C., and Richard Jansson. "Miniature Packaging of Electronics in Three Dimensional Form." Cambridge, Mass: MIT Instrumentation Laboratory, June 1959.

Hallion, Richard. *On the Frontier: Flight Research at Dryden, 1946–1981.* NASA History Series. Washington, D.C.: NASA, 1984.

Hallion, Richard. *Test Pilots: The Frontiersmen of Flight*, rev. ed. Washington, D.C.: Smithsonian Institution Press, 1988.

Hallion, Richard, and Michael H. Gorn. *On the Frontier: Experimental Flight at NASA Dryden.* Washington, D.C.: Smithsonian Institution Press, 2003.

Hamilton, Margaret. "The Heart and Soul of Apollo: Doing It Right the First Time." Paper presented at Military and Aerospace Applications of Programmable Logic Devices Conference, Washington, D.C., 2004.

Hand, James A. "Computer-Aided Inertial Platform Realignment in Manned Space Flight." Cambridge, Mass.: MIT Instrumentation Laboratory, 1968.

Hand, James A. "MIT's Role in Project Apollo, Vol. I: Project Management, Systems Development, Abstracts and Bibliography." Cambridge, Mass.: MIT Instrumentation Laboratory, 1971.

Hanley, David L., Jayne Partridge, and Eldon C. Hall. "The Application of Failure Analysis in Procuring and Screening of Integrated Circuits." Cambridge, Mass.: MIT Instrumentation Laboratory, 1965.

Hansen, James R. *Enchanted Rendezvous: John C. Houbolt and the Genesis of the Lunar-Orbit Rendezvous Concept.* Washington, D.C.: NASA, 1995.

Hansen, James R. *Engineer in Charge: A History of the Langley Aeronautical Laboratory, 1917–1958.* NASA History Series. Washington, D.C.: NASA Scientific and Technical Information Office, 1987.

Hansen, James R. *First Man: The Life of Neil A. Armstrong.* New York: Simon & Schuster, 2005.

Harland, David M. *How NASA Learned to Fly in Space: An Exciting Account of the Gemini Missions.* Burlington, Ont.: Apogee Books, 2004.

Heims, Steve J. *Constructing a Social Science for Postwar America: The Cybernetics Group, 1946–1953.* Cambridge, Mass.: MIT Press, 1993.

Heppenheimer, T. A. *History of the Space Shuttle.* Washington, D.C.: Smithsonian Institution Press, 2002.

Herfort, John. "MIT Computers 'Miraculous.'" *Boston Globe,* December 29, 1968: 3.

Hersch, Paul. "Engineers Reassessing Electronic Hardware in Light of Some Near Failures on Apollo 12." *IEEE Spectrum* (January 1970): 72–74.

Hoag, David G. "Apollo Navigation, Guidance, and Control Systems: A Progress Report." E-2411. Cambridge, Mass.: MIT Instrumentation Laboratory, February 1969.

Hoag, David G. "The Eagle Has Returned." In *Dedication Conference of the International Space Hall of Fame.* Alamogordo, N.M.: American Astronautical Society, 1976.

Hoag, David G. "Guidance, Navigation, and Control of Manned Lunar Landing." Cambridge, Mass.: MIT Instrumentation Laboratory, n.d.

Hoag, David G. "The History of Apollo On-Board Guidance and Navigation." P–357. Cambridge, Mass.: Charles Stark Draper Laboratories, September 1976.

Hoag, David G. "LEM Guidance Computer Programs to Be Supplied by MIT." Cambridge, Mass.: MIT Instrumentation Laboratory, 1965.

Hoey, Bob. "X-15 Ventral-Off." In Stoliker, Hoey, and Armstrong, eds., *Flight Testing at Edwards,* 155–158.

Holleman, E. C., N. A. Armstrong, and W. H. Andrews. "Utilization of the Pilot in the Launch and Injection of a Multistage Vehicle." In *28th Annual Meeting of the Institute of Aeronautical Sciences.* New York, N.Y., 1960.

Holley, M. D., W. L. Swingle, S. L. Bachman, C. J. LeBlanc, H. T. Howard, and H. M. Biggs. "Apollo Experience Report: Guidance and Control Systems: Primary Guidance, Navigation, and Control System Development." NASA TN D-8227. Houston, Tex.: Johnson Space Center, 1976.

Holliday, Will L., and Dale P. Hoffman. "Systems Approach to Flight Controls." *Astronautics* (May 1962): 36–37, 74–80.

Hong, Sungook. "Man and Machine in the 1960s." *Techne* 7, no. 3 (2004): 49–77.

Hopkins, Albert L. "A Fault-Tolerant Information Processing Concept for Space Vehicles." Cambridge, Mass.: MIT Instrumentation Laboratory, 1970.

Hopkins, Albert L. "A Fault-Tolerant Information Processing System for Advanced Control, Guidance, and Navigation." Cambridge, Mass.: Charles Stark Draper Laboratories, 1970.

Hopkins Jr., Albert L., Ramon Alonso, and Hugh Blair-Smith. "Logical Description for the Apollo Guidance Computer (AGC4)." Cambridge, Mass.: MIT Instrumentation Laboratory, 1963.

Horner, Richard. "Banquet Address before the first Annual Awards Banquet of the Society of Experimental Test Pilots." *SETP Quarterly Review* 2, no. 1 (Fall 1957): 1–10.

Hughes, Agatha C., and Thomas Parke Hughes, eds. *Systems, Experts, and Computers: The Systems Approach in Management and Engineering, World War II and After*. Cambridge, Mass.: MIT Press, 2000.

Hughes, Thomas Parke. *Elmer Sperry; Inventor and Engineer*. Baltimore: Johns Hopkins University Press, 1971.

Hughes, Thomas Parke. *Networks of Power: Electrification in Western Society, 1880–1930*. Baltimore: Johns Hopkins University Press, 1983.

Hughes, Thomas Parke. *Rescuing Prometheus*, 1st ed. New York: Pantheon Books, 1998.

Hutchins, Edwin. *Cognition in the Wild*. Cambridge, Mass.: MIT Press, 1995.

IBM Corporation. "Apollo Mission Simulator." Press release, n.d.

IBM Corporation. "IBM'S Role as NASA Prime Contractor in Apollo/Saturn Program." Company brochure, n.d.

Irwin, James B., and William A. Emerson. *To Rule the Night: The Discovery Voyage of Astronaut Jim Irwin*, 1st ed. Philadelphia: A. J. Holman Co., 1973.

Israel, Paul. *Edison: A Life of Invention*. New York: John Wiley, 1998.

Jenkins, Dennis R. *Hypersonics Before the Shuttle: A Concise History of the X-15 Research Airplane*. Monographs in Aerospace History No. 18. Washington, D.C.: NASA, 2000.

Jenkins, Dennis R. *Space Shuttle: The History of the National Space Transportation System: The First 100 Missions*, 3rd ed. Cape Canaveral, Fla.: D. R. Jenkins, 2001.

Jenkins, Dennis R., and Tony Landis. *Hypersonic: The Story of the North American X-15*. North Branch, Minn.: Specialty Press, 2003.

Jenkins, Morris V. "Software Compatibility with Lunar Mission Objectives." In Manned Space Flight Center, *Apollo Lunar Landing Mission Symposium*, 389–426. Houston, Tex.: NASA, 1966.

Johnson, Madeline S., and Donald R. Giller. "MIT's Role in Project Apollo, Vol. V: The Software Effort." Cambridge, Mass.: Charles Stark Draper Laboratories, 1971.

Johnson, Stephen B. *The Secret of Apollo: Systems Management in American and European Space Programs*. New Series in NASA History. Baltimore: Johns Hopkins University Press, 2002.

Johnson, Stephen B. "Samuel Philips and the Taming of Apollo." *Technology and Culture* 42 (October 2001): 685–709.

Johnston, Malcolm W. "A Manual LEM Backup Guidance System." Cambridge, Mass.: MIT Instrumentation Laboratory, 1964.

Kalman, R. E. "A New Approach to Linear Filtering and Prediction Problems," *Transactions of the ASME—Journal of Basic Engineering* 82 (1960): 34–45.

Kauffman, James Lee. *Selling Outer Space: Kennedy, the Media, and Funding for Project Apollo, 1961–1963*. Studies in Rhetoric and Communication. Tuscaloosa: University of Alabama Press, 1994.

Kaye, David N. "The Indispensable Men." *Electronic Design* (August 16, 1969): 40–42.

Keese, W. M., B. F. Leibowitz, W. J. Martin, I. D. Nehama, A. H. Scheinman, C. S. Sherrard, W. C. Dennis, S. E. Fleige, D. A. Jackson, and B. J. Thielen. "Management Procedures in Computer Programming for Apollo-Interim Report." Bellcomm, Inc., 1964.

Keller, W. F. "Study of Spacecraft Hover and Translation Modes above the Lunar Surface." Paper presented at American institute of Aeronautics and Astronautics, Annual Meeting, Washington, D.C., June 1964.

Kelly, Lloyd L., and Robert B. Parke. *The Pilot Maker.* New York: Grosset & Dunlap, 1970.

Kelly, Thomas J. *Moon Lander: How We Developed the Apollo Lunar Module.* Smithsonian History of Aviation and Spaceflight Series. Washington, D.C.: Smithsonian Institution Press, 2001.

Kennedy, John F. "Special Message to the Congress on Urgent National Needs," May 25, 1961. At http://history.nasa.gov/spdocs.html (accessed April 10, 2007).

Kennett, Lee B. *The First Air War, 1914–1918.* New York: Free Press, 1991.

Klass, Philip J. "Apollo Guidance Bidders Protest NASA Choice of Non-Profit Firm." *Aviation Week and Space Technology* (January 8, 1962): 44–45.

Klass, Philip J. "First Apollo Control Prototype Is Readied," *Aviation Week and Space Technology* (May 25, 1964): 91–105.

Kluever, E. E., D. Mallick, and G. Matranga. "Flight Results with a Non-Aerodynamic, Variable Stability, Flying Platform." *SETP Technical Review* 8, no. 2 (1966): 98–121.

Klumpp, Allan R. "A Manually Retargeted Automatic Descent and Landing System for LEM." Cambridge, Mass.: MIT Instrumentation Laboratory, 1966.

Kraft, Christopher C. *Flight: My Life in Mission Control.* New York: Dutton, 2001.

Kraft, Christopher C. "Some Operational Aspects of Project Mercury." *SETP Quarterly Review* 4, no. 2 (Winter 1960): 52–62.

Kramer, P. C., E. E. Aldrin, and W. E. Hayes. "Onboard Operations for Rendezvous." In Manned Spacecraft Center (U.S.), *Gemini Summary Conference*, 27–40.

Kranz, Gene. *Failure Is Not an Option: Mission Control from Mercury to Apollo 13 and Beyond.* New York: Simon & Schuster, 2000.

Laning, J. Hal, and N. Zierler. "A Program for Translation of Mathematical Equations for Whirlwind I." Cambridge, Mass.: MIT Instrumentation Laboratory, 1954.

Latour, Bruno. *Science in Action: How to Follow Scientists and Engineers through Society.* Cambridge, Mass.: Harvard University Press, 1988.

Launius, Roger D. *Apollo: A Retrospective Analysis.* Monographs in Aerospace History No. 3. Washington, D.C.: NASA, 1994.

Launius, Roger D. "Heroes in a Vacuum: The Apollo Astronaut as Cultural Icon." Paper presented at 43rd AIAA Aerospace Sciences Meeting and Exhibit. Reno, Nev., January 2005.

Launius, Roger D., ed. *Innovation and the Development of Flight,* 1st ed. College Station: Tex. A&M University Press, 1999.

Launius, Roger D. "Interpreting the Moon Landings: Project Apollo and the Historians," *History and Technology* 22, no. 3 (September 2006): 225–255.

Launius, Roger D., and Howard E. McCurdy, eds. *Spaceflight and the Myth of Presidential Leadership.* Urbana: University of Illinois Press, 1997.

Leary, William M. "The Search for an Instrument Landing System, 1918–48." In Launius, ed., *Innovation and the Development of Flight,* 80–99.

Lécuyer, Christophe. *Making Silicon Valley: Innovation and the Growth of High Tech, 1930–1970, Inside Technology.* Cambridge, Mass.: MIT Press, 2006.

Leondes, Cornelius T., R. W. Vance, and M. S. Agbabian. *Lunar Missions and Exploration.* New York: John Wiley and Sons, 1964.

Leveson, Nancy G. *Safeware: System Safety and Computers.* Reading, Mass.: Addison-Wesley, 1995.

Liebergot, Sy, with David M. Harland. *Apollo EECOM: Journey of a Lifetime.* Burlington, Ont.: Apogee Books, 2003.

Lindbergh, Charles A., and Fitzhugh Green. *We.* New York, London: G. P. Putnam's Sons, 1927.

Littleton, Orval P. "Apollo Experience Report—Guidance and Control Systems: Command and Service Module Stabilization and Control System." NASA TN D-7785. Houston, Tex.: Johnson Space Center, 1974.

Loftus, J. P. "Crew Tasks and Training." In *Apollo Lunar Landing Mission Symposium,* 461–528. Houston, Tex.: NASA, 1966.

Logsdon, John M. *The Decision to Go to the Moon: Project Apollo and the National Interest.* Cambridge, Mass.: MIT Press, 1970.

Logsdon, John M. "A Failure of National Leadership: Why No Replacement for the Space Shuttle?" In Dick and Launius, eds., *Critical Issues in the History of Spaceflight,* 269–7.

Logsdon, John M., ed. *Managing the Moon Program: Lessons Learned from Project Apollo.* Monographs in Aerospace History No. 14, Washington, D.C.: NASA, 1999.

Lovell, Jim, and Jeffrey Kluger. *Lost Moon: The Perilous Voyage of Apollo 13.* Boston: Houghton Mifflin, 1994.

MacKenzie, Donald A. *Inventing Accuracy: An Historical Sociology of Nuclear Missile Guidance.* Inside Technology. Cambridge, Mass.: MIT Press, 1990.

MacKenzie, Donald A. *Mechanizing Proof: Computing, Risk, and Trust.* Cambridge, Mass.: MIT Press, 2001.

Mailer, Norman. *Of a Fire on the Moon*, 1st ed. Boston: Little, Brown & Co. 1970.

Manned Spacecraft Center (U.S.). *Gemini Summary Conference; Technical Papers*. SP-138. Washington, D.C.: NASA, Scientific and Technical Information Division, 1967.

Manned Space Flight Center. "Apollo Lunar Landing Mission Symposium: Proceedings and Compilation of Papers," June 25–27, 1966. At http://klabs.org/history/ntrs_docs/manned/apollo/index.htm (accessed April 2007).

Manningham, Dan. "The Cockpit: A Brief History." *Business and Commercial Aviation* 80, no. 6 (June 1, 1997): 56.

Martin, Fred H. "Apollo 11: 25 Years Later." At http://www.hq.nasa.gov/office/pao/History/alsj/a11/a11.1201-fm.html (accessed January 2007).

Martin, Frederick H., and Richard H. Battin. "Computer-Controlled Steering of the Apollo Spacecraft." *J. Spacecraft* 5, no. 4 (1968): 400–407.

Matranga, G., W. H. Dana, and N. A. Armstrong. "Flight Simulated Off-the-Pad Escape and Landing Maneuver for a Vertically Launched Hypersonic Glider." Technical memorandum. Washington, D.C.: NASA, 1962.

Matranga, Gene J., Wayne Ottinger, and Calvin R. Jarvis. *Unconventional, Contrary, and Ugly: The Lunar Landing Research Vehicle*. Monographs in Aerospace History No. 35. Washington, D.C.: NASA, 2004.

Matranga, Gene J., and Joseph A. Walker. "Investigation of Terminal Lunar Landing with the Lunar Landing Research Vehicle." Paper presented at American Institute of Aeronautics and Astronautics Manned Space Flight Meeting, St. Louis, Mo., October 11–13, 1965. NASA-TM-X-74475.

Maynard, Owen. "General Mission Summary and Configuration Description." In Manned Space Flight Center, *Apollo Lunar Landing Mission Symposium*, 11–83. Houston, Tex.: NASA, 1966.

Mazlish, Bruce. *The Railroad and the Space Program: An Exploration in Historical Analogy, Technology, Space, and Society*. Cambridge, Mass.: MIT Press, 1965.

McCurdy, Howard E. *Inside NASA: High Technology and Organizational Change in the U.S. Space Program*. New Series in NASA History. Baltimore: Johns Hopkins University Press, 1993.

McCurdy, Howard E. *Space and the American Imagination*. Smithsonian History of Aviation Series. Washington, D.C.: Smithsonian Institution Press, 1997.

McCurdy, Howard E. "Observations on the Robotic versus Human Issue in Spaceflight." In Dick and Launius, eds., *Critical Issues in the History of Spaceflight*, 77–106.

McDivitt, James, and Neil Armstrong. "Gemini Manned Flight Program to Date." *SETP Proceedings* (1965): 134–156.

McDonnell Corporation. *NASA Project Gemini Familiarization Manual*, 1965. At ftp://sf.gds.tuwien.ac.at/2/m/ms/mscorbaddon/GeminiManualVol1Sec2.pdf (accessed April 10, 2007).

McDougall, Walter A. *The Heavens and the Earth: A Political History of the Space Age*. New York: Basic Books, 1985.

McElheny, Victor. "Aldrin Simplified Moon Procedure." *Boston Globe*, July 21, 1969: 16.

McElheny, Victor. "Man Returns to the Moon; Shepard, Mitchell Overcomes Computer Hitch; MIT Engineers Provide Bypass System." *Boston Globe*, February 5, 1971: 1.

McElheny, Victor. "MIT Unit Set to Orbit." *Boston Globe*, October 20, 1968: A1.

McElheny, Victor. "More Lunar Probes Urged by Group of U.S. Scientists." *Boston Globe*, October 31, 1969: 38.

McElheny, Victor. "Report Says NASA Concentrates on Machine, Ignores Man." *Boston Globe*, November 18, 1969.

McElheny, Victor. "Some Fear Showy Space Trips Will Stifle Progress." *Boston Sunday Globe*, November 9, 1969: 50.

McElheney, Victor. "Space Tests Pinpoint Lunar Module Bugs." *Boston Globe*, November 4, 1968: 60.

McElheny, Victor. "Little Errors Added Up to 4-Mi. Apollo Mistake." *Boston Globe*, August 13, 1969: 14.

McRuer, Duane T., Irving Louis Ashkenas, and Dunstan Graham. *Aircraft Dynamics and Automatic Control*. Princeton, N.J.: Princeton University Press, 1974.

Miller, John E., and Ain Laats. "Apollo Guidance and Control System Flight Experience." Cambridge, Mass.: MIT Instrumentation Laboratory, 1969.

Mindell, David A. *Between Human and Machine: Feedback, Control, and Computing before Cybernetics*. Johns Hopkins Studies in the History of Technology. Baltimore: Johns Hopkins University Press, 2002.

Mindell, David A. "Opening Black's Box: Rethinking Feedback's Myth of Origin." *Technology and Culture* 41 (July 2000): 405–434.

Mindell, David A. *War, Technology, and Experience Aboard the USS Monitor*. Baltimore: Johns Hopkins University Press, 2000.

MIT Instrumentation Laboratory. "Astronauts' Guidance and Navigation Course Notes." Cambridge, Mass.: MIT Instrumentation Laboratory, 1962.

MIT Instrumentation Laboratory. "Control of Flight Software Development Costs." Cambridge, Mass.: Charles Stark Draper Laboratory, 1971.

MIT Instrumentation Laboratory. "Guidance System Operations Plan, AS-278: Vol. 1, CM GNCS Operations." Cambridge, Mass.: MIT Instrumentation Laboratory, 1966.

MIT Instrumentation Laboratory. "Guidance System Operations Plan for Manned LM Earth Orbital and Lunar Missions Using Program Luminary 1E: Section 5, Guidance Equations." R-567. Cambridge, Mass.: MIT Instrumentation Laboratory, December 1971.

MIT Instrumentation Laboratory. "Space Navigation Guidance and Control." R-500. Cambridge, Mass.: MIT Instrumental Laboratory, 1965.

Mitchell, Edgar D., and Dwight Arnan Williams. *The Way of the Explorer: An Apollo Astronaut's Journey through the Material and Mystical Worlds*. New York: G. P. Putnam's Sons, 1996.

Morring, Frank. "A Base to Build On." *Aviation Week and Space Technology* 165, no. 23 (December 11, 2006): 24–26.

Mudgway, Douglas J. *Uplink-Downlink: A History of the NASA Deep Space Network, 1957–1997*. Washington, D.C.: NASA, 2001.

Mueller, George. "Final Report: Apollo Guidance Software Task Force." Washington, D.C.: NASA, 1968.

Mumford, Lewis. *Technics and Civilization*. New York: Harcourt, 1934.

Mumford, Lewis. *The Myth of the Machine: The Pentagon of Power*, 1st ed. New York: Harcourt Brace Jovanovich, 1970.

Mumford, Lewis. *The Myth of the Machine: Technics and Human Development*, 1st ed. New York: Harcourt, 1967.

Muntz, Charles A. "Users Guide to the Block Ii AGC/LGC Interpreter." Cambridge, Mass.: MIT Instrumentation Laboratory, 1965.

Murray, Arthur. "Pilot-Oriented Dyna-Soar Designs." *SETP Quarterly Review* 6, no. 3 (Summer 1963): 9–15.

Murray, Charles A., and Catherine Bly Cox. *Apollo, the Race to the Moon*. New York: Simon & Schuster, 1989.

Mutchler, J. V. "Apollo CMC/LGC Software Development Plan." Houston, Tex.: Manned Spacecraft Center, 1968.

NASA. "Proceedings of the F-8 Digital Fly-by-Wire and Supercritical Wing First Flight's 20th Anniversary Celebration." Edwards, Calif.: NASA Dryden Flight Research Center, Conference Publication 3256, May 27, 1992.

NASA. "Technical Information Summary AS–501 Apollo Saturn V Flight Vehicle." Huntsville, Ala.: Marshall Space Flight Center, 1967.

Nevins, James L. "Man-Machine Design for the Apollo Navigation, Guidance and Control System." E-2476. Cambridge, Mass.: MIT Instrumentation Laboratory, January 1967.

Nevins, James L. "Man-Machine Design for the Apollo Navigation, Guidance, and Control System—Revisited: Apollo, a Transition in the Art of Piloting a Vehicle." Cambridge, Mass.: Charles Stark Draper Laboratory, January 1970.

Nevins, J. L., and I. S. Johnson. "Man-Computer Interface for the Apollo Guidance, Navigation and Control System." Cambridge, Mass.: MIT Instrumentation Laboratory, 1967.

Nevins, J. L., I. S. Johnson, and T. B. Sheridan. "Man/Machine Allocation in the Apollo Navigation, Guidance, and Control System." E-2305. Cambridge, Mass.: MIT Instrumentation Laboratory, July 1968.

Nevins, J. L., E. A. Woodin, and R. W. Metzinger. "Man-Machine Simulations for the Apollo Navigation, Guidance, and Control System." E-2081. Cambridge, Mass.: MIT Instrumentation Laboratory, January 1967.

Noble, David F. *Forces of Production: A Social History of Industrial Automation*, 1st ed. New York: Knopf, 1984.

Noyce, Robert N. "Integrated Circuits in Military Equipment." *IEEE Spectrum* (June 1964): 71–72.

Nyquist, Harry. "Certain Factors Affecting Telegraph Speed." *Bell System Technical Journal* 3 (April 1924): 324–346.

Nyquist, Harry. "Certain Topics in Telegraph Transmission Theory." *Transactions of the AIEE* 47 (1928): 617–644.

O'Leary, Brian. *The Making of an Ex-Astronaut*. Boston: Houghton Mifflin, 1970.

Oxford University Press. *The Oxford English Dictionary* on compact disc, 2nd. ed. Oxford: Oxford University Press, 1992.

Partridge, Jayne, L. David Hanley, and Eldon C. Hall. "Progress Report on Attainable Reliability of Integrated Circuits for Systems Application." Cambridge, Mass.: MIT Instrumentation Laboratory, 1964.

Perrow, Charles. Normal Accidents: Living with High-Risk Technologies. Updated ed. Princeton, N.J.: Princeton University Press, 1999.

Phillips, William H. *Journey in Aeronautical Research: A Career at NASA Langley Research Center*. Monographs in Aerospace History No. 12. Washington, D.C.: NASA, 1998.

Phillips, William H. *Journey into Space Research: Continuation of a Career at NASA Langley Research Center*. Monographs in Aerospace History No. 40. Washington, D.C.: NASA, 2005.

Pinney, Benjamin W. "Projects, Management, and Protean Times: Engineering Enterprise in the United States, 1870–1960." Ph.D. dissertation, Massachusetts Institute of Technology Program in Science Technology and Society, 2001.

Pitts, John A. *The Human Factor, Biomedicine in the Manned Space Program to 1980*. SP-4213. Washington, D.C.: NASA Scientific and Technical Information Branch, 1985.

Pynchon, Thomas. *Gravity's Rainbow*. New York: Viking Press, 1973.

Pyne, Stephen J. "Seeking Newer Worlds: An Historical Context for Space Exploration." In Dick and Launius, eds., *Critical Issues in the History of Spaceflight*.

Quesada, E. R. "A Pilot's Philosophy for the Space Age." *SETP Quarterly Review* 4, no. 1 (Fall 1959): 3–6.

Rabinbach, Anson. *The Human Motor: Energy, Fatigue, and the Origins of Modernity*. New York: BasicBooks, 1990.

Ramo, Simon. "ICBM: Giant Step into Space," *Astronautics* (August 1957): 34–41.

Rankin, Daniel Allen. "A Model of the Cost of Software Development for the Apollo Spacecraft Computer." Massachusetts Institute of Technology Alfred P. Sloan School of Management. Master of Science thesis, 1972.

Recer, Paul. "They Feud Over Moon Flights." *Miami Herald*, October 12, 1969.

Reichart, Tony. "Son of Apollo: The Next Lunar Lander Will Be a Giant Leap Ahead of the First." *Air and Space* (April/May 2006): 21–27.

Ridenour, Louis N. *Radar System Engineering*, 1st ed. Radiation Laboratory Series (Massachusetts Institute of Technology) 1. New York: McGraw-Hill Book Co., 1947.

Roberts, J. O. "The Case against Automation in Manned Fighter Aircraft." *SETP Quarterly Review* 2, no. 3 (Fall 1957): 18–23.

Roland, Alex. "Barnstorming in Space: The Rise and Fall of the Romantic Era of Spaceflight, 1957–1986." In Byerly, ed., *Space Policy Reconsidered*, 33–52.

Roland, Alex. *Model Research: The National Advisory Committee for Aeronautics, 1915–1958*. NASA History Series. Washington, D.C.: NASA Scientific and Technical Information Office, 1985.

Roughton, Bert. "Jets with Byte Stuff Hovering on Horizon for Airline Industry." *Atlanta Journal and Constitution* (December 11, 1988): A1.

Rusnak, Walter. "Avionics Aspects of the Lunar Landing Research Vehicle/ Armstrong Accident LLRV Investigation Press Release." Bell Aerosystems Company, 1968.

Sapolsky, Harvey M. *The Polaris System Development: Bureaucratic and Programmatic Success in Government*. Cambridge, Mass.: Harvard University Press, 1972.

Sato, Yasushi. "Local Engineering in the Early American and Japanese Space Programs: Human Qualities in Grand System Building." Ph.D. dissertation, University of Pennsylvania, 2005.

Savage, Bernard I., and Alice Drake. "AGC4 Basic Training Manual. Volume I of II." Cambridge, Mass.: MIT Instrumentation Laboratory, 1967.

Schirra, Walter. "A Real Breakthrough—The Capsule Was All Mine." *Life* 26 (October 1962): 39.

Schirra, Wally, and Richard N. Billings. *Schirra's Space*. Annapolis, Md.: Naval Institute Press, 1995.

Scott, David. "The Apollo Guidance Computer: A User's View." Lecture at the Computer Museum, Boston, Mass., June 10, 1982. *Computer Museum Report* 2 (Fall 1982). Transcript of Scott remarks courtesy of Eldon Hall.

Scott, David, A. A. Leonov, and Christine Toomey. *Two Sides of the Moon*, 1st U.S. ed. New York: Thomas Dunne Books, 2004.

Seamans, Robert C. *Aiming at Targets: The Autobiography of Robert C. Seamans, Jr.* NASA History Series. Washington, D.C.: NASA History Office, 1996.

Seamans, Robert C. *Project Apollo: The Tough Decisions*. Washington, D.C.: NASA, 2005.

Sears, Norman E. "Technical Development Status of Apollo Guidance and Navigation." Cambridge, Mass.: MIT Instrumentation Laboratory, 1964.

See, Eliot. "Engineering and Operational Approaches for Projects Gemini and Apollo." *Society of Experimental Test Pilots 7th Annual Symposium Proceedings* (1963): 135–147.

Sewall, Tom. "Happy Day for MIT Experts." *Boston Herald*, December 28, 1968: 3.

Shepard, Alan B. *Training by Simulation*. Edwin A. Link Lecture Series. Washington, D.C.: Smithsonian Institution, 1965.

Shepard, Alan B., and Donald K. Slayton. *Moon Shot: The Inside Story of America's Race to the Moon*, 1st ed. Kansas City, Mo.: Turner Publications, 1994.

Siddiqi, Asif A. "American Space History: Legacies, Questions, and Opportunities for Future Research." In Dick and Launius, eds., *Critical Issues in the History of Spaceflight*, 433–480.

Siddiqi, Asif A. *Challenge to Apollo: The Soviet Union and the Space Race, 1945–1974*. NASA History Series. Washington, D.C.: NASA, NASA History Division, 2000.

Slayton, Donald K. "Operation Plan and Pilot Aspects of Project Mercury." *SETP Quarterly Review* 4, no. 2 (Winter 1960): 63–69.

Slayton, Donald K., and Michael Cassutt. *Deke!: U.S. Manned Space: From Mercury to the Shuttle*, 1st ed. New York: Forge, 1994.

Smith, Andrew. *Moondust: In Search of the Men Who Fell to Earth*, 1st U.S. ed. New York: Fourth Estate, 2005.

Smith, Michael. "Selling the Moon: The U.S. Manned Space Program and the Triumph of Commodity Scientism." In Fox and Lears, eds., *The Culture of Consumption*, 177–209.

Society of Experimental Test Pilots. *History of the First Twenty Years*. Covina, Calif.: Taylor Publishing Company, 1978.

Speer, Raymond. "Strict Control Kept Out Semiconductor Flaws." *Electronic Design* 17 (August 16, 1969): 29.

Stafford, Thomas P., and Michael Cassutt. *We Have Capture: Tom Stafford and the Space Race*. Washington, D.C.: Smithsonian Institution Press, 2002.

Steinhoff, Ernst A. *The Eagle Has Returned: Proceedings of the Dedication Conference of the International Space Hall of Fame*. Alamagordo, N. M.: American Astronautical Society, 1976.

Stengel, Robert F. "Manual Attitude Control of the Lunar Module." *Journal of Spacecraft and Rockets* 7, no. 8 (August 1970): 941–947.

Stillwell, Wendell H. *X-15 Research Results*. SP-60. Washington, D.C.: NASA Scientific and Technical Information Office, 1964.

Stoliker, Fred, Bob Hoey, and Johnny Armstrong, eds. *Flight Testing at Edwards: Flight Test Engineers' Stories 1946–1975*. Lancaster, Calif.: Flight Test Historical Foundation, 1996.

Strickland, Zack. "Series of Lunar Landings Simulated." *Aviation Week and Space Technology* (June 30, 1969): 55–59.

Sullivan, Madeline M. "Hybrid Simulation of the Apollo Guidance Navigation and Control System." R-525. Cambridge, Mass.: MIT Instrumentation Laboratory, December 1965.

Swenson, Lloyd S., James M. Grimwood, and Charles C. Alexander. *This New Ocean: a History of Project Mercury*. NASA History Series. SP-4201. Washington, D.C.: NASA, Scientific and Technical Information Division, Office of Technology Utilization, 1966.

Taylor, Lawrence W., Jr. "Recent X-15 Flight Test Experience with the Mh-96 Adaptive Control System." Paper presented at the Intercenter Technical Conference on Control, Guidance, and Navigation Research for Manned Lunar Missions, Ames Research Center, Moffett Field, Calif., July 24–25, 1962.

Taylor, Lawrence W., and George B. Merrick, "X-15 Stability Augmentation System." NASA TN D-1157, 1961.

Thomas, B. K. "Apollo 8 Proves Value of Onboard Control." *Aviation Week and Space Technology* (January 20, 1969): 40–46.

Thompson, Milton O. *At the Edge of Space: The X-15 Flight Program*. Washington, D.C.: Smithsonian Institution Press, 1992.

Thompson, Milton O. *Flight Research: Problems Encountered and What They Should Teach Us*. Monographs in Aerospace History No. 22. Washington, D.C.: NASA History Division, Office of Policy and Plans, 2000.

Thompson, Milton O., and James R. Welsh. "Flight Test Experience with Adaptive Control Systems." Paper presented at the AGARD Guidance and Control and Flight Mechanics Panels, Oslo, Norway, September 3–5, 1968.

Thompson, Neal. *Light This Candle: The Life and Times of Alan Shepard, America's First Spaceman*, 1st ed. New York: Crown Publishers, 2004.

Tindall, Howard W., Jr. "Techniques of Controlling the Trajectory." In *What Made Apollo a Success*. SP-287. Washington, D.C.: NASA Scientific and Technical Information Office.

Tomayko, James E. *Computers in Spaceflight: The NASA Experience*. Linthicum Heights, Md: NASA History Office, 1988.

Tomayko, James E. *Computers Take Flight: A History of NASA's Pioneering Digital Fly-by-Wire Project*. NASA History Series. Washington, D.C.: NASA, 2000.

Trageser, Milton B., and David G. Hoag. "Apollo Spacecraft Guidance System." Cambridge, Mass.: MIT Instrumentation Laboratory, 1965.

Turkle, Sherry. *Life on the Screen: Identity in the Age of the Internet.* New York: Simon & Schuster, 1995.

Turkle, Sherry. *The Second Self: Computers and the Human Spirit*, 20th anniversary ed. Cambridge, Mass.: MIT Press, 2005.

Turner, Fred. *From Counterculture to Cyberculture: Stewart Brand, the Whole Earth Network, and the Rise of Digital Utopianism.* Chicago: University of Chicago Press, 2006.

Vincenti, Walter G. *What Engineers Know and How They Know It: Analytical Studies from Aeronautical History*, Johns Hopkins Studies in the History of Technology. Baltimore: Johns Hopkins University Press, 1990.

Voas, Robert. "Manual Control of the Mercury Spacecraft." *Astronautics* (March 1962): 18.

Voas, Robert. "A Description of the Astronaut's Task in Project Mercury." *Human Factors* 3, no. 3 (September 1961): 149–165.

Von Braun, Wernher, "Address to the Society of Experimental Test Pilots." *SETP Newsletter* (August 1959): 3–9.

Von Braun, Wernher. "Space Flight: Past, Present, and Future." New York: Wings Club General Harold L. Harris "Sight" Lectures, 1968.

Von Braun, Wernher. *Space Frontier*, 1st ed. New York: Holt, 1967.

Von Braun, Wernher, and Cornelius Ryan. *Conquest of the Moon.* New York: Viking Press, 1953.

Vonbun, Friedrich O. "Ground Tracking of Apollo." *Astronautics and Aeronautics* 4 (May 1966): 104–115.

Vonnegut, Kurt. *Player Piano*, new ed. New York: Holt, 1966.

Walker, Chuck, and Joel Powell. *Atlas: The Ultimate Weapon: By Those Who Built It.* Burlington, Ont.: Apogee Books, 2005.

Walker, Joseph A. "Some Concepts of Pilot's Presentation." *SETP Quarterly Review* 4, no. 2 (Winter 1960): 90–96.

Walker, Joseph, and Joseph Weil. "The X-15 Program." Paper presented at Second AIAA-NASA Manned Space Flight Meeting, April 22–23, 1963, Dallas, Tex. NASA Dryden Archives.

Waltman, Gene L. *Black Magic and Gremlins: Analog Flight Simulations at NASA'S Flight Research Center.* Monographs in Aerospace History No. 20. SP-4520. Washington, D.C.: NASA, NASA History Division, Office of Policy and Plans, 2000.

Ward, John W. "The Meaning of Lindbergh's Flight." *American Quarterly* 10 (1958): 3–16.

Watkins, Billy. *Apollo Moon Missions: The Unsung Heroes.* Westport, Conn.: Praeger Publishers, 2006.

Weitekamp, Margaret A. *Right Stuff, Wrong Sex: America's First Women in Space Program, Gender Relations in the American Experience*. Baltimore: Johns Hopkins University Press, 2004.

Weitekamp, Margaret A. "Critical Theory as a Toolbox: Suggestions for Space History's Relationship to the History Subdisciplines." In Dick and Launius, eds., *Critical Issues in the History of Spaceflight*, 549–572.

White, Stanley C. "Human Factors and Bioastronautics: State of the Art 1960." *Astronautics* (1960): 35–36.

Widnall, William S. "Lunar Module Digital Autopilot." *Journal of Spacecraft and Rockets* 8, no. 1 (1971): 56–62.

Wiener, Norbert. *Cybernetics; or, Control and Communication in the Animal and the Machine*. New York: MIT Press, 1948.

Wiesner Committee. "Report to the President-Elect of the Ad Hoc Committee on Space," January 10, 1961. At http://www.hq.nasa.gov/office/pao/History/report61.html (accessed April 10, 2007).

Wohl, Robert. *A Passion for Wings: Aviation and the Western Imagination, 1908–1918*. New Haven: Yale University Press, 1994.

Wolfe, Tom. *The Right Stuff*. New York: Farrar, Straus & Giroux, 1979.

Wood, James. "Pilot Control of the X-20/Titan II Boost Profile." In *SETP 7th Annual Symposium Proceedings* (September 27–28, 1963): 165–175.

Woodling, C. H., S. Faber, J. J. Van Bockel, C. C. Olasky, W. K. Williams, J. L. C. Mire, and J. R. Homer. "Apollo Experience Report: Simulation of Manned Space Flight for Crew Training." Houston, Tex.: Manned Spacecraft Center, 1973.

Wright, Wilbur. "Some Aeronautical Experiments." Lecture delivered before the Western Society of Engineers, September 18, 1901, reprinted in Wright et al., *The Papers of Wilbur and Orville Wright*, 100.

Wright, Wilbur, Orville Wright, Octave Chanute, and Marvin Wilkes McFarland. *The Papers of Wilbur and Orville Wright: Including the Chanute-Wright Letters and Other Papers of Octave Chanute*. New York: McGraw-Hill, 2001.

X-15, feature film directed by Richard Donner, 107 minutes. Essex Productions, 1961.

Index

About the Cover Image

Computer reconstruction of Neil Armstrong's view out the Lunar Module on Apollo 11, 520 feet above the lunar surface just as he transferred from automatic control to semi-manual "attitude hold" (note his hand reaching for the switch), to fly the vehicle past West Crater (visible out the window) to a smooth area for landing. Note the landing point designator, the graded angles on the window that would guide Armstrong's eye to the computer's estimate of a landing spot, and the 1202 program alarm indications on the guidance computer display at lower right. The image was created according to the author's conception by digital artist John Knoll, with research input from Paul Fjeld. A small number of compromises were made in order to create the image: the viewpoint is actually about 18 inches behind where Armstrong's would have been; the view of West Crater is a few seconds out of sync with the mission timer and the events depicted inside the LM; the 1202 alarm, the last of them, occurred approximately 15 seconds earlier so it would most likely have been cleared from the display by this point; some of the checklists that the crew had arrayed around them may be missing; boulders around the crater are included from Armstrong's description.

The LM interior was modeled in AutoDesSys FormZ and LuxologyModo from a variety of reference sources, including NASA drawings, historical photographs, and photographs of the LM simulator at the Cradle of Aviation Museum in Long Island. The lunar surface was modeled in Luxology Modo using lunar orbiter photographs and the Apollo 11 powered descent film as reference. Textures were created in Adobe Photoshop. Final rendering was done in LuxologyModo.